Elemental Graphene Analogues

David J. Fisher

One of the greatest revolutions in materials science in recent years has been the literal renaissance of age-old materials in new and unexpected guises and possessing correspondingly astounding properties. There was once a time, for instance, when textbooks declared that only metals could offer any progress in superconduction. Since then, familiar perovskites – and even humble magnesium boride – have been recognised as being so-called ‘room-temperature’ superconductors. Carbon in particular has benefited from this revolution and has now found application as routinely deposited diamond coatings and as C60 ‘buckyballs’.

The most recent innovation has been the discovery and preparation of graphene; single-monolayer carbon having a remarkable strength. This success has naturally led researchers to ask whether other materials might also be prepared in an analogous monolayer form and offer similarly amazing properties.

The present monograph summarizes all of the work carried out on such monolayer materials up to the beginning of 2017, with attention being restricted to those, like graphene, being composed of a single element. Most of the work done so far on these ‘elemental graphene analogues’ has been theoretical, but the existing experimental data suggest that they may well become as useful as graphene.

Elemental Graphene Analogues

David J. Fisher

Published by **Materials Research Forum LLC**
Millersville, PA 17551, USA

Published as part of the book series
Materials Research Foundations
Volume 14 (2017)
ISSN 2471-8890 (Print)
ISSN 2471-8904 (Online)

Print ISBN 978-1-945291-30-2
ePDF ISBN 978-1-945291-31-9

Distributed worldwide by

Materials Research Forum LLC
105 Springdale Lane
Millersville, PA 17551
USA
http://www.mrforum.com

Printed in the United States of America
10 9 8 7 6 5 4 3 2 1

Table of Contents

Introduction

Silicene..1

1.1 Mechanical Properties 6

1.2 Preparation and Structure 14

1.3 Composites 17

1.4 Substrate Interaction 19

1.5 Thermoelectricity...40

1.6 Thermal Conductivity...46

1.7 Magnetoresistance 50

1.8 Valley and Spin Phenomena...53

1.9 Magnetic Properties..65

1.10 Optical Properties 70

1.11 Electrical Conductivity..73

1.11.1 Superconductivity...77

1.11.2 Hall Effect...78

1.11.3 Semiconductivity and Band Structure...80

1.12 Surface Interaction...107

1.13 Defects 122

1.13.1 Line Defects..123

1.13.2 Point Defects...124

1.13.3 Stone-Wales Defect 128

1.13.4 Miscellaneous 131

1.14 Diffusion Processes 132

1.15 Applications...133

1.15.1 Transistors....133
1.15.2 Spintronics....138
1.15.3 Environmental Protection....141
1.15.4 Sensors....141
1.15.5 Gas Purification....143
1.15.6 Hydrogen Storage....143
1.15.7 Energy Storage....145
1.15.8 Catalysis....147
Germanene....149
2.1 Preparation....149
2.2 Structure....152
2.3 Mechanical Properties....155
2.4 Semiconduction and Band Structure....156
2.5 Substrate Interaction....158
2.6 Surface Interaction....163
2.7 Defects....170
2.8 Thermal Conductivity....171
2.9 Thermo-Electricity....172
2.10 Optical Properties....173
2.11 Composites....174
2.12 Applications....175
2.12.1 Solar Energy....175
2.12.2 Energy Storage....176
2.12.3 Catalysis....176
2.12.4 Transistors....176

2.12.5 Optical Devices 178
2.12.6 Spintronics 178
Stanene (Tinene) 180
3.1 Structure 181
3.2 Preparation 182
3.3 Mechanical Properties 182
3.4 Substrate Interaction 183
3.5 Surface Interaction 184
3.6 Superconductivity 186
3.7 Defects 188
3.8 Diffusion 189
3.9 Magnetism 189
3.10 Band Structure 189
3.11 Thermal Conductivity 192
3.12 Applications 193
Antimonene 195
4.1 Preparation 196
4.2 Composites 197
4.3 Substrate Interaction 198
4.4 Surface Interaction 199
4.5 Defects 199
4.6 Thermal Conductivity 200
4.7 Optical Properties 201
4.8 Semiconduction 201
4.9 Applications 201

Indiene......203
Arsenene......204
6.1 Mechanical Properties......205
6.2 Semiconduction......205
6.3 Composites......208
6.4 Surface Interaction......209
6.5 Magnetism......211
6.6 Thermal Conductivity......212
Phosphorene......213
7.1 Preparation......213
7.2 Structure......216
7.3 Mechanical Properties......217
7.4 Diffusion......221
7.5 Magnetism......222
7.6 Semiconduction and Band Structure......226
7.7 Surface Interaction......241
7.8 Substrate Interaction......246
7.9 Optical Properties......247
7.10 Defects......249
7.11 Composites......254
7.12 Superconductivity......256
7.13 Thermal Conductivity......257
7.14 Applications......259
7.14.1 Solar Power......259
7.14.2 Sensors......259

7.14.3 Catalysis 261
7.14.4 Energy Storage 262
7.14.5 Energy Harvesting 263
7.14.6 Thermoelectricity 264
7.14.7 Electronics 265
7.14.8 Biological 270
Bismuthene 271
Borophene 272
9.1 Structure 272
9.2 Defects 273
9.3 Mechanical Properties 273
9.4 Composites 275
9.5 Superconductivity 276
9.6 Magnetism 276
9.7 Applications 276
9.7.1 Energy Storage 276
9.7.2 Hydrogen Storage 278
9.7.3 Sensors 278
9.7.4 Catalysis 279
9.7.5 Biological 279
References 280
Keywords 365

Introduction

Graphene has entered the public consciousness as few other 'non high-street' materials have done in the recent past; even being mentioned in the TV comedy series, *The Big Bang Theory*[1]. Engineers already dream[2] of using super-strength graphene to realize science-fiction writer Arthur C.Clarke's 'space elevator'[3]. Much of the initial public interest in graphene was probably piqued by the fact that its discovery earned a Nobel prize for Andre Geim and Konstantin Novoselov.

But scientists never rest on their laurels, and it was only natural to seek other analogous materials which might offer similarly spectacular properties. In barely more than half a decade, this search has already yielded a large amount of information; albeit much of it based upon computer modelling rather than experimental investigation. Density-functional theory computations in particular are a powerful tool for investigating the electronic structure (especially the ground state) of nanomaterials. By using this method, many graphene-like materials have been explored, and amazing properties have been revealed.[4]

Monolayer materials in general have been among the 'hottest' topics in condensed-matter physics ever since the experimental fabrication of graphene. The new classes of monolayer materials to be manufactured experimentally include the group-IV elements (yielding 'silicene', 'germanene' and 'stanene [tinene]'), the group-V elements (yielding 'phosphorene') and transition-metal dichalcogenides. The group-IV monolayers are predicted to be topological insulators[5], while the transition-metal dichalcogenides are potentially useful for valleytronics. Monolayer materials can be used as field-effect transistors and are expected to be a key feature of future nano-electronic devices.

Opto-electronic applications require materials having suitable band-gaps and high mobilities. The broad range of band-gaps and high mobilities offered by two-dimensional semiconductors composed of monolayers of group-15 elements (phosphorene, arsenene, antimonene, etc.) thus offers great promise. The calculated binding energies and phonon band dispersions of these allotropes reflect thermodynamic stability. The energy band-gaps range from 0.36 to 2.62eV, and phosphorene, arsenene and bismuthene in particular have carrier mobilities which are as high as several thousand cm^2/Vs.[6]

This work is summarised in the present book. The contents are however restricted to analogues consisting of a single element, thus being closest to graphene in nature, and so the transition-metal dichalcogenides will not be covered here.

Chapter 1

Silicene

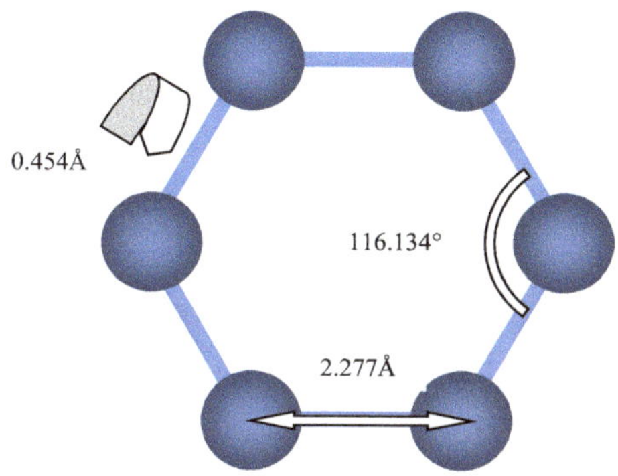

The advent of graphene ushered-in a new era in materials science. Graphene is a two-dimensional planar honeycomb array of carbon atoms in sp^2-hybridized states. A natural question to ask was whether other elements of the group-IV elements of the periodic table, such as silicon and germanium, could also form graphene-like structures. Silicene was theoretically predicted in 1994 and has been created experimentally much more recently, in the face of some skepticism[7]. Like graphene, silicene exhibits electronic and mechanical properties that can be exploited in nano-electronic applications. That is, silicene shares many of the intriguing properties of graphene; such as so-called Dirac electronic dispersion. The different structure, compared to that of graphene, also offers the ability to open a band-gap in the presence of an electric field, or when deposited onto a substrate. These are important properties from the viewpoint of digital electronics applications. Experimental evidence nevertheless indicates that silicene is very different to graphene in terms of its stability, atomic structure, electronic properties and processing. Some of these differences impair the use of silicene for practical application. It is necessary to consider the tendency of silicene to exhibit multiple structural forms, and the role played by strong hybridization with substrates with regard to the electronic band structure of silicene.

There was once doubt over whether graphene could be prepared, because its existence contradicted the Landau-Peierls-Mermin-Wagner predictions that there could exist no stable flat form of such crystals[8]. The so-called flat shape of graphene arises due to microscopic buckling at the finest interatomic scale. There were similar misgivings over the feasibility of silicene and other analogues.

As long ago as 1994, the possibility of corrugation in the silicon analogue of graphite had already been theoretically examined. There was however very little study of silicene before 2009, when silicene with a low-buckle structure was proved to be dynamically stable using *ab initio* calculations. The buckling amounts to 0.454Å, as indicated in the figure at the head of this chapter. In spite of the buckled geometry, silicene exhibits most of the good electronic properties of planar graphene: such as a high Fermi velocity, a Dirac cone (figure 1) and carrier mobility. Silicene moreover offers some distinct advantages. There is better tunability of the band gap, which will facilitate the preparation of an effective room-temperature field effect transistor. It has a much stronger spin-orbit coupling which thus promises the occurrence of a quantum spin Hall effect at experimentally useful temperatures. Silicene permits easier valley polarization, and this aids valleytronics experimentation.

It can be shown however that silicene and other two-dimensional crystals are stable due to transverse short-range displacements of certain atoms. The distortions are small and form various patterns. As the temperature decreases, two transitions - disorder to order and order to disorder - can occur. The ordered state of graphene takes the form of stripes, where carbon atoms are shifted regularly with respect to the plane. The flat graphene, silicene and germanene planes look like microscopic so-called washboards with a wavelength of a few interatomic spacings. Because of an up-down asymmetry in such two-dimensional crystals, deposited on a substrate, a mini band-gap can appear. Mini-gap formation is related to the buckling and to interaction with the substrate.

Molecular dynamics simulations, using the Lennard-Jones potential, of a two-dimensional array of silicon atoms at various temperatures and densities shows that the radial distribution function does not change as the parameters change, and resembles the corresponding (111) surface of the face-centered cubic structure. The liquid phase appears at very high temperatures, suggesting that the system is very stable in the solid phase.[9]

For silicon, which is usually sp^3-hybridized, silicene is an unusual and rare structure. Silicene was theoretically proposed and its structure was subsequently calculated as a possible candidate for nanoribbons of silicon grown onto the anisotropic Ag(110) surface. Since 2012, monolayer silicene sheets with various superstructures have been synthesized on many substrates, including Ag(111), Ir(111), $ZrB_2(00{\bullet}1)$, ZrC(111) and MoS_2. Multilayer sheets have also been grown onto Ag(111). A silicene field effect transistor has recently been fabricated which exhibits the predicted ambipolar Dirac charge transport and which will hopefully form the basis of a silicene-based nano-electronics industry.

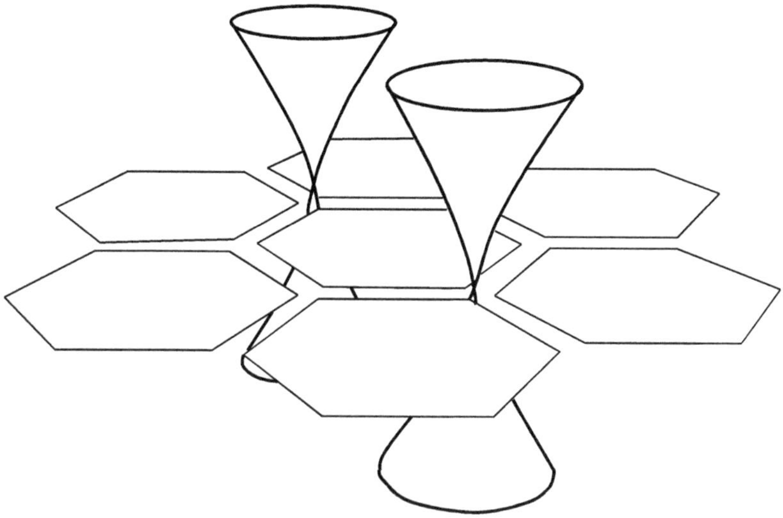

Figure 1. Generic Dirac Cone. Depicted here for graphene, but it is a common feature of the electronic structure of 2-dimensional materials. The electrons here behave like massless Dirac particles which appear in the electronic band structure as gapless excitations with a linear dispersion.

The physical and chemical properties of this peculiar form of silicon, demonstrating the presence of π and π^* bands and giving the so-called Dirac cones at the K corners of the Brillouin zone, later revealed the sp^2-like nature of the valence orbitals of the Si-Si bonds and a strong resistance to oxygen. The two-dimensional structure is also the basis for a variety of potentially useful chemical and physical applications. As noted above, the material has been prepared via the epitaxial growth of silicon in the form of stripes on Ag(001), of ribbons on Ag(110) and of sheets on Ag(111). The nano-ribbons observed on Ag(110) are found, using high definition scanning tunnelling microscopy and density functional theory calculations, to consist of an arched honeycomb structure. Angle-resolved photo-emission experiments performed on these ribbons, along their length, reveal a band structure which is analogous to the Dirac cones of graphene. Unlike silicon surfaces, which react strongly with oxygen, the silicene nano-ribbons are resistant to oxygen attack.

Theoretical studies of stand-alone silicene sheets and nano-ribbons, using tight-binding and density functional theory calculations show that, unlike graphene, silicene sheets are stable only if a small (0.44) degree of buckling is present. The electronic properties of silicene nanoribbons and sheets resemble those of graphene, with massless Dirac fermions that carry charge, and a quantum-spin Hall effect. It is expected that silicene will be quite easy to incorporate into existing silicon-based electronics. Silicene is symmetrically buckled in each of its six-membered units, and this buckling is periodically repeated across the surface. The Raman spectra of silicene clusters, calculated using first-principles density functional theory methods, show that the presence of metal clusters over the silicene units affects the intensity of the buckling modes; which can be enhanced by increasing the number of atoms in the clusters.[10]

The symmetrical buckling of each of the six-membered rings of silicene distinguishes it from graphene and imparts various interesting properties. The pseudo-Jahn-Teller distortion breaks the symmetry and leads to the buckling. In graphene, the two sub-lattice structures are equivalent and therefore do not allow for the opening of the band-gap by an external electric field. Silicene's stronger spin-orbit coupling, compared to that of graphene promises far-reaching applications in spintronic devices, given the electric field- and exchange field-tunable topological properties of silicene.[11]

The origin of the periodic buckled structure of silicene is ultimately due to the pseudo Jahn-Teller instability on each of its planar six-membered rings; in turn attributed to a coupling of the planar D_{6h} ground state with the first b_{2g} excited state, via a b_{2g} vibrational mode. By explicitly calculating the vibronic coupling-constants, by means of a complete study of the pseudo Jahn-Teller effect, it can be shown that the use of vibronic coupling of the ground state, with only one excited state, to explain the planar instability is inconsistent with linear multi-level pseudo Jahn-Teller effect theory. It can also be shown that, in order to ensure consistency, the pseudo Jahn-Teller model should include the next excited state, which is symmetrically coupled to the puckering mode. This can all be shown by means of an analysis of the vibronic instability of the ground state of hexasilabenzene; the basic silicon hydrogenated hexagonal ring-unit which defines silicene.[12]

Because silicon prefers sp^3 hybridization over sp^2, hydrogenation is much easier in silicene. The hydrogenation of silicene to form silicane opens the band-gap and increases the puckering angle. However, silicane will not be extensively considered here because it does not count as an *elemental* analogue of graphene. Lithiation can potentially suppress the pseudo-Jahn-Teller distortion in silicene and could therefore flatten the silicene structure while opening the band-gap. When silicene sheets and nanoribbons are prepared as deposits on substrates such as silver, diboride thin films and iridium, the supporting substrate critically controls the electronic properties of silicone. Matching with an appropriate support may be critical in practical applications of silicene.

When zig-zag edge nanoribbons, doped with a single carbon chain, are studied using a first-principles projector augmented wave potential within the density functional theory framework, it is found that the carbon chain is close to being straight; thus resulting in a transverse contraction near to the carbon chain and thus of the ribbon width. The C-Si and Si-H bonds are typically ionic, while the C-H bond is covalent. Zig-zag nanoribbons doped with a single carbon chain are all metallic, regardless of the position of the carbon chain. These results can be explained in terms of electronegativity differences and the binding force of electrons; due to the atomic-radius difference between the elements.[13]

Generalized gradient approximations of H-terminated nanoribbons with zig-zag or armchair edges, using the first-principles projector-augmented wave potential method within the density functional theory framework, show that the length of the Si-H bond is always 1.50Å. The edge Si-Si bonds are shorter than inner ones of identical orientation; implying a contraction relaxation of the edge silicon atoms. An edge state appears at the Fermi level in broader zig-zag nanoribbons, but does not appear in all armchair nanoribbons due to their dimer Si-Si bond at the edge. With increasing width of armchair nanoribbons, the direct band-gaps exhibit an oscillation behavior and a periodic feature, $\Delta_{3n} > \Delta_{3n+1} > \Delta_{3n+2}$, for a certain integer, n. Charge-density contour analysis shows that the Si-H bond is an ionic bond, due to the relatively larger electronegativity of the hydrogen atom. All types of Si-Si bond display however typical covalent bonding features, although their strength depends upon not only the bond-orientation but also upon the bond position. The larger the deviation of the Si-Si bond orientation from the nanoribbon axis, and the closer the Si-Si bond to the nanoribbon edge, the stronger is the strength of the Si-Si bond.[14] The contraction of the nanoribbon occurs mainly in the width-direction; especially the near edge. Another reason may be a contribution arising from the terminated hydrogen atoms.

Recent first-principles calculations have led to the proposal of a new class of material: so-called Janus silicene. This is silicene which is asymmetrically functionalized with hydrogen and halogen atoms. The formation energies and phonon dispersions indicate that all Janus silicene systems will exhibit a good kinetic stability. When compared with silicane, any Janus silicene system is a direct band-gap semiconductor. The band-gap of Janus silicene can take any value between 1.91 and 2.66eV upon carefully tuning the chemical nature of the adatoms. Elastic straining can further reduce the band gap to 1.11eV at a biaxial tensile strain of up to 10%.[15] Such a material has potential application in opto-electronics; exhibiting as it does a very wide spectral range and high stability under ambient conditions.

1.1 Mechanical Properties

Because it is the mechanical properties of graphene which have aroused the greatest interest, it is perhaps logical to begin here with the mechanical properties of silicene, bearing in mind that most of the results in this field as a whole are calculated and simulated rather than experimentally measured.

Molecular dynamics simulations of the mechanical properties of silicene under uniaxial tensile deformation indicate that the fracture strength and fracture strain of silicene are much higher than those of bulk silicon, even though the Young's modulus of silicene is lower than that of bulk silicon. An increase in temperature significantly decreases the fracture strength and fracture strain of silicene, while an increase in strain-rate slightly enhances them. Brittle fracture behavior is predicted by the simulations[16]. Molecular dynamics simulations of the mechanical properties of polycrystalline silicene involves firstly an annealing process which helps to construct a more realistic modelling structure for the polycrystalline material. A more stable structure is then formed due to the breaking and re-formation of bonds between atoms on the grain boundaries. As the grain size decreases, the efficiency of the annealing process, quantified in terms of the energy change, increases. Biaxial tensile tests of the annealed samples then explore the relationship between the grain size, and properties such as the in-plane stiffness, fracture strength and fracture strain. These indicate that, as the grain size decreases, the fracture strain increases while the fracture strength exhibits the opposite trend. The decreasing fracture strength can be partly attributed to the weakening effect which arises from the increasing area density of defects, which acts as a reservoir of stress-concentrating sites on the grain boundary[17]. The observed crack localization, crack propagation and fracture strength are easily explainable using a defect pile-up model.

Molecular dynamics simulations of hydrogen-functionalized silicene nanosheets (silicane) show that the mechanical properties of the nanosheets are degraded by the functionalizing. Upon comparing the mechanical properties of armchair and zig-zag silicene and silicane nanosheets, it is found that armchair nanosheets have a higher Young's modulus, fracture strength and fracture strain than do zig-zag silicene and silicane nanosheets of the same dimension. As to the fracture pattern of silicane nanosheets, brittle behavior is observed for both armchair and zig-zag forms.[18] The mechanical properties of both armchair and zig-zag silicane exhibit an inverse dependence upon temperature.

The mechanical properties of silicene nanostructures subject to tensile loading have been studied via molecular dynamics simulation. The maximum in-plane stress and the corresponding critical strain of the armchair and the zig-zag silicene sheets at 300K are 8.85 and 10.62, and 0.187 and 0.244N/m, respectively. The in-plane stresses of the silicene sheet in the armchair direction at 300, 400, 500 and 600K are 8.85, 8.50, 8.26 and 7.79N/m, respectively. The in-plane stresses of the silicene sheet in the zig-zag direction at 300, 400, 500 and 600K are 10.62, 9.92, 9.64 and 9.27N/m, respectively.[19] The silicene sheet yielded in the zig-zag direction compared with tensile loading in the armchair direction. Wrinklons and waves were observed at the shear band across the center-zone of the silicene sheet. Another density functional theory analysis of the stress-strain relationship of low-buckle silicene under equiaxial tensile strain, and uniaxial tensile strain along armchair and zig-zag directions, predicts that the ideal strengths for equiaxial tension and armchair uniaxial tensions are 7.59 and 6.76N/m, respectively. In the case of zig-zag uniaxial tension, there exist two ideal strengths (5.26 and 5.29N/m) due to a phase transition of silicene from the original low-buckle structure to a true planar structure.

The influence of defects upon the mechanical properties and failure behavior of silicene sheets, when investigated using molecular mechanics and molecular dynamics methods, is such that the intrinsic strength of the sheets decreases with increasing linear density of vacancies, increasing width ratio of cracks and increasing inflection angle of grain boundaries. The elastic properties of sheets are affected not only by defects but also by their corrugated structure. The fracture failure of sheets with defects usually starts from Si-Si bonds located at the defect edge.[20] The stretching meanwhile tunes the electronic structure of the sheets.

Analytical expressions for the central, k_0, and non-central, k_1, force constants of two-dimensional (graphene, silicene) and three-dimensional (diamond, silicon) structures have been obtained within the previously proposed model of the binding energy of carbon atoms in graphene. The Kleinman internal displacement parameter of the two-dimensional structure has been determined and this shows that the ratio, k_0/k_1, depends only upon the dimension of the structure.[21]

Fully atomistic first-principles molecular dynamics methods predict an elastic stiffness of 50.44N/m for the zig-zag direction and 62.31N/m for the armchair direction; with an ultimate strength of 5.85N/m and an ultimate strain of the order of 18% for monolayer silicene. A weak directional dependence is observed. A predicted effective bending stiffness of 38.63eV/unit-width indicates that its corrugated structure increases the bending rigidity, compared to the similar graphene system[22].

Density functional based tight-binding and reactive molecular dynamics studies of the mechanical properties of suspended single-layer silicene yield Young's modulus values for armchair and zig-zag membranes upon fitting the linear sections of predicted stress-strain curves. Very small differences are found between the values for membranes having differing edge terminations. For armchair membranes, values of 43N/m (0.043TPa nm) and 62.7N/m (0.0627TPanm) are found using various techniques. For zig-zag membranes, values of 43N/m (0.43TPanm) and 63.4N/m (0.0634TPanm) are found, depending upon the technique[23].

Finite-difference calculations of phonon dispersion under three types of tension predict that the failure mechanisms under armchair and zig-zag uniaxial tension are elastic instability. Phonon instabilities occur near to the center of the Brillouin zone and the phonon soft modes for armchair and zig-zag uniaxial tension are longitudinal acoustical modes along the pulling direction, as in graphene. The failure mechanism under equiaxial tension is attributed only to elastic instability, unlike the case of graphene. The phonon instability is dictated by an out-of-plane acoustic mode rather than by the K_1 mode, as in graphene. Phonon behavior also becomes unstable in the planar structure of silicene transformed by zig-zag uniaxial tension. Uniaxial tension along armchair or zig-zag directions cannot open a gap in silicene. On the other hand, it is feasible to tune the gaps introduced by spin-orbital coupling via these two types of tension[24].

Density functional theory analysis of the mechanical stabilities of planar and low-buckled honeycomb monolayer structures of silicon under large strains shows that both types can sustain large strains (≥0.15) under armchair, zig-zag and biaxial deformation. It is found that knowledge of the third-, fourth- and fifth-order elastic constants is essential for the accurate modelling of mechanical properties under strains greater than 0.03, 0.06 and 0.08, respectively. The second-order elastic constants, including the in-plane stiffness, are predicted to increase monotonically with pressure while the Poisson ratio monotonically decreases with increasing pressure. Results on the positive ultimate strengths and strains, second-order elastic constants and the in-plane Young's modulus indicate that both of the above forms of silicon are mechanically stable[25].

First-principles calculations predict that the in-plane stiffness of silicene is much lower than that of graphene. The yield strain of silicene under uniform expansion under ideal conditions is about 20%. Homogeneous strain can introduce a semimetal-metal transition. The semimetallic state of silicene, in which the Dirac cone is located at the Fermi level, can persist only up to tensile strains of 7%, with an almost invariable Fermi velocity. At larger strains, silicene changes into a conventional metal. The work function changes significantly under biaxial strain. Calculations show that strain-tuning is important for nano-electronic applications. The armchair Young's modulus is estimated to be 63.00GPa.nm, while the zig-zag Young's modulus is 51.00GPa.nm.[26] Others have estimated the armchair Young's modulus to be 61.33GPa.nm.[27] Further estimates are shown in table 1.[28] When first-principles and tight-binding calculations are used to investigate the structures of silicene bilayers under isotropic tensile strain, it is found that the strain induces several barrier-less phase transitions. Following the phase transitions, the bilayer structures become planar, similar to AA-stacking graphene bilayers; but combined with strong covalent interlayer bonds. The tight-binding results demonstrate that the silicene bilayer is characterized by an intralayer sp^2 hybridization and an interlayer sp^1 hybridization. With increasing strain, the electronic properties of silicene bilayers change from semiconducting to metallic.[29]

Table 1
Mechanical properties of silicene

Material	Property	Value
armchair	Young's modulus	61.7GPa.nm
zig-zag	Young's modulus	59GPa.nm
armchair	Poisson ratio	0.29
zig-zag	Poisson ratio	0.33
armchair	UTS	7.2GPa.nm
zig-zag	UTS	6.0GPa.nm
armchair	ε_{UTS}	0.175
zig-zag	ε_{UTS}	0.19

Ab initio density functional theory calculations of the properties of multi-layers with various stacking configurations reveal the evolution of those properties as the number, n, of layers ranges from 1 to 10. A monolayer possesses properties which are similar to those of graphene, but the geometrical and electronic properties of multi-layers are markedly different from those of graphene multilayers. Strong interlayer covalent bonding exists between the layers in multilayers of silicene, unlike the weak van der Waals bonding which exists between graphene layers. The interlayer bonding strongly affects the geometrical and electronic structures of the multilayers. As in the case of graphene bilayers, silicene with two different stacking configurations, AA and AB, exhibits linear and parabolic dispersions, respectively, around the Fermi level. Again unlike the case of graphene, the dispersion curves for bilayers of silicene are shifted in the band diagram; a behavior which is attributed to the strong interlayer bonding present. When n is greater than 3, the geometrical and electronic properties of multilayers with four different stacking configurations, AAAA, AABB, ABAB and ABC, can be considered. Results based upon cohesive energy show that all of these multilayers are energetically stable. The three stacking configurations, AAAA, AABB and ABC, which have a tetrahedral coordination, exhibit higher cohesive energies than does the Bernal, ABAB, stacking configuration. This further differs from the case of multilayers of graphene, where ABAB is the lowest-energy configuration. Bands near to the Fermi level in the lower-energy stacking configurations, AAAA, AABB and ABC, correspond to the surface atoms and these surface states are responsible for the semi-metallic nature of these multilayers.[30]

Molecular dynamic simulations of silicene nanoribbons of various widths, under 5 or 10% uniaxial strain at 1 and 300K, predict that they are very ductile, possess considerable toughness and exhibit a very long plastic range before fracture. Under uniaxial strain, the nanoribbon structure also gradually changes from two-dimensional to one-dimensional[31].

Molecular dynamics simulations, in canonical ensemble, of large deformations of two-dimensional silicene nanosheet under uniaxial and biaxial tension predict that the Young's and bulk moduli, and ultimate tensile stress, are lower than those of graphene. The ultimate strain is higher than that of graphene for armchair silicene, unlike the zig-zag form. The Poisson ratio of silicene is also predicted to be greater than that of the carbon counterpart; due to the longer Si-Si bond-lengths and the low buckled honeycomb structure. The bulk modulus is strongly size-dependent and decreases with increasing length of the nanosheet. Under large deformations, the formation of topological defects and silicon chains are predicted. Silicene is moreover noticeably weaker than graphene in the zig-zag direction[32].

Molecular dynamics simulation, using the environment-dependent interatomic potential to describe the interaction of silicon atoms and predict the Young's modulus of defect-free and defective silicene nanoribbons as a function of length and temperature, shows that the moduli of pristine and defective nanoribbons increase with ribbon length in both chirality directions. The Young's modulus of defective nanoribbons exhibits a complex dependence upon the combinations of vacancies. With respect to temperature, the Young's modulus of nanoribbons, with and without vacancy defects, exhibits a non-linear behavior which might well be tailored for a given length and chirality[33].

When the mechanical properties of silicene are investigated using *ab initio* calculations and molecular dynamics simulations using various empirical potentials, the simulation results show that the Young's modulus calculated for bulk silicene is consistent with the *ab initio* calculations. The chirality has a significant effect upon the critical strain and stress of bulk silicene under uniaxial tension. The Young's modulus also depends strongly upon the chirality and size of a nanoribbon, due to edge-effects[34].

Ab initio calculations of the effects of biaxial tensile strain predict that, up to a strain of 5, the Dirac cone remains essentially at the Fermi level while higher strains induce hole-doped Dirac states because of weakened Si-Si bonding. The lattice is expected to be stable up to a strain of 17. Buckling decreases with strain of up to 10, and then increases again; accompanied by band-gap variation. The strain-dependence is similar to that of graphene. Values of the Grüneisen parameter were also calculated (table 2).[35]

Table 2
Grüneisen parameter of silicene as a function of strain

Strain (%)	γ_G
5	1.64
10	1.62
15	1.54
20	1.34
25	1.42

First-principles density functional theory calculations predict that buckled single-layer silicene can transform into planar hexagonal silicene at a critical tensile strain of 0.20. Phonon dispersion analysis suggests that planar hexagonal silicene is stable under tension. The Poisson ratio exhibits a marked strong anisotropy: it increases when stretched in the zig-zag direction, but decreases when strained in the armchair direction[36]. When stretched in the former direction, the Poisson ratio of silicene might attain 0.62.

A hierarchical first-principles investigation of the entangled effects of lattice dimensionality and bond characteristics in lattice dynamics shows that bond-bending contributes negatively to the Grüneisen constant, γ, resulting in a negative acoustic γ. Bond-stretching similarly contributes positively to the Grüneisen constant, resulting in a positive optical γ. Layer-thickening caused by chemical functionalization tends to increase the acoustic γ, due to the increased bond-stretching effect. Bond-weakening caused by chemical functionalization tends to decrease the optical γ, due to the decreased bond-stretching effect. Excitation of the negative-γ modes results in a negative thermal expansion, while mode excitation and thermal expansion compete with each other thermomechanically.[37]

During the synthesis of silicene, structural defects such as edge cracks are likely to be generated and affect its properties. Molecular dynamics simulations of armchair nanoribbons with edge cracks show that the mechanical properties are impaired by edge cracks. Both pristine nanoribbons, and those with edge cracks, exhibit brittle fracture. The crack length plays an important role in determining the critical strain and the fracture strength. Investigations of the effect of strain-rate and temperature upon the mechanical properties of nanoribbon with edge cracks show that an increasing strain-rate increases the critical strain and fracture strength, but decreases the Young's modulus[38]. Low strain-rates also change the expanded directions of cracks. Increasing the temperature can also significantly impair the mechanical properties of cracked nanoribbons.

The presence of a substrate can be expected to have an effect upon the properties. First-principles calculations of the stability and electronic structure of competing silicene phases under in-plane compressive stress, when either free-standing or deposited onto ZrB_2 (00•1) surfaces, reveal a particular (√3 x √3)-reconstruction to be stable on ZrB_2 (00•1) under epitaxial conditions. Unlike the planar and buckled forms of free-standing silicene, all but one of the silicon atoms per hexagon of this planar-like phase, reside in a single plane. Without a substrate, and for a wide range of strains, this phase is energetically less favorable than the buckled one. It is calculated to represent the ground state on the ZrB_2 (00•1) surface, and the atomic positions are found to be determined by interactions with the nearest-neighbor Zr atoms competing with Si-Si bonding interactions imposed by the constraint of the honeycomb lattice[39]. The preparation of silicene-free ZrB_2(00•1) thin films, grown onto Si(111) by Ar^+ ion bombardment, permits study of the spontaneous formation of silicene on their surfaces. Imaging of the ZrB_2(00•1) surfaces using scanning tunnelling microscopy reveals the structures of Zr-terminated and B-terminated ZrB_2(00•1) created by the bombardment.[40] The spontaneous formation of a continuous silicene sheet on a sputtering-induced disordered ZrB_2 surface demonstrates that silicene does not require an atomically-flat crystalline template in order to be stabilized.

In a similar manner, adsorption can be expected to have an effect. First-principles plane-wave pseudopotential method based on the density functional theory, applied to the structure and stability of lithium-adsorbed silicene under biaxial strain show that it largely maintains its original configuration when tensile and compressive strains are applied. The silicene plane bulges towards the lithium atom however when a large compressive strain is applied and the total energy of the corresponding system becomes distinctly lower[41].

Going one stage further, actual reaction with gases can be expected to have an effect. First-principles calculations of fully-oxidized silicene predict that the zig-zag ether-like conformation is the most energetically favorable structure. These nanosheets have interesting elastic characteristics and even exhibit an auxetic behavior, with a negative Poisson ratio. Following oxidation, the semi-metallic nanosheets transform into semiconductors having narrow direct band-gaps. Due to the anisotropic mechanical and electronic properties, the material possesses a high axial intrinsic charge mobility of up to $10^4 cm^2/Vs$; comparable to that of graphene nanoribbons[42].

First-principles investigation of halogenated silicene, SiX (X = F, Cl, Br or I), predicts that such materials exhibit an enhanced stability as compared with silicene and possess a tunable direct gap with a small carrier effective mass. The halogen-dependence of the energy gap can be clearly understood in terms of variations in buckling and bond-energy perturbation based upon orbital hybridization. A negative Poisson ratio is predicted in the case of fluorinated silicene with a boat structure[43].

Although the present book limits itself to discussing elemental analogues, it is nevertheless interesting to see how the properties of silicene and graphene 'blend into' one another. The Harrison bonding-orbital method and the Keating model, when applied to the concentration dependences of the elastic constants of the two-dimensional Si_xC_{1-x} system, predict that the central and non-central force constants and the Grüneisen parameter all exhibit a non-linear behavior over the transition from graphene to silicene. Short-range repulsion plays a non-trivial role in this. It is also found that the elastic constants and Young's modulus change almost linearly over the transition from graphene to silicene[44].

First-principles calculations of the atomic structures and non-linear properties of single-layer graphene, bilayer graphene, single layer silicene and bilayer silicene under equiaxial tension and uniaxial tensions along armchair and zig-zag directions show that there exists a weak Van der Waals interaction between the two layers of bilayer graphene, and that the interlayer distance is not variable with strain for the three types of tension. On the other hand, the interlayer of bilayer silicene involves a covalent bond interaction and the distance decreases with increasing strain for all three types of tension. A continuum description of the elastic response is achieved by expanding the elastic-strain energy-density as a Taylor series in strain; truncated after the third-order term. The in-plane second- and third-order elastic constants of bilayer graphene and bilayer silicene are obtained by fitting the strain energy density versus Lagrangian strain relationships. The results predict that the in-plane stiffnesses of bilayer graphene and bilayer silicene become slightly greater than those of their single-layer counterparts. In spite of the interlayer Si-Si covalent bond between two layers of bilayer silicene, its stiffness is still much lower than that of bilayer graphene and single-layer graphene. The Poisson ratios of bilayer graphene and bilayer silicene remain essentially unchanged as compared with their single-layer counterparts[45].

An important point, to be re-iterated in later sections, is that deformation affects the electronic properties. When stretching in the elastic range, the electronic and magnetic properties can be strongly modified. In particular, the band-gap of a specific armchair nanoribbon is closed under strain and the highest valence and the lowest conduction bands are linearized. In this way, massless Dirac fermion behavior can be realised; even in a semiconducting nanoribbon. Under plastic deformation, the honeycomb structure changes irreversibly and creates a number of new structures and functionalities.[46] By analogy with graphene, cage-like structures and even suspended atomic chains, could be created between two honeycomb flakes.

1.2 Preparation and Structure

As mentioned elsewhere, strong doubts have been raised concerning the possibility of synthesizing silicene on metallic substrates. This is because of the non-negligible interaction between silicon and metal atoms. In order to avoid such growth problems, silicon has been directly deposited onto a chemically inert graphite substrate at room temperature. Atomic force microscopy, scanning tunnelling microscopy and *ab initio* molecular dynamics simulations, reveal the growth of silicon nanosheets in which the substrate/silicon interaction is minimized. Scanning tunnelling microscopic measurements clearly reveal the atomically resolved unit cell and the small buckling of the silicene honeycomb structure. Similarly to the carbon atoms in graphene, each of the silicon atoms has 3 nearest-neighbors and 6 second-nearest neighbors; thus demonstrating its dominant sp^2 configuration.[47] These scanning tunnelling spectroscopic investigations confirm the metallic character of the deposited silicene, and band-structure calculations also indicate the presence of a Dirac cone.

Silicene, like graphene, is now successfully fabricated by growing it epitaxially onto various substrates[48]. Graphene is grown on Ru(00•1) and Pt(111), silicene is grown on Ag(111) and Ir(111) and honeycomb hafnium is grown on Ir(111). Epitaxy on a transition metal substrate is a promising method for producing various two-dimensional atomic crystals. This method is particularly valuable in the case of two-dimensional materials that do not exist in three-dimensional form, such as silicene[49]. The existence of the honeycomb crystal structure[50] and Dirac cone is crucial to observation of the intrinsic properties of silicene. *Ab initio* calculations and two-dimensional nucleation theory studies of the structural evolution of planar silicon clusters, and the nucleation of silicene in the initial stages of its epitaxial growth on Ag(111) surfaces, show that ground-state Si_N clusters ($1 \leq N \leq 25$) on the Ag(111) surface undergo transformation from non-hexagonal plane structures to fully hexagonal ones at N = 22. This is a crucial step in growing high-quality silicene nanosheets. When compared with graphene nucleation on transition-metal surfaces, the low diffusion barrier to silicon atoms and the low nucleation barrier are responsible for the rapid nucleation of silicene on Ag(111) surfaces. These calculations demonstrated that silicene should be synthesized at about 500K in order to reduce the defect density[51]. Similar studies of silicon clusters ($N \leq 24$) on Ag(111) surfaces had shown that, unlike dome-shaped graphene clusters, silicon clusters prefer nearly flat structures with little buckling and are more stable than directly deposited three-dimensional free-standing silicon clusters on silver surfaces. A p-d hybridization between silver and silicon was revealed, as well as sp^2 characteristics in Si_N on Ag(111). Molecular dynamics simulations indicated a high thermal stability of silicene on Ag(111) surfaces, in contrast to that on Rh(111).[52] Experimental techniques, combined with *ab initio* density functional theory calculations, can describe the two-dimensional Si/Ag(111) system in terms of a sp^2-sp^3 form of silicon, characterized by a vertically distorted honeycomb lattice maintained by the constraint imposed by the substrate. The Raman spectrum reflects the multi-hybridized nature of the two-dimensional silicon nanosheets which results from a buckling-induced distortion of a purely sp^2-hybridized structure. The vibrational and electronic properties of the two-dimensional silicon nanosheets are closely related to the buckling arrangement.[53]

In an innovative scheme, first-principles density functional theory has been used to model the evolution of the structural and electronic properties in going from SiC sheet to silicene. The planar configurations of Si-C monolayer systems are essentially retained, apart from an increase in the buckling of the planar structure as the substitution ratio of silicon increases. The band-gap of the Si-C monolayer system gradually decreases as the substitution ratio of silicon atoms increases from 0 to 100%. The energy and type of the band-gap are closely related to the substitution ratio of silicon atoms and to the degree of order of Si-C. Further analysis of the density of states reveals an orbital contribution of silicon and carbon atoms near to the Fermi level.[54]

The structure and thermodynamic properties of free-standing silicene have been studied by means of computer simulation: structures are obtained by cooling from two-dimensional liquid silicon via molecular dynamics simulations using the Stillinger-Weber interatomic potential. The temperature dependences of the total energy, heat capacity, mean ring size and mean coordination number show that silicenization of two-dimensional liquid silicon exhibits a first-order behavior. The evolution of the radial distribution function upon cooling from the melt also shows that solidification occurs in the system. The final configuration of the silicene is then analyzed in terms of coordination, bond-angle, interatomic distance and ring distributions or a distribution of buckling. Two-dimensional visualization of the atomic configurations clearly demonstrates that silicene which is obtained naturally by cooling from the melt exhibits various previously-unreported behaviors. These include the formation of polycrystalline silicene having clear grain boundaries containing defects such as vacancies, Stone-Wales defects or skew rings and multi-membered rings. The atoms of the silicene are mainly incorporated into six-fold rings; forming a buckled honeycomb structure. Buckling is not unique to all of the atoms in the models, although most atoms exhibit the buckling of the most stable low-buckling silicene reported. The buckling distribution is broad and symmetrical.[55]

Armchair- and zigzag-edge nanoribbons, terminated with oxygen or hydroxyl, are revealed - using first-principles methods – to have edges which are rippled in the case of oxygen termination. On one edge, the neighboring Si-O bonds move uniformly right or left from the silicene plane. On one edge of armchair oxygen-terminated ribbons, the neighboring Si-O bonds move right and left, respectively, to result in larger ripple amplitudes. The influence of OH-termination upon the edge is comparatively small, and leads to smaller rippled edges. Electronic structure calculations show that the p_x electrons of oxygen on the rippled edges of oxygen-terminated zig-zag ribbons sp^3-hybridize with the edge silicon atoms to form one more band. The band-gaps of armchair ribbons terminated with oxygen or OH also obey three-family behavior, due to quantum confinement and the effect of the edges. In OH-terminated armchair ribbons, taking account of the new atom chains formed by the hydrogen bonds of the neighboring OH molecules, the band-gaps obey the same hierarchy, $\Delta_{3p}>\Delta_{3p-1}>\Delta_{3p-2}$, as in oxygen-terminated ones.[56]

Also observed have been high aspect-ratio perfectly straight aligned nanoribbons with a pyramidal cross-section. They are multistacks of silicene and, in angle-resolved photo-emission, exhibit a cone-like dispersion of their φ and φ* bands at the $\bar{X}$ point of their one-dimensional Brillouin zone. The Fermi velocity is about 1.3 x 106m/s.[57]

The atomic structure of the multilayer silicene grown on the Ag(111) single crystal surface has been investigated by using low-energy electron diffraction and scanning tunnelling microscopy. The intensity of the low-energy electron diffraction spot is measured as a function of the incident electron energy (I-V curve) and the I-V curve is analyzed using dynamic low-energy electron diffraction theory. The Si(111)(√3×√3)-Ag model closely reproduces the I-V curve, whereas models consisting of the honeycomb structure do not. The bias dependence of the scanning tunnelling microscopic image of multilayer silicene agrees with that of the Si(111)(√3×√3)-Ag reconstructed surface.[58] It is concluded that the multilayer silicene, grown on Ag(111), is identical to the Si(111)(√3×√3)-Ag reconstructed structure.

1.3 Composites

Superlattices can be constructed by alternately stacking graphene and hexagonal silicene in three possible ways: top-, bridge- and hollow-. Top-stacking is the most stable pattern. Although free-standing graphene and silicene are both semi-metals, the results suggest that the graphene and silicene layers in the superlattice both exhibit metallic electronic properties, due to a small amount of charge-transfer from the graphene to the silicene layers. Density functional theory calculations show that weak van der Waals interactions dominate between silicene and graphene, with their intrinsic electronic properties preserved. In particular, interlayer interactions in hybrid silicene/graphene nanocomposites induce tunable p-type and n-type doping of silicene and graphene, respectively, showing that their dopant carrier concentrations can be modulated via the interfacial spacing.[59] The Dirac point of graphene is, moreover, folded to the Γ-point of the superlattice instead of the K-point in the isolated graphene. Such a change in the Dirac point of graphene could lead to a significant change in the transportation properties of the graphene layer. The band-structure and the charge-transfer indicate that the interaction between the stacking sheets in the graphene/silicene superlattice amounts to more than just a van der Waals interaction.[60] Study of the electronic transport properties of such a graphene-silicene bilayer system, using density-functional theory combined with a non-equilibrium Green's function formalism shows that, depending upon the energy of the electrons, transmission can be larger in the combined system, as compared to the sum of the transmissions of separate graphene and silicene monolayers. This effect is related to the increased electron density-of-states in bilayer samples. At some energies the electronic states become localized in one of the layers, resulting in suppression of the electron transmission. The effect of an applied voltage upon the transmission becomes more pronounced in the layered sample, as compared to graphene, due to the larger variation in the electrostatic potential profile.[61] Again using density functional theory, study of the electronic properties of a graphene-silicene bilayer shows that a single layer of silicene binds to the graphene layer with an adhesion energy of about 25meV/atom. This adhesion energy between the two layers closely obeys the well-known $-1/z^2$ dispersion energy law found for two infinite parallel plates. In small flakes of graphene-silicene bilayers with hydrogenated edges, negative charge is transferred from the graphene layer to the silicene layer; producing a permanent and a switchable polar bilayer. In an infinite graphene-silicene bilayer, the negative charge is transferred from the silicene layer to the graphene layer. The graphene-silicene bilayer is expected to be a good candidate material for producing a nanocapacitor possessing piezoelectric capabilities. The permanent dipole of the bilayer can be tuned by applying a perpendicular external electric field.[62]

Electronic structure calculations based upon density functional theory reveal the presence of a noticeable interlayer van der Waal interaction in hybrid bilayers consisting of silicene and a SiC monolayer sheet, with a binding energy of 45meV/atom. This interlayer interaction also leads to significant changes in the nature of the chemical bonding, thus provoking a curving of the previously planar SiC sheet. Modification of the charge distribution at the interface then produces a dissimilar chemical environment for the two sub-lattices of silicene, and opens up the band-gap in the heterogeneous bilayer system. The application of an external electric field linearly modulates the band-gap of the silicene/SiC bilayer, thus indicating its potential usefulness in nanoscale devices.[63]

In silicene/silicane nanoribbons, where silicane is fully-hydrogenated silicene, the minimum energy barrier to the diffusion of hydrogen atoms at the zig-zag and armchair interfaces is 1.54 and 1.47eV, respectively. The diffusion of hydrogen atoms at both interfaces is always endothermic. The minimum diffusion energy barriers for one and two hydrogen atoms on pristine silicene are 0.73 and 0.87eV, respectively. The thermal stability of the composite nanoribbons can therefore be markedly enhanced by increasing the hydrogen diffusion barriers via silicene/silicane interface-engineering. The zig-zag composite nanoribbons remain metallic whereas the armchair composites are semiconducting. The silicene nanoribbons partly determine the metallic or semiconducting nature of the composite.[64]

Density functional theory and empirical molecular dynamics investigations of interlayer interaction, electronic structure and thermal transport of a bilayer heterostructure consisting of silicene and hexagonal boron nitride show that the two layers interact weakly via non-covalent binding. The Dirac cone of silicene is thus preserved, with the Dirac cone point being located exactly at the Fermi level, and with only a small number of electrons being transferred from the hexagonal boron nitride to silicene; suggesting that silicene dominates the electronic transport. Molecular dynamics results demonstrate that the heat current along hexagonal boron nitride is six times that along silicene; suggesting that hexagonal boron nitride dominates thermal transport. This decoupled role of hexagonal boron nitride and silicene with regard to thermal and electronic transport suggests that the BN-silicene bilayer heterostructure is favourable for thermoelectric applications.[65]

High-resolution angle-resolved photo-emission spectroscopy of $CaSi_2$, which has an ABC stacking-sequence of both calcium and silicene layers along the [00•1] crystal axis, reveals - in the band structure - massless Dirac-cone π-electron dispersions at the K(H) point in the Brillouin zone, together with ω-band dispersions. The Dirac point is located about 2eV from the Fermi level, suggesting substantial charge-transfer from calcium atoms to silicene layers. These observations indicate that a sp^2 bonding-framework holds in $CaSi_2$ and produces the massless Dirac-cone state at the K-point in spite of the strongly buckled structure of the silicene layer.[66]

Pseudomorphic lateral heterostructures can be envisaged which are based upon ribbons of silicene and germanene and are the 2-dimensional analogues of conventional 3-dimensional silicon/germanium superlattices and quantum wells. In spite of the considerable (about 4%) lattice mismatch between free-standing silicene and germanene, *ab initio* simulations predict that, for striped 2-dimensional lateral heterostructures made by alternating silicene and germanene ribbons of constant width, the silicene/germanene junction remains pseudomorphic. That is, it maintains lattice-matched edges up to critical ribbon widths which approach some tens of nm.[67] Such critical widths are an order of magnitude greater than the critical thickness measured in 3-dimensional pseudomorphic silicon/germanium heterostructures and the resolution of state-of-the-art lithography; thus permitting the possibility of lithography-patterned silicene/germanene junctions.

1.4 Substrate Interaction

Silicene is predicted to contain massless Dirac fermions and to exhibit an experimentally accessible quantum spin Hall effect. Such properties are not seen in two-dimensional silicon honeycomb lattices prepared on metallic substrates, where the crystal and hybrid electronic structures of the so-called epitaxial silicene phases are strongly affected by the substrate and are thus different to those predicted for isolated two-dimensional structures. The prospect of exploiting such low-dimensional silicon π materials makes the development of silicene an important challenge.[68]

Epitaxial silicene on atomically flat ultra-thin Ag(111) films, prepared by low-temperature deposition onto silicon substrates followed by annealing at room temperature, exhibits patchy (4 x 4), (√13 x √13), (√7 x √7) and (2√3 x 2√3) domains in regions that are a monolayer lower than pristine Ag. The post-deposition annealing temperature plays an important role in determining the resulting morphology.[69] On first-layer silicene containing the above domains, two types of (√3 x √3) silicene domain, (√3 x √3)-α and –β, form in the second layer. A bias voltage-independent Moiré pattern is observed for (√3 x √3)-α silicene and a bias voltage-dependent standing wave pattern is observed for (√3 x √3)-β silicene. The latter silicene has a dispersion relationship that can be fitted using a linear function, although the resultant Fermi velocity is twice as large as the expected value.[70] Scanning tunnelling microscopy and spectroscopy show that the coupling strength of (√3 x √3)-silicene with a Ag(111) substrate is variable in different regions. This helps to explain the structure and electronic properties of silicene.[71] Applying the same geometrical approach for the (2√3 x 2√3)R30° structure, to the (√13 x √13)R13.9° type-II structure, an atomic model for the silicene layer has been proposed which is based upon a periodic relaxation of the strain epitaxy. This creates periodic arrangements of perfect areas of (√13 x √13)R13.9° type-II structure, surrounded by defected areas.[72] A detailed analysis of published experimental results, obtained using scanning tunnelling microscopy and low-energy electron diffraction, indicates good agreement with this geometrical model.

Low-temperature scanning tunnelling spectroscopy and Raman spectroscopy of the electronic and vibrational properties of silicene on Ag(111), at coverages of up to one monolayer at 230 to 250C, reveals the strong contribution made by silver states to the measured density of states around the Fermi level.[73] The Raman spectra evolve as a function of the sub-monolayer coverage, giving rise - at one-monolayer coverage - to peaks that are characteristic of chemical bonds with distorted sp^3 hybrid orbitals. Such properties account for the electronic transparency of the silicene/Ag(111) interface.

First-principles calculations were used to investigate systematically various phases of silicene observed on Ag(111) substrates: $(2 \times 2)_{silicene}/(7 \times 7)_{Ag(111)}$, $(7 \times 7)_{silicene}/(23 \times 23)_{Ag(111)}$, $(7 \times 7)_{silicene}/(13 \times 13)_{Ag(111)}$. The original Dirac cone of silicene is 1.5 to 1.7eV below the Fermi level, and is destroyed by band hybridization between silicene and silver in all of the examined phases. Silicene synthesized on Ag(111) substrates could thus not preserve its electronic properties.[74]

Grazing-incidence X-ray diffraction measurements and density functional theory calculations have been used to determine precisely the atomic positions of the silicon and silver atoms which form the (4x4)Si/Ag(111) silicene reconstruction. Remarkable agreement is found between the experimental structure factors and the theoretically predicted ones. The results confirm the previously proposed honeycomb structure, with the buckling of the silicon atoms being equal to 0.77Å.[75] The silver substrate atoms are also relaxed, leading to a non-negligible elastic deformation energy of the substrate which is equal to $43mJ/m^2$.

Four coincidence lattices, (7 x 7) on (13 x 13), (3 x 3) on (4 x 4), (2 x 2) on (7 x 7) and (7 x 7) on (23 x 23) are related to silicon coverage, biaxial strain and other conditions. Scanning tunnelling microscopic images, calculated for the optimized atomic geometries, agree with those observed experimentally. Destruction of the original honeycomb symmetry, and the strong adsorbate-substrate interaction, significantly influence the electronic structure. Peeled-off silicene sheets show conical linear bands with small gaps. The band edges of the (3 x 3) on (4 x 4) geometry cannot be explained in terms of gap-opening between Dirac cones, for symmetry reasons.[76]

Synchrotron-based soft X-ray emission and absorption spectroscopy, and first-principles full-potential density functional theory simulations predict a novel tristable AA-stacked bilayer structure that can explain the (√3 x √3)R30° honeycomb topography that is commonly observed in scanning tunnelling microscopy and non-contact atomic force microscopy. The electronic structure of this epitaxial bilayer is similar to that of epitaxial monolayers on Ag(111). That is, it is metallic and exhibits significant interaction with the underlying substrate. Soft X-ray spectroscopy instead suggests that, during multilayer growth, a majority of the epitaxial silicon reverts to a bulk-like state.[77] The experimental data cannot be explained by proposed models which consider only the buckling of silicon atoms. Multi-layer silicene on Ag(111) is actually a thin film of bulk-like silicon, terminated by a Si(111)-(3 x 3)-Ag surface.[78] The (√3 x √3)R30 silicene honeycomb monolayer on Ag(111) undergoes a phase transition to two types of mirror-symmetric boundary-separated rhombic phase below 40K. First-principles calculations reveal that weak interactions between silicene and Ag(111) drive the unusual spontaneous buckling in monolayer silicene; forming two energy-degenerate and mirror-symmetric (√3 x √3)R30 rhombic phases in which a linear band dispersion near to the Dirac point and a significant gap opening (150meV) at the Dirac point appear. The low transition barrier between these two phases permits them to interchange via a dynamic flip-flop motion, thus resulting in the (√3 x √3)R30 honeycomb structure observed at high temperatures.[79] For silicene film consisting of a few layers, synchrotron radiation photo-electron spectroscopy reveals a cone-like dispersion at the Brillouin zone centre due to band-folding. The π*- and π-states meet at about 0.25eV below the Fermi level, providing clear evidence of the presence of gap-less Dirac fermions[80]; silicene had long been predicted theoretically to be a buckled honeycomb arrangement having an electronic dispersion which resembles that of relativistic Dirac fermions[81]. Convincing evidence arose from both structural and electronic properties of epitaxial sheets on a Ag(111) substrate.

Silicene nanoribbons grown onto a Ag(110) substrate, when studied using reflection electron energy-loss spectroscopy as a function of the electron-beam incidence angle, α, show that spectra taken at the silicon K-absorption edge (1.840keV) reveal the presence of two distinct loss structures which are attributed to the transitions, 1s→ π and 1s→ σ, according to their intensity dependence upon α. Such behavior, when compared with than of carbon, indicates sp^2-like hybridization of silicon valence orbitals in silicene nanoribbons just as it does for the carbon atomic bonds of graphene.[82]

First-principles simulations predict that silicene is a gap-less semiconductor, due to sp^2-hybridization of the silicon atoms. The weak overlapping of their $3p_z$ orbitals leads however to a very reactive surface and results in a more energetically stable semiconducting surface upon the adsorption of chemical species. It is therefore further predicted that silicene, when inserted into a graphite-like lattice such as an ultra-thin AlN stack preserves its sp^2-hydridization and therefore its graphene-like electronic properties.[83] The application of a perpendicular electric field can drive silicene into a gap-less state, characterized by two nearly fully spin-polarized Dirac cones, due to both relatively large spin-orbital interactions and to inversion symmetry breaking. It can be argued that, since intervalley scattering from non-magnetic impurities is highly suppressed by time-reversal symmetry, the behavior should be effectively single Dirac cone-like. Numerical calculations demonstrate that there is no significant back-scattering from a single impurity that is non-magnetic and unit-cell uniform; indicating a stable delocalized state. This view is further confirmed by a scaling of the conductance for disordered systems using the same type of impurities.[84]

The use of angle-resolved photo-emission spectroscopy to map the entire Brillouin zone of (√3 x √3)R30° reconstructed epitaxial multilayer silicene islands, growing on top of the first (3 x 3) reconstructed silicene wetting layer on Ag(111) substrates, reveals V-shaped linear dispersions. These are related to the π- and π*-bands of massless quasiparticles in multilayer silicene at the Brillouin zone centre and at all of the centres of the (√3 x √3)R30° Brillouin zones in the extended scheme, due to folding of the Dirac cones at the K and K̄' points of the (1 x 1) silicene Brillouin zone. The obtained Fermi velocity, of about 3 x 10^5m/s is deemed to be highly promising for potential silicene-based devices.[85]

The discovery of (4 x 4) silicene formation on Ag(111) raises the question of whether silicene maintains its Dirac fermion character, as does graphene, on a supporting substrate. Angle-resolved photo-emission spectroscopy and density-functional theory calculations show that the Dirac cone fades out.[86] The special coupling between Dirac fermion and lattice vibrations (i.e. electron-phonon coupling) in silicene layers on an Ag(111) surface, when probed using *in situ* Raman spectroscopy, is found to be significantly modulated by tensile strains resulting from the lattice mismatch between the silicene and the substrate and by charge-doping from the substrate. Special phonon modes corresponding to two-dimensional electron-gas scattering at the edge sites of the silicene have been identified.[87]

The two-dimensional epitaxial silicon layers, (4 x 4), (√13 x √13) and (2√3 x 2√3), exhibit differing electronic and vibrational properties.[88] Peaks in the experimental Raman spectra are attributed to the vibrational modes of the silicene layers. The positions and behaviors of the Raman peaks with respect to the excitation energy are a fundamental tool for investigating and distinguishing the various silicene phases on Ag(111).[89]

In (4 x 4) silicene, the silicon atoms are arranged so as to form a buckled honeycomb structure wherein six silicon atoms out of eighteen silicon atoms in the unit cell are displaced vertically[90]. The displacements lead to a vertical shift of the substrate silver atoms, thus indicating non-negligible coupling at the interface between the silicene layer and the substrate. The interface coupling significantly modifies the electronic structure of (4 x 4) silicene. No Landau level sequences are observed using scanning tunnelling spectroscopy with magnetic fields applied perpendicularly to the sample surface. Density functional theory calculations show that the π and π^* bands derived from the silicon $3p_z$ are hybridized with the silver electronic states, leading to a drastic modification of the band structure and an absence of Dirac fermion features together with two-dimensionality of the electronic states. The strong coupling at the interface causes symmetry-breaking of the (4 x 4) silicene and to a consequent disappearance of Dirac fermion features.[91] The interaction between the silicene and the underlying silver substrate is strong, and therefore silicene on silver surfaces loses its Dirac fermion character.[92] High-resolution images of silicene grown on Ag(111) reveal five phases: (4 x 4)-α, 4 x 4-β, -α, -β and -γ. For each of the phases.[93]

When silicene sheets are grown onto a close-packed surface, Ag(111), by the direct condensation of a silicon atomic flux onto the single-crystal substrate under ultra-high vacuum conditions, a highly-ordered silicon structure arranged as a honeycomb lattice is formed which in fact consists of two sub-lattices which occupy positions of different heights (by 0.02nm); indicating possible sp^2-sp^3 hybridization.[94] When silicene is grown onto Ag(111) substrates, it has a nearest-neighbour distance of about 0.23nm; in agreement with the figure predicted theoretically for free-standing silicene. The above work however reported the observation of silicene sheet which is strongly compressed: giving a Si-Si distance of only 0.19nm. This raised the question of whether silicene can support a large compressive strain. The apparent contradiction was explained here by a careful analysis which revealed that the pure Ag(111) surface can mimic a honeycomb structure, and thus easily be taken to be a strained silicene layer. It was concluded that there is no evidence for the existence of such strongly compressed silicene layers.[95]

Low-energy electron microscopy permits direct visualization of the formation and stability of silicene layers on a Ag(111) substrate. Theoretical calculations question the stability of this graphene analogue, in that silicene layers are intrinsically unstable with respect to the formation of sp^3-like hybridized bulk-like silicon. Irreversible formation of the latter structure is triggered by thermal silicon adatoms that are created by the silicene layer itself. The same instability prevents the formation of a fully closed silicene layer, or thicker bilayers. This was once thought to make the large-scale fabrication of silicene layers on silver substrates unlikely[96]. Low-energy electron diffraction, low-temperature scanning tunnelling microscopy and density functional theory calculations show that the structure of silicene is controlled by the substrate temperature during deposition. It is possible to synthesize bilayer silicene by growing silicene on silicene/Ag(111) substrates[97].

Silicene nanosheets grown epitaxially onto Ag(111) have, after carefully tuning the growth parameters, complex two-dimensional silicon structures, with corrugated domains. Local modifications of the density of states are observed throughout the reconstructed silicene domains and these are attributed to symmetry-breaking induced by interactions with the silver lattice; as in the case of graphene.[98]

Early scanning tunnelling microscopy and *ab initio* calculations based upon density functional theory were used to study self-aligned silicon nanoribbons, on Ag(110), which had honeycomb graphene-like structure. The silicon honeycomb structure was clearly observed by scanning tunnelling microscopy, while the density functional theory calculations confirmed that the silicon atoms spontaneously adopted the new silicon structure.[99] Straight 1.6nm-wide silicene nanoribbons on Ag(110), arranged in a one-dimensional grating with a pitch of 2nm, had a honeycomb geometry. Angle-resolved photo-emission revealed quantum confined electronic states of one-dimensional nature. The silicon band dispersion along the direction of the nanoribbons suggested a behavior which was analogous to the Dirac cones of graphene on various substrates.[100]

Scanning tunnelling microscopy was used to make real-time studies of the growth of silicene on Ag(111) at 200 to 533K. The growth mode depends strongly upon the deposition temperature. At 300K and above, formation of silicene results from exchanges between surface silver atoms, and silicon atoms which are inserted into the substrate top layer. Density functional theory calculations confirm that silicon insertion is thermodynamically favored. At deposition temperatures greater than 400K, ordered structures are observed as soon as the silicene domains are sufficiently large. With increasing coverage, disordered and dotted phases progressively transform into stable (13 x 13)R13.9° and (4 x 4) structures[101]. The growth of silicene can be accompanied by the release of about 65% of the surface silver atoms from the silicon-covered areas. Silicon islands develop on the silver terraces and silicon strips at the silver step edges, progressively forming the above ordered (4 x 4), (13 x 13) R13.9° and dotted phases. The displaced silver atoms group together so as to develop additional bare silver terraces growing round the silicon islands from the pristine silver step edges. This indicates a strong interaction between silicon and silver atoms, with a large modification of the silver substrate beneath the surface layer. This contradicts the model of a silicene layer interacting weakly with the unreconstructed silver substrate, and strongly suggests that the structure of silicene on Ag(111) corresponds either to a Si-Ag surface alloy or to a silicon plane covered with silver atoms.[102] In the case of the (13 x 13)R13.9° superstructure grown on Ag(111), scanning tunnelling microscopy reveals two co-existing atomic arrangements having with two different orientations relative to the Ag(111) surface. Density functional theory calculations, with and without including van der Waals interactions, indicate corrugated silicon atomic positions for both orientations; implying a significant interaction with the Ag(111) surface. The electronic structures of the silicene and Ag(111) surfaces are such that new interface states emerge close to the Fermi level.[103]

Investigation of the silicene structure forming the (4 x 4) superstructure on Ag(111), using low-energy electron diffraction, shows that the silicon atoms form a buckled honeycomb structure which fully matches the structure as optimized using density functional theory calculations. The Si-Si bond-lengths range from 2.29 to 2.31Å; close to the upper limit of Si-Si double bonds, and shorter than the bond length in bulk diamond. Buckling of the silicene layer causes displacement, of silver atoms in the first substrate layer, by 0.3Å perpendicularly to the surface.[104] Rocking-curve analysis shows that silicene has a buckled structure, with a spacing of 0.83Å between the top and bottom silicon layers. The distance between the bottom silicon layer in the silicene and the first silver layer is 2.14Å. These results agree with the theoretically predicted values to within an error of ±0.05Å.[105]

On Ag(100), Ag(110) and Ag(111), when various structures are obtained merely by varying the silicon coverage and/or its atomic arrangement, all of the structures show that silicene is buckled, with a Si-Si nearest-neighbor distance of between 2.28 and 2.5Å. Due to the buckling of the silicene sheet, the apparent (lateral) Si-Si distance can be as low as 1.89Å. It is also found that, for a given coverage and symmetry, various scanning tunnelling microscopic images can be observed which correspond to structures that differ only by a translation.[106]

The first non-contact atomic force microscopic study of silicene grown on a silver surface, obtained by combining non-contact atomic force microscopy with scanning tunnelling microscopy, revealed both ($\sqrt{13}$ x $\sqrt{13}$)R13.9° and (4 x 4) superstructures. In the case of the widely observed (4 x 4) structure, the non-contact atomic force microscopic image is very similar to that recorded using scanning tunnelling microscopy. The structure which is resolved by non-contact atomic force microscopy is compatible with only one of two silicon atoms being visible; indicating marked buckling of the silicene honeycomb layer.[107]

Total-energy electronic structure calculations, within density-functional theory, of ultra-thin atomic layers of silicon on Ag(111) surfaces reveal several distinct stable silicene structures, ($\sqrt{3}$ x $\sqrt{3}$), (3 x 3) and ($\sqrt{7}$ x $\sqrt{7}$), as the silicon thickness increases from a monolayer to a quadruple layer. The calculated transition barriers imply the occurrence of flip-flop motion at low temperatures. Stable (2 x 1) π-bonded chain and (7 x 7) dimer-adatom-stacking fault Si(111)-surface structures on Ag(111) clearly show a cross-over of silicene and silicon structures for multilayer silicon on silver surfaces. There is also an absence of Dirac states in multilayer silicene on Ag(111), due to covalent interactions of the silicene/Ag interface and Si/Si interlayer. A new state is instead found, near to the Fermi level, composed of π-orbitals located on the surface layer of (√3 x √3) multilayer silicene. This satisfies hexagonal symmetry and exhibits linear energy dispersion. By examining the electronic properties of (2 x 1) π-bonded chain structures, it is found the surface-related π-states of multilayer silicon structures are sufficiently robust on silver surfaces.[108] At multi-layer coverage, only one structural phase has been reported: (√3 x√3)R30°, but with various domain orientations resulting from the structural properties of the initial wetting monolayer.[109] *Ab initio* density functional theory, assuming van der Waals interactions for the formation of a silicene bilayer on the Ag(111) surface, shows that (23 x 23)R30° bilayer silicene on a (19 x 19)R23.4 silver substrate is the most stable. Calculated scanning tunnelling microscopic images indicate instead a (3 x 3)R30 symmetry. This translational symmetry, the associated lattice spacing and the height of the top-most monolayer agree with experimental findings. The band structure of the complete adsorbate system exhibits conical linear bands near to the Fermi level, due to hybridization of the adsorbate and substrate states.[110] The (3 x 3)R30 reconstruction of multilayers grown on Ag(111), studied using low-energy electron diffraction, is shown to be induced by a surfactant layer of silver atoms during growth and is not due to pristine Si. The silicon film exhibits the optical characteristics of bulk cubic diamond-like silicon, thus indicating that it is not formed of any other allotropic phase of silicon which does not exist in nature.[111]

When grown on Ag(111), silicene is characterized by a number of critical parameters: the substrate temperature plays a key role in determining the morphology during growth, but an unexpectedly important role is also played by the post-deposition annealing temperature. This governs the self-organization of silicene domains, even in the sub-monolayer coverage regime, and the consecutive transitions between silicene having differing periodicities. These temperature-driven phase transitions can be used to select the desired majority silicene phase.[112] The deposition of one silicon monolayer onto a (111) silver substrate at 150 to 300C gives rise to a mixture of (4 x 4), (2√3 x 2√3)R30° and (√13 x √13)R13.9° superstructures whose ratios again depend strongly upon the substrate temperature. Detailed analysis of low-energy electron diffraction patterns and scanning tunnelling microscopic images shows that all these superstructures are generated by a quasi-identical silicon single layer having a honeycomb structure with various rotations relative to the substrate. The scanning tunnelling microscopic images are explained in terms of the positions of the silicon atoms relative to the silver atoms. A complete analysis of all possible rotations of the silicene layer predicts the existence of a (√7 x √7)R19.1° superstructure.[113]

Closer study of the (2√3 x 2√3)R30° phase reveals an unconventional rotated ("2√3 x 2√3") pattern yielding a complicated set of split diffraction spots. Scanning tunnelling microscopy results disclose a Ag(111) surface that is homogeneously covered by ("2√3 x 2√3") silicene having an additional quasi-periodic long-range ordered superstructure. The homogeneous coverage by ("2√3 x 2√3") silicene permits an angle-resolved photo-electron spectroscopic study which describes a silicene band structure in unprecedented detail.[114] On the basis of a detailed re-analysis of experimental results for the (2√3 x 2√3)R30° structure, obtained using scanning tunnelling microscopy and low-energy electron diffraction, it has lately been proposed that a silicene layer consists of periodic arrangements of perfect areas of (2√3 x 2√3)R30°, surrounded by defected areas. A generalization of the model explains experimental observations such as deviation of the average direction, Moiré patterns and an apparent overall disorder.[115]

For the (3 x 3) reconstructed first silicene layer, non-contact atomic force microscopy indicates the same lateral arrangement of silicene atoms as does scanning tunnelling microscopy. Both non-contact atomic force microscopy and scanning tunnelling microscopy furnish a unifying interpretation of the second layer (√3 x √3)R±30 structure. There is support for the possible existence of less-stable ~2%-stressed (√7 x √7)R±19.1 rotated silicene domains in the first layer.[116]

First-principles investigations and scanning tunnelling microscopy show that (presumably) van der Waals packed multilayered silicene sheets transform spontaneously into diamond-structured bulk silicon film due to strong interlayer coupling. In contrast to the marked surface reconstruction on conventional Si(111), multilayered silicene prepared by bottom-up epitaxy on Ag(111) exhibits a nearly ideal flat surface with only weak buckling. Without the use of silver surfactants, (√3 x √3) honeycomb patterns emerge as a result of the dynamic fluctuations of mirror-symmetric rhombic phases; similar to monolayer silicene. The weak relaxation permits the existence of novel surface states with Dirac linear dispersion.[117]

The electronic structure of (3 x 3) reconstructed silicene on (4 x 4) Ag(111) is such that new states emerge due to strong hybridization between silicene and Ag. It is possible to distinguish them clearly from states arising from the Dirac cone of free-standing silicene or from the sp bands of bulk Ag. Silicene atoms contribute to the Fermi level, leading to similar scanning tunnelling microscopy patterns as observed below or above the Fermi level.[118] Study of (3 x 3) silicene on the (4 x 4) Ag(111) surface shows that no true covalent bonds are formed between silicene and the silver surface, but there is an overlap between the charge densities of the lower silicon atoms and the nearest silver atoms. Charge difference calculations show that clear charge-reorganisation takes place upon bringing together the silicene and the silver substrate. Bader charge calculations indicate that the upper silicon atoms are slightly positively charged, while the silver surface plane carries a negative charge. An electrostatic interaction thus exists between the upper silicon atoms and the underlying silver atoms, offering a possible explanation for silver buckling.[119] Total-energy electronic-structure calculations of monolayer and bilayer silicene on Ag(111) surfaces suggest that the experimentally observed (3 x 3) form, which is usually argued to be a monolayer, is really a bilayer. This conclusion is based upon accurate density-functional calculations in which the local density approximation, the generalized-gradient approximation and the van der Waals density-functional approximations to the exchange-correlation energy are all carefully examined. It is found that structural tristability exists for (3 x 3) bilayer silicene. The calculated energy barriers between the three stable structures range from 7 to 9meV per silicon atom; indicating possible flip-flop changes between them. Flip-flop motion between two of the three structures produces honeycomb structures in scanning tunnelling microscopic images. Changing between all three produces the (1 x 1) structure. Electron states, which effectively obey the Dirac equation in free-standing silicene, couple with substrate silver orbitals due to bond formation, and shift downwards deep into the valence bands. This is a common feature of all stable or metastable silicene layers on a Ag(111) substrate.[120] Some first-principles calculations show that (3 x 3) reconstructed silicene is made up of dumb-bell units of silicon atoms, arranged in a honeycomb pattern. This model gives excellent agreement with experimentally reported lattice constants and scanning tunnelling microscopic images. The (3 x 3) reconstruction is governed mainly by interaction between silicon atoms, with a weak influence arising from the silver substrate.[121]

Silicene loses its Dirac fermion characteristics, due to substrate-induced symmetry breaking, when synthesized on Ag(111). No Landau-level sequences appear in the tunnelling spectra in a magnetic field, and density functional theory calculations show that the band structure is greatly modified by hybridization between the silicon and silver atoms. This was the first direct demonstration of a lack of Dirac fermions in a single-layer honeycomb lattice, due to appreciable symmetry breaking.[122] Dirac electrons are also absent, near to the Fermi level in all of the observed stable structures, due to buckling of the silicon monolayer and mixing between silicon and silver orbitals. This again was the first theoretical investigation that confirmed the absence of Dirac electrons in silicene, due to strong interactions with the substrate. It was proposed that a BN substrate or an hydrogen-processed silicon surface were good candidates for preserving Dirac electrons in silicene.[123] Total-energy electronic-structure calculations of silicene on Ag(111), hexagonal BN and hydrogen-processed Si(111) surfaces show that, on the commonly-used Ag(111) surfaces, there are several stable and metastable structures with (4 x 4), (√13 x √13) and (2√3 x 2√3) periodicities with respect to the (1 x 1) Ag(111) lateral cell within a total-energy difference of 70meV per silicon atom. In all of the stable and metastable structures, the silicene is so buckled that the π-state re-hybridizes with the σ-state, leading to the π+ state, and the linear energy dispersion peculiar to Dirac electrons then disappears in several cases which are associated with opening of the energy gap. The substantial mixing of the π+ state thus generated, with the states of the silver atoms in the substrate, converts the π+ state to the mixed π+ state and makes the state shift downwards or upwards; eventually annihilating Dirac electrons near to the Fermi level. The absence of Dirac electrons, caused in this way, is common to all of the stable and metastable structures of silicene on Ag(111). The interaction between the π+ and substrate orbitals should be weak enough to preserve Dirac electrons and also be enough to keep the system stable. Hexagonal BN and hydrogen-processed Si(111) surfaces instead are such that silicene on those substrates is sufficiently stable; with the binding energy being comparable to, or twice that, of graphite and enough to preserve Dirac electrons near to the Fermi level.[124]

Low-temperature scanning tunnelling microscopy and spectroscopy studies of the (√3 x √3)R30 phase of silicene on Ag(111) surface reveal the existence of Dirac fermion chirality via the observation of -1.5 and -1.0 power-law decays of quasi-particle interference patterns. Unlike the trigonal warping of the Dirac cone in graphene, it is found that the Dirac cone of silicene is hexagonally warped. This is further confirmed by density functional calculations, and is attributed to the unique superstructure of silicene. The (√3 x √3)R30 phase is an ideal system within which to investigate the unique Dirac fermion properties of silicene.[125]

By unfolding the band structure in the Brillouin zone of a super-cell to that of a primitive cell, followed by projecting onto silver and silicene sub-systems, it is demonstrated that the Dirac cone in silicene on Ag(111) is destroyed. The linear dispersions observed in angular-resolved photo-emission spectroscopy and scanning tunnelling spectroscopy come from the silver substrate and not from silicene.[126]

Five distinct epitaxial silicene morphologies on Ag(111), when examined using density functional theory and soft X-ray spectroscopy at the silicon $L_{2,3}$ edge, imply that hybridization with the Ag(111) substrate causes the monolayers to become strongly metallic. The results further imply that epitaxial silicene on Ag(111) does not possess the

Dirac cone electronic structure that is characteristic of free-standing silicene and graphene sheets.[127]

Silicene clearly forms a variety of phases on Ag(111) substrate, but their structures are still not fully understood.[128] The real-time formation of silicon single layers on Ag(110) and Ag(111) has recently been investigated using a combination of scanning tunnelling microscopy, grazing-incidence X-ray diffraction, optical differential reflectance spectroscopy and density functional theory. This showed that the reactivity of the silver substrate with the silicon atoms is important. The formation of silicon nanoribbons on Ag(110) is accompanied by missing-row reconstruction of the silver surface, while the removed silver atoms form additional islands. The mechanism of silicon island formation on Ag(111) involves the insertion of single silicon atoms into the silver top layer, and exchange with silver atoms. The released silver atoms then diffuse to step edges and form new silver fingers. This strong interaction between silicon and silver atoms, and the possibility of alloying, calls into question the true structures of the various silicon phases which are observed on Ag(110) and Ag(111) and are usually attributed to silicene single layers.[129] In other recent research, the local structure around the silicon atoms of silicene, deposited onto Ag(110) and Ag(111), has been determined using extended X-ray absorption fine-structure spectroscopy at the silicon K-edge. The results show that the silicon atoms are not situated in a flat honeycomb network that is locally buckled, but in a structure which mimics the double Si(111)-plane of crystalline silicon; having almost identical first and second interatomic distances (2.35, 3.83Å) on a regularly buckled geometry. The results also indicate that the silver atoms are at a well-defined distance from the silicon ones.[130]

In spite of the extensive research on silicene nanoribbons, their atomic structure is still not fully understood. Recent research suggests that the structure of silicene nanoribbons on Ag(110) is that of an armchair having reconstructed edges. Pronounced quantum-well states are observed which are due to the confinement of quasi-particles perpendicular to the nanoribbon and can be explained using the elementary one-dimensional particle-in-a-box model.[131]

Angle-resolved photo-emission measurements of silicene(3x3)/Ag(111), reveal the presence of six pairs of Dirac cones, located on the edges of the first Brillouin zone of Ag(111). This is in sharp contrast to the expected six Dirac cones centered at the K points of the primary silicene(1x1) Brillouin zone. Analysis shows clearly that this unusual Dirac cone structure is not tied to pristine silicene alone, but originates from the combined effects of the silicene(3x3) and the Ag(111) substrate.[132] This thus reveals an unique type of Dirac cone which is generated by the interaction of two different constituents. The observation of Dirac cones in silicene/Ag(111) thus offers an unique opportunity to investigate unusual quantum phenomena and to explore applications which are based upon 2-dimensional silicon systems.

The substrate effect upon the electronic properties of silicene is important for its practical applications. First-principles calculations for silicene on two representative inert substrates, hexagonal boron nitride monolayer and SiC(00•1) show that the silicene-substrate interaction energies range from 0.067 to 0.089eV per silicon atom; typical of van der Waals interaction. The characteristic Dirac cone is preserved in silicene on a h-BN monolayer or on hydrogenated silicon-terminated SiC(00•1). Silicene becomes metallic when it is placed on an hydrogenated carbon-terminated SiC(00•1) surface. This is explained in terms of the work functions for the silicene and the substrates.[133] First-principles calculations predict that hexagonal boron nitride with a flat surface is an ideal substrate for silicene. Van der Waals interactions hold the silicene and h-BN together, forming silicene/BN moiré superstructures. The latter open a band-gap of about 30meV at the Dirac point of silicene at equilibrium distance. The band-gap is almost independent of the rotation angle between the two lattices, but can be tuned by changing the interlayer spacing. The high Fermi velocity of silicene is preserved in these superstructures.[134]

Studies of ballistic electron transport through silicene barriers of width, d, and height, U, in the presence of an exchange field, M, and a normal electric field, E_z, show that away from the Dirac point the spin- and valley-resolved conductances, as functions of U, exhibit resonances. When close to it however there is a pronounced dip that can be transformed into a transport gap by varying E_z. The charge conductance, g_c, changes from an oscillatory to a monotonically decreasing function of d beyond a critical value of E_z, and this can be used to enable electric-field controlled switching. The field, M, also splits each resonance of g_c into two spin-resolved peaks. The spin polarization near to the Dirac point increases with E_z or M and becomes almost perfect above certain values of these parameters; a similar result holding true for double barriers.[135]

First-principles calculations predict the stability of silicene, as a buckled honeycomb lattice, on passivated substrates of group-IV (111)-(1 x 1) surfaces. The weak van-der-Waals interaction between the silicene and these substrates does not prevent its linear bands forming Dirac cones at the Brillouin zone corners. Only very small fundamental gaps are opened around the Fermi level.[136]

Although Dirac-like electronic properties are predicted for silicene, some experiments reveal substrate-induced complex buckling that could destroy such features. First-principles calculations suggested that gallium sulfide nanosheet would be an appropriate substrate for silicene, which can form commensurate Si/GaS heterosheets that are analogous to graphene/BN systems. In these heterosheets, the characteristic Dirac-like band feature of linear dispersions is retained and, due to the intrinsic electric field, a sizable band-gap is opened at the Dirac point, which could be further tuned by use of a bias voltage or straining.[137] First-principles methods and density-functional theory studies of silicene on GaAs(111)-(1 x 1) reveal stable structures on both the arsenic- and gallium-terminated GaAs surfaces. The structures have a honeycomb-hexagonal geometry. The cohesive energies of silicene on both the arsenic and gallium sides of GaAs surfaces are comparable to those of their bulk structures to those on Ag(111) substrates. The corresponding binding energies range from 0.56 to 1.37eV per silicon atom. This is 10 times higher than the usual van der Waals interaction, reflecting a covalent interaction between the silicene and GaAs surfaces. Band-structure calculations show that such a covalent interaction leads to an absence of Dirac electrons for silicene on GaAs surfaces. Intercalated hydrogen atoms chemically bond to the GaAs surface, and the silicene shifts upwards from the GaAs surface by 2.50 to 3.86Å where a covalent van-der-Waals interaction transition occurs between silicene and the GaAs surface. The distances between silicene and hydrogen atoms are 30 and 8% larger than the atomic-radius sum of silicon and hydrogen on arsenic-terminated and gallium-terminated GaAs surfaces, respectively. The interaction between silicene and hydrogen on the arsenic-terminated GaAs surface is thus obviously weaker than the typical covalent interaction. On the gallium-terminated GaAs surface, it is comparable to the typical covalent interaction. This difference is attributed to the difference in electronegativity between arsenic and gallium atoms. Hydrogen intercalation recovers the Dirac electrons on arsenic-terminated GaAs(111), due to the weaker Si-H interaction. It does not do so on gallium-terminated GaAs(111), due to the stronger Si-H interaction.[138]

First-principles calculations of a superlattice of silicene and hexagonal boron nitride demonstrate that the interaction between the layers of the superlattice is very small. Quasi free-standing silicene can therefore be achieved. In particular, the Dirac cone of silicene is preserved. Due to the wide band-gap of hexagonal boron nitride, the superlattice offers the same characteristic physical phenomena as free-standing silicene. The combined effect of intrinsic spin-orbit coupling, and an external electric field, induces a transition from a semimetal to a topological insulator and thence to a band insulator[139].

Calculations of the electronic structure of silicene, and of the stability of its weakly buckled honeycomb lattice, in an external electric field oriented perpendicular to the monolayer show that the electric field produces a tunable band-gap in the Dirac-type electronic spectrum; but the gap is suppressed, by a factor of about 8, by the high polarizability of the system. Under low electric fields, interplay between the tunable band-gap - specific to electrons on a honeycomb lattice - and Kane-Mele spin-orbit coupling induces a transition from a topological to a band insulator. At much higher electric fields, the material becomes a semimetal.[140]

The band-gap of silicene first closes and then opens again in an increasing electric field, indicating a topological phase transition from the quantum spin Hall state to the band insulator state. Due to the relatively large atomic spacing of silicene, which reduces the band-width, the electron-electron interaction is quite strong and cannot be ignored.

A so-called silicene nanodisk is silicene with a closed edge. In the case of the analogous graphene nanodisk, the number of zero-energy modes is given by the lower bound on the inequality for a bipartite system, $\xi_0 \geq 2|N_A - N_B|$; where ξ_0 is the number of zero-energy modes and $N_{A/B}$ is the number of sites in the A/B sub-lattice. They are standing-wave states. In the case of a silicene nanodisk however, they propagate around the nanodisk as a helical current because silicene is a topological insulator due to the spin-orbit interaction. The counting-rule of zero-energy states and a low-energy theory of zig-zag triangular silicene have been deduced. The validity of a bulk-edge correspondence in a nanodisk with a rough edge has been revealed by calculating the local probability amplitude and current.[141]

In order to study how non-local Coulomb interactions affect the phase diagram of correlated electron materials, a correlated lattice model with non-local interactions can be approximated by using an effective Hubbard model with only on-site interactions, U*. The effective model is defined by the Peierls-Feynman-Bogoliubov variational principle. The local part of the interaction, U, is reduced according to: $U^* = U - \bar{V}$, where $\bar{V}$ is a weighted average of non-local interactions. It can then be shown that the non-local Coulomb interaction can decrease the effective local interaction by more than a factor of two over a wide doping range.[142] The Hubbard interaction, intrinsic spin orbital coupling and electric field can be taken into consideration by using a tight-binding model, with the phase diagram of silicene being carefully investigated at the mean field level. When the magnitudes of the two mass terms, produced by the Hubbard interaction and electric potential, are close the intrinsic spin orbital coupling flips the sign of the mass term at either K or K′ for one spin and leads to the emergence of the spin-polarized quantum anomalous Hall state.[143]

First-principles calculations of the stability of silicene layers as buckled honeycomb lattices on chlorine-passivated Si(111) and clean $CaF_2(111)$ surfaces show that van der Waals interactions between silicene and the inert substrate stabilize the adsorbate system without destroying the silicon p_z-derived linear bands which form Dirac cones at the Brillouin zone corners. Only small gaps (3 and 52meV) are opened up.[144]

First-principles calculations of the energetics and electronic properties of silicene and $MoSe_2$ heterobilayers show that the silicene is bound to the $MoSe_2$ substrate with a binding energy of -0.56eV per silicon atom; indicating weak interaction between the two layers. An almost linear band dispersion behavior in the silicene, with a sizable band-gap, is obtained due to the variation of on-site energy induced by the $MoSe_2$ substrate. The band-gaps and electron effective mass of such heterobilayers can be effectively tuned by using the interlayer spacing, an external electric field or straining.[145]

The electronic properties of silicene on $NiSi_2$(111) and on hydrogenated $NiSi_2$(111) (H:$NiSi_2$), as well as hydrogenated silicene (H:silicene) on $NiSi_2$(111), when deduced from first principles, are such that hydrogenation of $NiSi_2$ lowers the binding energy between silicene and the substrate, resulting in partial decoupling of those properties. Relaxed silicene on H:$NiSi_2$ exhibits a small band-gap opening of 0.14eV. Silicene on H:$NiSi_2$ has a calculated electron effective mass of $0.08m_0$, and a Fermi velocity of 0.39 x 10^6m/s; values which are similar to those for free-standing silicene. The H:silicene on $NiSi_2$ retains its indirect band structure and density-of-states, as compared to free-standing H:silicene. The band-gap of H:silicene on $NiSi_2$ is 1.97eV; similar to the free-standing H:silicene band-gap of 1.99eV. Hydrogenation may therefore be a viable method for decoupling a silicene layer from a $NiSi_2$(111) substrate and tuning its electronic properties[146].

The interaction of silicene with the hybrid substrate, β-Si_3N_4(00•1)/Si(111), having an exposed β-Si_3N_4(00•1) surface, has been studied by means of density functional calculations; including van der Waals interactions. It was demonstrated that electron transfer occurs from the substrate to the silicene layer, and that the electron transfer can be modulated by doping the substrate. The β-Si_3N_4/silicene interaction appears to be sufficiently strong to ensure an adequate adsorption stability, and the electronic states of the substrate and adsorbate remain decoupled; permitting potential exploitation of the special electron-mobility properties of the silicene layer.[147]

First-principles simulations of the interaction of silicene with (00•1)ZnS surfaces show that charge-transfer at the silicene/(00•1)ZnS interface leads to opening of an indirect energy band-gap of about 0.7eV in the silicene. The nature of the band-gap, indirect or direct, and its magnitude can be controlled by means of an external electric field. The energy-gap is predicted to become direct for electric fields greater than about 0.5V/Å, and then to decrease almost linearly with increasing electric field.[148]

The study of (√3 x √3)-reconstructed epitaxial silicene, especially the case of silicene on ZrB_2(00•1) thin films, shows that the structural and electronic properties are highly interlinked and leave their signature on the chemical states of individual silicon atoms as revealed in core-level photo-electron spectra as well as in the valence electronic structure and low-energy interband transitions.[149] Analysis of high-resolution silicon 2p photo-electron and near-edge X-ray absorption fine structure spectra shows that core-level excitations of epitaxial silicene occur on ZrB_2(00•1) thin films, and the data are characteristically different to those from sp^3-hybridized silicon. The lower silicon 2p binding energies and the low onset in the near-edge X-ray absorption fine structure, as well as the occurrence of satellite features in the core-level spectra, are attributed to screening by low-energy valence electrons and interband transitions between π-bands, respectively. Analysis of the observed silicon 2p intensities, related to chemically distinct silicon atoms, indicates the presence of at least one previously unidentified component. Its presence suggests that the observation of stress-related stripe domains in scanning tunnelling microscopy images is intimately linked to the relaxation of silicon atoms away from energetically unfavourable positions.[150] First-principles calculations reveal the origin of the large-scale stripe pattern of epitaxial silicene on the ZrB_2(00•1) surface, as observed by scanning tunnelling microscopy. In the absence of stripes, the (3 x 3)-reconstructed atom-thick silicon layer exhibits a so-called zero-frequency phonon instability at the M-point. In order to avoid a divergent response, the relevant phonon mode then triggers the spontaneous formation of a new phase having a particular stripe pattern; offering a means for both decreasing the atomic surface density and the total energy of silicene on that substrate. This mechanism manages epitaxial strain and is expected to be more common in two-dimensional epitaxial materials which exhibit a small lattice mismatch with the substrate.[151]

Silicene forms an ordered phase on ZrC(111)-(2 x 2), and phonon dispersion relationships were measured here for the first time using high-resolution electron energy loss spectroscopy. The atomic structure and phonon dispersions were calculated using *ab initio* density functional theory. Nearly all of the surface-localized phonon modes were accurately reproduced and the dipole-active phonon modes in particular corresponded closely to the peaks in the specular high-resolution electron energy loss spectrum. The optimized atomic structure was one of distorted silicene on Zr-terminated ZrC(111), in which the ($\sqrt{3}$ x $\sqrt{3}$) unit cell of Si_6 coincides with the (2 x 2) unit of the substrate ZrC(111). In the unit cell, one silicon atom sits on the on-top site of the Zr substrate. Three atoms are placed at the bridge sites of the substrate, and two other atoms are located on the 3-fold hollow sites above the third layer (C) and above the fourth layer (Zr). This structure is observed on several metal substrates.[152]

The phonon dispersion of silicene on ZrB_2(00•1), measured using high-resolution electron energy loss spectroscopy and compared with *ab initio* density functional theory calculations, shows that, for silicene having the periodicity of the substrate, the most stable silicene structure, which is similar to the so-called planar-like model, closely reproduces the observed phonon modes. The calculated electronic structure suggested that silicene is metallic on $ZrB_2(00\bullet1)$ and semiconducting on ZrC(111).[153]

It is challenging to reproduce the angle-resolved photo-electron spectra of epitaxial silicene by means of first-principles calculations and to resolve issues which are related to the structural configuration of silicene on $ZrB_2(00\bullet1)$ surfaces. By representing the band structure in a large Brillouin zone associated with a single silicon atom, it is found that the imaginary part of the one-particle Green's function models the spectral weight observed in angle-resolved photo-electron spectra. By varying the in-plane lattice constant, the results of density functional theory calculations and angle-resolved photo-electron spectra converge to the so-called planar-like phase and predict the orbital character of electronic states in the vicinity of the Fermi level.[154]

Low-temperature scanning tunnelling spectroscopy and density functional theory calculations reveal the electronic properties of epitaxial silicene on $ZrB_2(00\bullet1)$ thin film, grown onto Si(111). The positions of the silicon atoms, and thus the localization of the valence and conduction states, are deducted by comparing the spectra with calculated local densities of state. A strong contribution of the p_z-orbitals of specific atoms, to those states, indicates the π-character of the conduction and valence bands. A clear correlation between hybridization of the orbitals of the silicon atoms and buckling is revealed.[155]

The evolution of the electronic structure and the structural stability of epitaxial silicene on $ZrB_2(00\bullet1)$ thin films, exposed to potassium atoms, was studied using angle-resolved photo-electron spectroscopy and low-energy electron diffraction. Potassium adsorption leads to charge donation to the silicene lattice, accompanied by partial filling of a formerly unoccupied π^*-band and by increasing hybridization between the diboride surface state and the lower branch of the back-folded π-band. The results confirm that, before potassium adsorption, the interactions at the silicene/substrate interface are rather weak.[156]

The predictive power of first-principles calculations, based upon van der Waals density functional theory, is limited with regard to silicene on a MoS_2 substrate. Only one functional gives reasonable agreement with experimental results on structural properties. When using other functionals, the interlayer interaction is underestimated or the charge redistribution at the interface is not described correctly, so that the predicted electronic structure is qualitatively incorrect.[157]

Experimental studies have reported that several types of silicon monolayer structure form on metal surfaces. These typically have the topology of a honeycomb bonding network, but differ in terms of corrugation and surface coverage. Using first-principles calculations, atomic-scale mechanisms have been identified which underlie the appearance of various configurations as the coverage increases during silicon deposition on silver. The key factor is that any extra silicon adatoms which land on preformed silicene films can be incorporated into the honeycomb network and form bonds with underlying silver atoms. The corrugation profile then changes, giving rise to various overlayer geometries. The same set of mechanisms controls the appearance of silicene films on an iridium substrate.[158]

According to first-principles calculations the absence of the Dirac cone, due to strong band-hybridization, is a common feature of epitaxial silicene on metal substrates such as iridium, copper, magnesium, gold, platinum, aluminium and Ag. The destroyed Dirac cone can be effectively restored with linear or parabolic dispersion by intercalating alkali metal atoms between the silicene and the metal substrate. The properties of silicene can thus be studied without its further transfer from the metal substrate.[159] When silicene is deposited onto metals, the general approach is that (111) surfaces are used in the case of face-centered cubic crystals, while (00•1) surfaces are used in the case of hexagonal crystals. The Ca(111)-(1 x 1) substrate is found to be the most promising candidate. Silicene adsorption on calcium-functionalized Si(111)-(1 x 1) and (2 x 1) surfaces has also been studied. The (1 x 1) substrates lead to overlayer silicene of hexagonal symmetry, and Dirac cones. The Dirac points are however below the Fermi level, and small energy gaps are opened up. Strong lattice relaxation occurs in the case of (2 x 1) surfaces and only rudimentary conical linear bands remain visible.[160] The electronic properties of epitaxially grown silicene on metal surfaces are influenced by the strong silicene-metal interactions. This has prompted studies of the growth of multi-layer silicene.[161]

First-principles molecular dynamics calculations show that a honeycomb silicene monolayer is stable on the Al(111)-(4 x 4) surface and has a binding energy which is comparable to that on the Ag(111)-(4 x 4) surface. A new ordered silicene monolayer structure, consisting of 3-, 4-, 5- and 6-sided polygons can also form on the Al(111) surface. This new silicene monolayer, so-called polygonal silicene, has an almost flat monolayer structure and exhibits a free electron-like density of states thus indicating high electronic conduction. This result suggests that silicene can take monolayer forms, other than a honeycomb one, by varying the choice of substrate.[162]

Low-energy electron diffraction and scanning tunnelling microscopy show that silicene, fabricated on Ir(111) substrates, has a ($\sqrt{7}$ x $\sqrt{7}$) superstructure with respect to the substrate lattice that coincides with the ($\sqrt{3}$ x $\sqrt{3}$) superlattice of silicene. First-principles calculations confirm that this is a $(\sqrt{3} \times \sqrt{3})_{silicene}/(\sqrt{7} \times \sqrt{7})_{Ir(111)}$ having a buckled configuration. The calculated electron localization function shows that the silicon adlayer on the Ir(111) substrate has two-dimensional continuity. This provides a method for fabricating high-quality silicene, and helps to explain the formation of buckled silicene sheets[163]. The buckled $(\sqrt{3} \times \sqrt{3})_{silicene}/(\sqrt{7} \times \sqrt{7})_{Ir(111)}$ configuration for silicene on Ir(111) and its electronic structure suggest that, for silicene-graphene layered structures, a silicene layer can be constructed beneath a graphene layer by intercalation. This offers the possibility of integrating silicene with graphene for applications in nano electronics.[164] Density functional theory shows that Ir(111) preserves the two-dimensional character of silicene but significantly distorts its structure away from the trigonal one expected for isolated silicene. The electronic structure of silicene is strongly hybridized with that of Ir(111), so that silicene on Ir(111) loses its linear band dispersion around the Fermi level; giving rise to a metallic band structure. Silicene/Ir(111) exhibits however a hidden linear-dispersive band which is related to the linear-dispersive conduction band of isolated silicene.[165]

Quantum size effects are useful for controlling interactions between silicene and a substrate, and permit the growth of silicene having tailored electronic properties. In the case of silicene on ultra-thin Pb(111) layers, the binding energy and the Dirac bands can easily be tuned via the quantum well states of the substrate. A novel mechanism can also protect the Dirac electrons from the influence of the substrate. This involves a special arrangement of some of the silicon atoms in silicene.[166] First-principles density functional theory calculations of silicene deposited onto a Pb(111) surface indicate several possible superstructures, with differing scanning tunnelling microscopy topographic images. All of the structures feature a low binding energy and a very small charge transfer, and thus interact weakly with the substrate. Linear band dispersion around the K-points of the Brillouin zone therefore survives and the bands have mainly the 3p character of silicene, with very little contribution from the 6p states of Pb. This suggests that lead may be the best metal substrate for silicene.[167]

Ab initio density functional theory calculations of the geometrical and electronic structures of silicene on Cu(111), covered with a monolayer of hexagonal boron nitride, show that silicene with a regularly buckled configuration is stabilized on h-BN layers which are stacked commensurately with the Cu(111) substrate. The electronic band structure projected to the silicon $3p_z$ orbital clearly exhibits a band crossing similar to a Dirac cone emerging in the band structure of free-standing buckled silicene. This contrasts with silicene on Cu(111), where the Dirac fermion features disappear completely due to strong interactions at the interface. The h-BN monolayer prevents silicene from interacting with the underlying Cu(111) substrate.[168]

It has indeed been challenging to obtain free-standing silicene by using mechanical exfoliation, as in the case of graphene. Fully atomistic simulations, with an effective potential for a silver substrate derived from first-principles calculations, were used to investigate possible means for peeling silicene by mechanical force alone. It was found that the peeling direction is critical for successfully exfoliating silicene, and that peeling at a 45° angle to the substrate is the most efficient method for detaching silicene[169].

The free-standing silicene-like structures obtained using metallic substrates tend to be very different to standalone buckled silicene. For instance, the Dirac cone is destroyed due to lattice distortion, and interaction with the substrate. When using graphene bilayers as a scaffold, the buckled hexagonal arrangement of silicene between the graphene layers is very similar to that theoretically predicted for standalone buckled silicene. The confinement between the flat graphene layers results in a planar clustering of silicon atoms with little buckling, which is energetically unfavorable in a vacuum[170]. It is very weakly van der Waals coupled to the graphene layers, with a graphite-like interlayer distance of 3.42Å and with no lattice distortion. These stacked layers are stable at well above room temperature[171].

When density functional theory is used to study the adsorption of a model organic molecule, benzene, on (3 x 3) silicene on a (4 x 4) Ag(111) surface, the calculations show that dispersion interactions have to be taken into account in order to describe the system properly. The adsorption energy is calculated by means of semi-empirical dispersion-corrected density functional theory. Charge-density and electron-localization function maps indicate that the molecule is chemisorbed on the silicene surface by means of two Si-C covalent bonds. In agreement with charge density difference calculations, two C-C double bonds are formed in the benzene molecule, which adopts a butterfly configuration. The silicene lattice is slightly deformed due to the benzene adsorption, but the Si-Si distance remains the same as in bare silicene on Ag(111). Charge-transfer from the top silicon atoms to both the molecules and the silver substrates occurs. Covalent functionalization of silicene is possible.[172]

The possibility of forming a two-dimensional ordered structure by the deposition of silicon onto metal surfaces was investigated by studying the atomic and electronic structures of the Si-(19 x 19)R23.4°/Pt(111) surface reconstruction using a set of experimental surface-science techniques and theoretical calculations. The theory gave very good agreement with experimental results and decisively confirmed that this phase is a surface alloy consisting of Si_3Pt tetramers arranged in a form which resembles a twisted Kagome lattice. This result was interpreted as meaning that the formation of silicene or germanene on Pt(111) and other transition-metal surfaces was unlikely.[173]

Using density functional theory calculations of the interaction between phenyl-modified silicene nanosheets, adhesion-energy curves between the nanosheets have been compared for various van der Waals density functionals and the density functional theory D2 Grimme method. It is shown that there is a weak attraction between the sheets for small separations, and that it is stronger when van der Waals forces are included. Without the latter forces, the interaction is negligible and occurs at a much larger separation. This highlights the need to include such forces when modelling nanosheets. Of the van der Waals methods, the so-called optB88 functional gives the highest interaction energy while the Grimme functional gives the lowest. The separation at which the nanosheets adhere more strongly varies from 10.04 to 11.24Å, as measured by the distance between the silicene layers. When the modified nanosheets are brought to separations as small as some 8Å, the phenyl groups on the bottom of one nanosheet can fit between the phenyl groups on the top of the adjacent nanosheet. This permits some degree of π-π interaction between the phenyl groups. The band-gap can be modified by compressing the nanosheets together while retaining a small attraction between them. There is also a change from a direct, to an indirect, band-gap.[174]

General guidelines have been proposed for choosing substrates which stabilize the monolayer honeycomb structure of silicene while simultaneously preserving the Dirac states. In addition to having a strong binding energy to the monolayer, a suitable substrate should be a large-gap semiconductor possessing a work function such that the Dirac point lies in the gap and far from the substrate states when their bands align. The concept has been illustrated by performing first-principles calculations for silicene on the Al-terminated (00•1) surface of Al_2O_3. Overlaid monolayers on Al-terminated Al_2O_3(00•1) retain the main structural profile of the low-buckle honeycomb structure via a binding energy which is comparable to that between silicene and Ag(111). An unfolded band structure, derived using the k-projection method, reveals that a gapped Dirac cone is formed at the K point due to structural distortion and interaction with the substrate.[175] A gap of 0.4eV for supported silicene suggests that it may have potential applications in nano-electronics.

1.5 Thermoelectricity

First-principles density functional and linear response studies have been made of two-dimensional silicon structures, together with nanoribbons of various widths. Also investigated was the possibility of nano-structuring the systems by mixing silicon and germanium[176]. It is found that the room-temperature figure-of-merit of these systems could attain 2.5.

Non-equilibrium Green's function method investigations of the thermoelectric performance of zig-zag and armchair silicene nanoribbons with central or edge defects show that, for perfect silicene nanoribbons, the maximum thermoelectric figure of merit,

ZT_M, decreases monotonically with increasing width while the phononic thermal conductance increases linearly. For various types of edges and defects, with increasing defect numbers in the longitudinal direction, ZT_M increases monotonically while the phononic thermal conductance decreases. Compared with zig-zag silicene nanoribbons, defected armchair silicene nanoribbons possess a higher thermoelectric performance. This is due to a higher Seebeck coefficient and a lower thermal conductance. An approximately 2.5 times enhancement in thermoelectric figure of merit values is obtained in armchair silicene nanoribbons with edge defects[177]. First-principles investigation of the thermospin properties of a two-probe model based upon zigzag-edge nanoribbons, as compared with odd-width ribbons, shows that the spin Seebeck coefficient of even-width ribbons is clearly higher at room temperature. This can be attributed to a nearly perfect symmetry of the linear conductance gap having a different spin index with respect to the Fermi level induced by the differing parity of the wave functions. The corresponding charge Seebeck coefficient is almost zero so that, when a thermal bias is present in even-width ribbons, an almost pure spin current is obtained. At the same time, the spin polarization of the current approaches infinity.[178]

Density functional theory, combined with non-equilibrium Green's functions, has been used to study crystal reconstruction in the presence of monovacancies, divacancies and linear vacancies in a zig-zag silicene nanoribbon having transversal symmetry. A 5-atom ring is formed in monovacancies, a 5-8-5 ring structure is formed in divacancies and a 8-4-8-4 ring structure is formed in linear vacancies. The linear conductance becomes strongly spin-dependent when the transversal symmetry is broken by vacancies; especially when they are located on the ribbon edges. The giant magnetoresistance can be smeared out by asymmetrical vacancies, and a single spin negative differential resistance can appear in the presence of linear vacancies and asymmetrical monovacancies or divacancies. A strong spin Seebeck effect is expected to appear at room temperature in zig-zag silicene nanoribbons with linear vacancies.[179]

Theoretical studies of the spin thermoelectric properties of zig-zag nanoribbons of two-dimensional hexagonal crystals, with respect to the thermoelectric properties of quasi-stable magnetic states and the effect of Coulomb and spin-orbit interactions on the topological edge states and transitions between topological insulator and conventional gap insulator states, show that the interplay of both interactions also has a significant effect upon the transport and thermoelectric characteristics of the nanoribbons. The spin-orbit interaction also determines the in-plane magnetic easy-axis. Nanoribbons with ferromagnetic alignment of the edge moments exhibit spin thermoelectricity in addition to the conventional phenomenon[180].

Electrons in two-dimensional crystals having a honeycomb lattice-structure possess a valley degree of freedom in addition to charge and spin. In the case of valley-resolved thermoelectric transport through a magnetic silicene superlattice, since spin is coupled to the valley, this permits the coexistence of the insulating transmission gap of one valley and the metallic resonant band of the other. This results in a strong valley polarization. The latter oscillates with the barrier strength; with its magnitude greatly enhanced by the superlattice structure. Controllable fully valley polarized transport, and an on/off switching effect in the conductance spectra, are also obtained. The spin- and valley-dependent thermopowers can be controlled by the on-site potential difference between the A and B sub-lattices and the Fermi energy, and are enhanced by the superlattice structure.[181]

Ab initio calculations of the thermoelectric and spin thermoelectric properties of silicene nanoribbons having bare, mono-hydrogenated and di-hydrogenated edges show that asymmetrical structures, in which one edge is bare or di-hydrogenated and the other is mono-hydrogenated, have a ferromagnetic ground state and exhibit remarkable conventional and spin thermoelectric properties. Strong enhancement of both the conventional and spin thermoelectric efficiencies results from a very specific band structure of these nanoribbons, where one spin channel is blocked due to an energy gap while the other spin channel is highly conductive. Nanoribbons with one edge bare or di-hydrogenated and the other di-hydrogenated, are antiferromagnetic in the ground state. The corresponding spin channels are thus equivalent, and only conventional thermoelectric effects can occur in these nanoribbons.[182]

The valley and spin resolved thermoelectric transport in a normal/ferromagnetic/normal silicene junction is such that, due to coupling between the valley and spin degrees of freedom, thermally-induced pure valley and spin currents can be found. The magnitudes and signs of these currents can be manipulated by adjusting the ferromagnetic exchange field and local external electric field. Fully valley- and/or spin-polarized currents are also found. In a similar manner to the currents arising from the band structure symmetry, tunable pure spin and/or valley thermopowers with a zero charge counterpart are generated.[183]

In silicene nanoribbons doped with aluminium and P, and with both antiparallel and parallel orientations of the edge magnetic moments it is found that, in the former case, suitably arranged impurities can lead to a net magnetic moment and therefore also to spin-dependent transport and spin thermoelectric effects. In the latter case, spin thermoelectric effects also occur in the absence of impurities. *Ab initio* calculations, based upon density functional theory, of the spin density distribution and transport parameters show that the spin thermopower can be considerably enhanced by the impurities. Such enhancement appears in certain regions of chemical potential.[184] *Ab initio* numerical methods based upon density functional theory, when applied to silicene nanoribbons doped with cobalt adatoms in the edge positions, show that the adatoms considerably influence the corresponding electronic band structure, resulting in two different magnetic states of the nanoribbons. One of these is ferrimagnetic with semiconducting transport properties, and the other is ferromagnetic and of metallic character. In the former state, the magnetic moments at the two edges are antiparallel while, in the latter case, they are parallel. Numerical calculations show that an appreciable spin thermopower can occur in the ferrimagnetic state due to a non-zero band-gap. In the ferromagnetic state, both the conventional and spin thermopowers are quite moderate. Overall, the ferrimagnetic state arising due to the presence of impurities at the edge positions appears to be the most relevant from the point of view of spin thermoelectric phenomena.[185] Non-equilibrium Green's function methods and non-equilibrium molecular dynamics simulations show that silicene nanoribbons are structurally stable if the edge atoms are passivated by hydrogen. Also, those having armchair edges usually exhibit a much better thermoelectric performance than do their zig-zag counterparts. The room-temperature figure-of-merit of armchair silicene nanoribbons exhibits a width-dependent oscillating decay, whereas it decreases slowly with increasing ribbon-width in zig-zag nanoribbons. There is also a strong temperature dependence of the thermoelectric performance of these nanoribbons[186]. These calculations indicate that, after optimizing the doping level and applied temperature, the figure-of-merit of the nanoribbons could attain 4.9.

A hydrogen-terminated zig-zag silicene nanoribbon heterojunction can be prepared and, by applying a temperature difference between the source and the drain, spin-up and spin-down currents can be generated which flow in opposite directions with almost equal magnitudes. This shows that the thermal spin current dominates carrier transport, while the thermal electron current is much suppressed. By modulating the temperature, a pure thermal spin current can be achieved. A thermoelectric rectifier and a negative differential thermoelectric resistance can also be obtained in the thermal electron current. A thermal magnetoresistance which can attain infinity is also obtained. By analyzing the spin-dependent transport characteristics, a phase diagram depicting various spin caloritronic phenomena can be constructed (figures 2 and 3).[187]

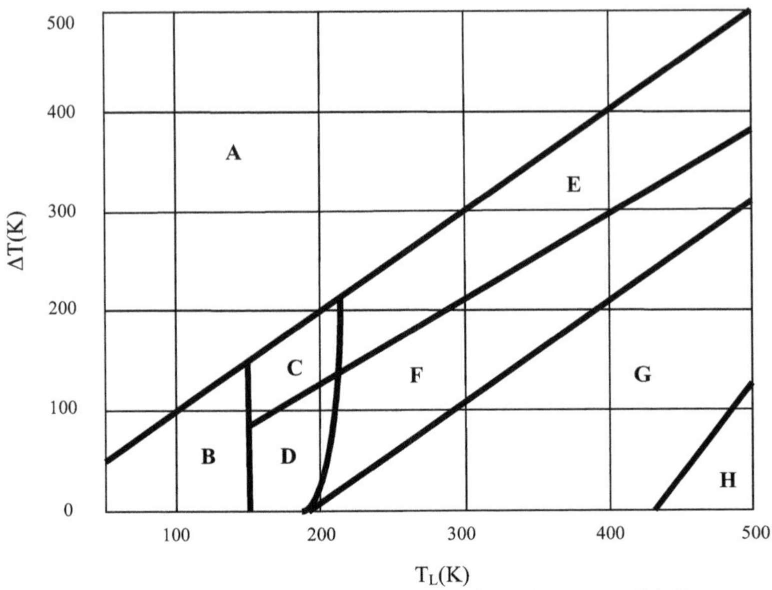

Figure 2. Phase diagram for total spin (I_s) and total current (I_c) flows in a 4SiZNR-H/4SiZNR-H_2 zig-zag silicene nanoribbon heterojunction, A: non-existent, B: $I_s = 0$, $I_c = 0$, C: $I_s > 0$, $I_c < 0$ with thermal rectification, D: $I_s > 0$, $I_c < 0$ opposite flow directions, E: $I_s > 0$, $I_c > 0$ with thermal rectification, F: $I_s > 0$, $I_c > 0$, same flow direction with negative differential thermal resistance effect in I_c, G: $I_s > 0$, $I_c > 0$, flow in same direction, H: $I_s > 0$, $I_c > 0$, same flow direction with negative differential thermal resistance effect in I_c.

The thermoelectric performance of silicene nanoribbons can be markedly improved by introducing nanopores and by tuning the Fermi energy. It is predicted that electronic thermoelectric figures-of-merit, ZT_e, of up to 160 are achievable, provided that the Fermi energy is located some 100meV above the charge neutrality point. Inclusion of the effect of phonons yields a full figure-of-merit, ZT, of 3.5. The sign of the thermopower S can be varied so as to yield values as high as +/-500aμV/K. One method of tuning the Fermi energy is to dope the silicene with a strong electron donor or a strong electron acceptor. Adsorbed layers of the former type can increase ZT_e to 3.1, and this is insensitive to temperatures of 100 to 400K.[188]

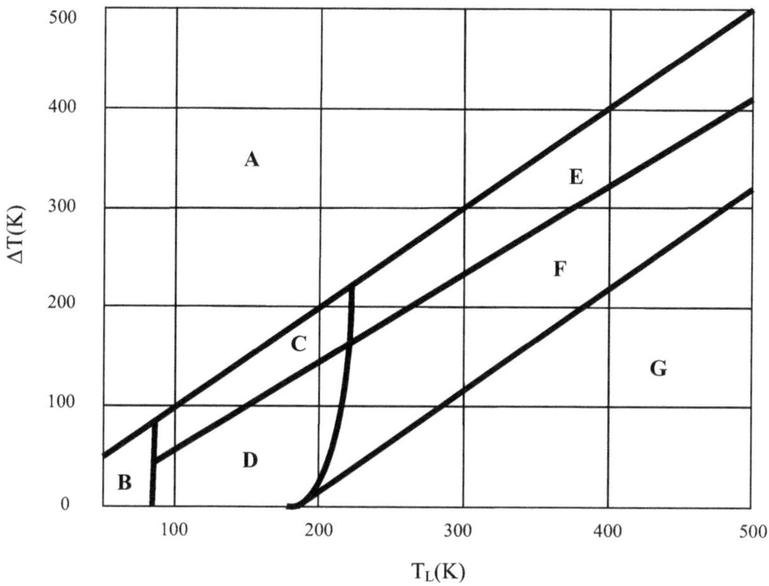

Figure 3. Phase diagram for total spin (I_s) and total current (I_c) flows in a 7SiZNR-7/4SiZNR-H_2 zig-zag silicene nanoribbon heterojunction, A: non-existent, B: $I_s = 0$, $I_c = 0$, C: $I_s > 0$, $I_c < 0$ with thermal rectification, D: $I_s > 0$, $I_c < 0$ opposite flow directions, E: $I_s > 0$, $I_c > 0$ with thermal rectification, F: $I_s > 0$, $I_c > 0$, same flow direction with negative differential thermal resistance effect in I_c, G: $I_s > 0$, $I_c > 0$, flow in same direction.

Thermoelectric effects in nanoribbons have been analyzed theoretically and numerically by using a model which includes a staggered exchange field and an external electric field that is normal to the atomic plane. Both fields separately open a gap in the edge states and therefore lead to non-zero thermopower in the vicinity of the gap edges. The interplay of the two fields leads to a spin-dependent gap and thus gives rise to a spin thermopower in addition to the conventional Seebeck effect. The Coulomb interaction, in the form of the Hubbard term in the mean-field approximation, leads to an antiparallel configuration of the edge magnetic moments in the ground state. The Coulomb interaction greatly modifies the topological properties, and therefore their transport and thermoelectric properties.[189]

Using the non-equilibrium Green's function method, an investigation of the transport properties of a graphene/silicene/graphene heterojunction with zigzag-edge nanoribbons – with modulation of the effective spin-orbit coupling and potential energy - demonstrates that both the transmission coefficient and the charge Seebeck coefficient exhibit oscillatory behavior and can be effectively modulated by effective spin-orbit coupling and the potential energy. The even-odd difference in transport and thermoelectric properties disappears in the graphene/silicene/graphene heterojunction.[190]

1.6 Thermal Conductivity

Equilibrium molecular dynamic simulations of the phonon thermal conductivity show that the in-plane thermal conductivity of silicene sheet is about an order of magnitude lower than that of bulk silicon. Vacancy defects significantly decrease the thermal conductivity, due to phonon-defect scattering. Phonon transport in silicene sheet is markedly affected by the vacancy concentration, vacancy size and vacancy boundary shape.[191]

The thermal conductivity of monolayer silicene at 300K is found, using first-principles methods, to be 9.4W/mK. The out-of-plane vibration contributes less than 10% of the overall thermal conductivity; differing from the results of similar studies of graphene. The difference is attributed to the presence of a small amount of buckling, which breaks the reflectional symmetry of the structure. The flexural modes are not purely out-of-plane vibrations and involve strong scattering with other modes.[192]

The dumb-bell structure of 2-dimensional group-IV material offers alternatives to growing thin films for various applications where thermal properties are important. The lattice thermal conductivity of low-buckle and dumb-bell silicene can be obtained by using first-principles calculations and the Boltzmann transport equation for phonons. For low-buckle silicene, the calculated lattice thermal conductivity with naturally occurring isotope concentrations is 27.72W/mK. For dumb-bell silicene, the calculated value is 2.86W/mK.[193] The thermal conductivity for dumb-bell silicene is much lower than low-buckle silicene, due to stronger phonon scattering. Molecular dynamics simulations of the effect of isotopes on thermal transport across silicene and graphene, using structures about 140nm-long and 4nm-wide, reveal the phonon spectra and the variation in thermal conductivity. This shows that out-of-plane modes are delocalized relative to in-plane counterparts. The absolute thermal conductivity reductions are more marked in graphene than in silicene.[194] The phonon transport in silicene is very different to that in graphene. The thermal conductivity of silicene exhibits a logarithmic increase with respect to sample size, due to the small scattering-rates of acoustic in-plane phonon modes. That of graphene is finite. The linear dispersion of the acoustic out-of-plane phonon modes, which is caused by the buckled structure, makes the long-wavelength longitudinal acoustic phonon modes in silicene less efficiently scattered as in graphene. In graphene, most of the heat is carried by the acoustic out-of-plane phonon modes. The same phonon modes in silicene make only an approximately 10% contribution to the total thermal conductivity. This can also be attributed to the buckled structure.[195] Density functional theory calculations, combined with a full-band Monte Carlo analysis of electron-phonon interaction and related transport properties, confirm that the out-of-plane acoustic phonon mode may play a dominant role; unlike the case of graphene. The small energy of this phonon mode, which arises from the weak sp^2 π-bonding between silicon atoms, contributes to the high scattering-rate and to an appreciable degradation of electron transport.[196]

When nano- to micron-sized monolayers of graphene and silicene were modelled using molecular dynamics, the graphene was modelled by using an optimum parameterization of the Tersoff potential while the silicene was modelled using parameterizations of the Tersoff potential for silicon. As the lengths of both materials increased, the corresponding thermal conductivities also increased; such that graphene had a far higher thermal conductivity than silicene at all length-scales. Armchair and zig-zag chiralities of either graphene or silicene had no significant effect on the thermal conductivities, given that the monolayer materials were modelled as being of infinite width. Graphene exhibited a significantly higher thermal conductivity than did silicene for every length-scale and chirality. This was attributed to the higher phonon group velocities of the dominant acoustic modes in graphene.[197] When short strips of single-layered material consisting of either graphene or silicene were modelled in the same way, the results indicated that both materials exhibited increasing thermal conductivity as the length increased; with the graphene exhibiting far more rapid increases. Armchair and zig-zag forms of silicene exhibited significant differences in thermal conductivity. This was not true of graphene. The latter had a significantly higher thermal conductivity than did silicene for every length-scale and chirality. This was attributed to the fewer excitable phonon frequencies, as suggested by the vibrational density of states.[198]

Using first-principles calculations and self-consistent solutions of the linearized Boltzmann-Peierls equation for phonon transport within a three-phonon scattering framework, the lattice thermal conductivities of free-standing silicene have been characterized for various isotropic tensile strains and temperatures. A marked size-dependence of the lattice thermal conductivity on tensile strain is observed, in that it diverges with increasing system-size. The intrinsic room-temperature lattice thermal conductivity of unstrained silicene converges, with system size, to give 19.34W/mK at 178nm. The room-temperature lattice thermal conductivity of strained silicene becomes as high as that of bulk silicon at 84μm; thus offering the possibility of using strain to manipulate the lattice thermal conductivity. The relative contribution to intrinsic lattice thermal conductivity, arising from out-of-plane acoustic modes, is greatest for unstrained silicene: about 39% at room temperature. The single-mode relaxation-time approximation, useful for bulk silicon, does not suitably describe phonon thermal transport within the temperature range considered. In large samples of silicene, the lattice thermal conductivity increases with tensile strain, peaks at about 7% strain and then decreases with further straining.[199]

Similar methods, involving solution of the Boltzmann transport equation, with interatomic force constants extracted from first-principles calculations, show that the thermal conductivity decreases only slightly with increased tensile strain in bulk materials whereas the thermal conductivity of silicene can increase markedly with strain. Depending upon the size, the maximum thermal conductivity of strained silicene can be a few times higher than in the unstrained case. This unusual strain dependence is attributed mainly to the sharp increase in acoustic phonon lifetime, probably arising from the flattening of the buckled silicene structure during stretching. This behavior is of course unique to silicene among other common two-dimensional materials.[200] No available interatomic potential can precisely characterize silicene. Optimization of the Stillinger-Weber potential parameters, specifically for a single-layer silicon sheet, can accurately reproduce the lightly buckled structure of silicene and the full phonon dispersion curves obtained from *ab initio* calculations. Performing equilibrium and non-equilibrium molecular dynamics simulations and anharmonic lattice dynamics calculations, using the new potential, show that the three methods consistently predict an extremely low thermal conductivity for silicene, plus a short phonon mean-free path. By qualifying the relative contributions of lattice vibrations in various directions, it was found that the longitudinal phonon modes dominated thermal transport in silicene.[201]

On the basis of first-principles calculations, it is proposed that doping with germanium may offer an effective means of tailoring the thermal conductivity of silicene. The electronic transport properties are not affected by germanium doping, while the room-temperature thermal conductivity is reduced by 62% at a dopant concentration of 6%. The depression of phonon transport is attributed to low-frequency phonon softening and enhanced phonon scattering by germanium doping.[202] Molecular dynamics simulation of the thermal conductivity of silicene nanosheets shows that the thermal conductivity of pristine samples is 25 to 30W/mK and exhibits anisotropy. Isotopic doping is, as above, effective in reducing the thermal conductivity. When silicene nanosheets are randomly doped with ^{30}Si to the extent of 50%, a maximum reduction of about 20% is observed. This reduction can be enhanced if the dopant is arranged in a superlattice pattern. The thermal conductivity of such superlattice-structured silicene nanosheets changes non-monotonically as the thickness of the lattice layers decreases. A competition between interface scattering and confinement effects of the phonon modes is responsible for this non-monotonic behaviour.[203] Other studies also show that isotopic doping results in a U-shaped change in the thermal conductivity for isotope concentrations ranging from 0 to 100% and, again, ordered doping (isotope superlattice) leads to a much larger reduction in thermal conductivity than does random doping.[204]

Non-equilibrium molecular dynamics simulations of thermal transport in a single-layer silicene sheet under uniaxial stretching show that, unlike graphene, it exhibits an anomalous thermal response to tensile strain. The thermal conductivity of sheet and nanoribbons first increases appreciably with applied tensile strain, rather than decreasing, and then fluctuates around a high plateau. By deducing the relative contributions made by various phonon polarizations, it is shown that the phonon transport is dominated by out-of-plane flexural modes, similarly to graphene. The markedly different behavior of silicene is then attributed to the interplay of two competing mechanisms determining heat conduction in stretched silicene sheet. One of these is uniaxial stretching modulation in the longitudinal direction, which significantly depresses the phonon group velocities of longitudinal and transverse modes and hinders heat conduction. The other is phonon stiffening in the flexural modes, which counteracts the above phonon softening effect and facilitates thermal transport. The abnormal behavior of silicene sheet is further attributed to the unique deformation characteristics of the hexagonal lattice[205].

Molecular dynamics simulations of the thermal conductivity of free-standing silicene and of silicene supported on an amorphous silicon dioxide substrate, at 300 to 900K, show that the thermal conductivity decreases with increasing temperature. The presence of the SiO_2 substrate leads to a reduction of up to 78% in the thermal conductivity at 300K. When the silicene is located on the SiO_2 substrate, the phonon relaxation times are decreased, from between 1 and 13ps, to less than 1ps. The phonon mean free paths are reduced, from between 10 and 120nm, to between 0 and 20nm. The thermal conductivities of free-standing and supported silicene are due mainly (>85%) to the longitudinal and transverse acoustic phonons, while the out-of-plane acoustic phonons make a contribution of less than 3%.[206]

Van der Waals heterostructures, prepared by stacking layers of isolated two-dimensional crystals such as graphene and silicene, are promising nanomaterials for the development of multifunctional devices. Molecular dynamic simulations of the interface thermal conductance and interface thermal rectification of hybrid graphene/silicene monolayers show that the thermal conductance not only increases with temperature, but also with monolayer length for a given temperature, before reaching a saturation value. At 300K, the saturated value is about 250MW/m^2K. The interface thermal rectification decreases with increasing monolayer length and temperature. Both the thermal conductance and interface thermal rectification are markedly affected by a tensile strain, imposed on silicene along the interface direction, but are almost independent of the heat flux. There is a critical value (42GW/m^2) of the latter flux, above which low-frequency kinetic waves are excited and provide an additional channel for heat transport.[207]

Other simulations show that the thermal conductance of pristine graphene/silicene bilayers at room temperature is 11.74MW/m^2K when the heat transfers from graphene to silicene, and is 9.52MW/m^2K for the opposite direction. The thermal conductance increases monotonically with temperature and interface coupling strength. Hydrogenation of graphene is effective in enhancing thermal conductance if optimum hydrogenation is employed. By changing the hydrogen coverage, the thermal conductance can be maximized so as to be five times higher than that of pristine graphene/silicene.[208]

The interfacial thermal transport across silicene and substrates such as crystalline silicon, amorphous silicon, crystalline silica and amorphous silica, as examined using classical molecular dynamics simulations and a transient pulsed-heating technique to characterize the interfacial thermal resistance shows that the interfacial thermal resistance between silicene and substrates decreases by some 40% between 100 and 400K. This is attributed to enhanced phonon coupling arising from the anharmonicity effect. Contrary to the traditional view, that amorphous structures tend to have poor thermal transport capabilities due to their disordered atomic configuration, it is calculated that amorphous silicon and silica substrates facilitate interfacial thermal transport, as compared with their crystalline counterparts. The coupling effect arising from the substrates can improve the interface thermal transport by up to 43.5%, for coupling strengths ranging from 1.0 to 2.0.[209]

1.7 Magnetoresistance

First-principles quantum transport calculations predict a giant magnetoresistance to exist in zig-zag silicene nanoribbons connecting two semi-infinite silicene electrodes via switching of the edge spin direction of the zig-zag silicene nanoribbons. The spin-filter efficiency of both antiferromagnetic and ferromagnetic zig-zag nanoribbons is sign-changeable with the bias voltage.[210] First-principles calculations of the transport properties of zig-zag silicene nanoribbons also indicate symmetry-dependent transport properties which are similar to those of zig-zag graphene nanoribbons, although the σ mirror plane is absent. Even-N and odd-N zig-zag nanoribbons have very different current-voltage relationships; attributed to the differing parity of their π and π bands under a c_2 symmetry operation with respect to the center axis. A magnetoresistance effect is observed in even-N zig-zag nanoribbons, and the order can reach 1,000,000.[211]

Density functional theory and non-equilibrium Green's function studies of the adsorption of titanium atoms on zig-zag silicene nanoribbons show that the titanium adatoms prefer sites inside the nanoribbons, rather than on the edges. Two neighboring adatoms are attractively coupled, and prefer to be adsorbed on the same side. The giant magnetoresistance in nanoribbons of even width is usually greatly reduced, except in symmetric adsorption cases. Strong spin negative differential resistance phenomena are observed, and a pure spin current can be produced by a temperature gradient in certain cases.[212]

Non-equilibrium Green's function and density functional theory investigations have been of the spin-dependent thermoelectric transport properties of zigzag-edged silicene nanoribbons, doped with Al-P bonded pairs at various edge positions. In the ferromagnetic configuration, strong quantum destructive interference effects between the localized states induced by the Al-P bonded pair and the side quantum states results in the appearance of spin-dependent transmission dips near to the Fermi level. This leads to simultaneous enhancement of the spin-filter efficiency and spin Seebeck coefficient at the Fermi level, while their signs depend upon the doping positions. In the antiferromagnetic configuration, spin-dependent transmission peaks with ordinary Lorentzian shapes near to the Fermi level can be introduced by the Al-P bonded pair. A pure spin current in doped antiferromagnetic zigzag-edged nanoribbons can be achieved by varying the temperature. In this case, the spin-filter efficiency can reach infinity, while the thermal magnetoresistance between the ferromagnetic and antiferromagnetic configurations can also reach infinity.[213]

Non-equilibrium Green's function and spin-polarized density functional theory studies of the spin-dependent electronic transport properties of zig-zag silicene nanoribbons with asymmetrical edge hydrogenation show that there exists a nearly 100% bipolar spin-filtering behavior in the zig-zag nanoribbon-based devices with an antiparallel spin configuration. A rectifying behavior and a giant magnetoresistance are also found.[214]

Investigation of charge conductance and spin and valley polarization, together with tunnelling magnetoresistance, in silicene junctions composed of normal silicene and ferromagnetic silicene reveals distinct features of the conductances for parallel and antiparallel spin configurations and of the tunnelling magnetoresistance when the ferromagnetic/normal/ferromagnetic junction is tuned by an external electric field. The behavior of the charge conductance and valley and spin polarizations can be analyzed in terms of the independent conductances of the differing spins at the two valleys and the band structure of ferromagnetic silicene. The conductances are affected by the vanishing of the propagating states at one or the other valley. Contrary to the case of graphene, the band structures at the two valleys are independently affected by the spin in the ferromagnetic regions and lead to non-zero and, for certain parameter regimes, pure valley and spin polarizations which can be tuned by an external electric field. The oscillatory behavior of the tunnelling magnetoresistance with respect to the strength of the barrier potential (spin-independent or spin-dependent) in the normal silicene region is such that - in some parameter regimes - the tunnelling magnetoresistance can even go from positive to negative as a function of the external electric field.[215]

The band structure in the presence of perpendicular electric and magnetic fields, taking account of the effects of off-resonant light, is such that - contrary to graphene - the light leads to a single Dirac cone state and thus to unusual plateaux and magnetotransport properties.[216]

Valley and spin-resolved transport through ferromagnetic/ferromagnetic/ferromagnetic silicene junctions is such that, for a normal silicene junction with zero exchange field in the presence of photo-irradiation, 100% valley/spin polarization is observed by modulating the energy. By enhancing the strength of the photo-irradiation or the gate voltage in the middle layer, a 100% spin and large valley polarization are found. If the staggered exchange field is considered, instead of photo-irradiation, fully polarized valley/spin transport is demonstrated upon tuning the energy. By tuning the staggered exchange field or the gate voltage in the middle layer, a 100% valley and large spin polarization are observed. A large negative or positive magnetoresistance can be obtained in a ferromagnetic/normal/ferromagnetic junction, depending upon the polarizing direction of the circularly polarized light.[217]

Two-dimensional topological insulators exhibiting helical edge states have been subjected to global and local manipulation at the edges by exposing them to a charge-density wave order, a superconductor, an out-of-plane antiferromagnetic or an in-plane antiferromagnetic field. These perturbations affect the helical edge states in differing fashions. Consequently, quantum spin-Hall effect can be realized without edge states. Such edge manipulations lead to promising applications such as a giant magnetoresistance and a perfect spin filter.[218]

Non-equilibrium Green's function and density functional theory studies, of substitutionally edge-doped zig-zag silicene nanoribbons in both the ferromagnetic and antiferromagnetic states, have been made of the doping effects of elements from groups III and V, and with a parallel or antiparallel magnetic configuration of the two electrodes. By switching the external magnetic field on and off, the metallic ferromagnetic zig-zag nanoribbons can be converted into insulating antiferromagnetic zig-zag nanoribbons. In the ferromagnetic state, even- and odd-width zig-zag nanoribbons exhibit markedly different magnetoresistances. In an odd-width edge-doped zig-zag nanoribbon, a large magnetoresistance occurs as compared with that in a pristine zig-zag nanoribbon. This situation is reversed in the case of even-width zig-zag nanoribbons. These results arise due to a marked change in the conductance in the antiparallel configuration.[219]

A theoretical study has been made of ballistic electron transport in silicene, with manipulation of a pair of ferromagnetic gates, and transport properties have been calculated by using the standard transfer matrix method for parallel and antiparallel magnetization configurations. Due to the stray field-induced wave-vector filtering effect, a marked difference in configuration-dependent transport gives rise to a giant tunnelling magnetoresistance. Together with the buckled structure of silicene and its electrically tunable energy gap, the magnetoresistance can be efficiently modulated by the externally-tunable stray field, electrostatic potential and staggered sub-lattice potential.[220]

1.8 Valley and Spin Phenomena

Valley-based electronics, or 'valleytronics', is a key feature of the next generation of electronic devices[221]. The valley degree of freedom is a common feature of the honeycomb lattice. The topological phases of silicene can be investigated by introducing differing exchange fields on the A and B sites. There then appears a wide variety of topologically protected states, each having a characteristic spin-valley structure. The single Dirac-cone semimetal, for example, is a state where one gap is closed while the other three gaps are open; thus avoiding the Nielsen-Ninomiya fermion-doubling problem. There is also a hybrid topological insulator, the so-called quantum spin-quantum anomalous Hall insulator, where a quantum anomalous Hall effect occurs at one valley and a quantum spin Hall effect occurs at the other. Single-valley semi-metals appear along the phase boundary, where only one of the two valleys is gap-less with degenerate spins. The semi-metals are also topologically protected because they appear at the interface of different topological insulators. Such spin-valley-dependent phenomena can be observed via optical absorption or edge modes.

The topological properties of a silicene superstructure, generated by an external periodic potential, are such that the superstructure is a quantum-spin Hall insulator if it is topologically connected to silicene. It is notable that two non-equivalent K and K′ points in the silicene Brillouin zone are identical in certain superstructures. In such cases, two Dirac cones co-exist at the same Dirac point in momentum space and are hexagonally warped by the Coulomb interaction. A numerical analysis of a (3 x 3) superstructure on the (4 x 4) structure of a silver substrate shows that it is a quantum-spin Hall insulator, that there exists no topological phase transition due to an external electric field and that hexagonal warping occurs in the band structure.[222]

Symmetry considerations, and use of the tight-binding method in combination with first-principles calculations, have been used to derive systematically the low-energy effective Hamiltonian involving spin-orbit coupling. This Hamiltonian is very general because it applies not only to silicene, but also to other low-buckled analogues of graphene, such as germanene and stanene; as well as to graphene - when the structure returns to a planar geometry. The effective Hamiltonian is itself the analogue of the graphene quantum spin Hall effect Hamiltonian. As in the graphene model, the effective spin-orbit coupling in a low-buckled geometry opens up a gap at the Dirac points and establishes the quantum spin Hall effect. The effective spin-orbit coupling comprises to first-order the atomic intrinsic spin-orbit coupling strength, ξ_0, while this leading-order contribution of spin-orbit coupling vanishes in the planar structure. Silicene, as well as germanene and stanene, therefore has a much larger gap, opened by the effective spin-orbit coupling at the Dirac points, than does graphene; due to the low-buckled geometry and the larger atomic intrinsic spin-orbit coupling strength. The more buckled the structure, the greater is the gap. The quantum spin Hall effect can be observed, in low-buckled silicon, germanium and tin systems, within experimentally accessible temperature ranges. In addition, the Rashba spin-orbit coupling in silicene is intrinsic due to its own low-buckled geometry; which vanishes at the Dirac point, K, while it has a non-zero value upon deviation from the K-point. The quantum spin Hall effect in silicene is therefore robust with respect to the intrinsic Rashba spin-orbit coupling.[223]

Recalling that silicene is a two-dimensional topological insulator, due to its quite large spin-orbital interaction and that it has a buckled lattice structure which permits control of the effective mass of Dirac electrons using a perpendicular electric field, the use of a spatially alternating electric field to generate multiple topologically-protected interface states has been proposed. When the valley-dependent electron mass (which defines the Chern number of insulating bulk material) changes sign, or discontinues due to spatial variation of the electric field, multiple topologically-protected interface states appear in the insulating bulk material. The topologically-protected interface states arise from the valleys, and sustain dissipation-less valley or spin-valley-dependent currents. These are immune to both valley-conservation and spin-observation scattering. The coupling among topologically-protected interface states, due to spatial electron tunnelling, excites the topological interface states. The existence or not of an excitation gap depends upon the parity of the topologically-protected interface state number.[224]

A valley polarized quantum Hall effect and topological insulator phase transition are identified. The Kubo formalism is used to discuss the Hall conductivity and analyze the longitudinal conductivity for elastic impurity scattering in the first Born approximation. The combination of an electric field with intrinsic spin orbit interaction leads to quantum phase transitions at the charge neutrality point and provides a tool for experimentally tuning the topological state[225]. Silicene is a suitable model system for exploring the spin and valley phenomena which are not accessible in graphene, due to the small spin orbit interaction.[226]

The spin-orbit torque in two-dimensional hexagonal crystals has two components: a field-like term due to an inverse spin galvanic effect, and an anti-damping torque which arises from Berry curvature in mixed spin-k space. In the presence of a staggered potential and exchange field, the valley degeneracy can be lifted and a valley-dependent Berry curvature is obtained. This leads to a tunable anti-damping torque by controlling the valley degree of freedom. The valley imbalance can be made as high as 100% by tuning the bias voltage or magnetization angle.[227] The intrinsic spin and valley Hall conductivities of two-dimensional crystals are also such that the type of insulating phase of the system (topological insulator or band insulator) can be tuned by means of a perpendicular electric field. At a finite frequency, the transverse edge to which carriers of a particular spin and valley label flow can be controlled so that accumulation of a particular combination of spin and valley index can be ensured. The direction of flow depends upon the type of insulating phase. The magnitude of the Hall conductivity response is increased from the DC values at certain incident photon frequencies associated with the onset of interband transitions.[228]

The reduced silicene/germanene single-particle Hamiltonian in a buckled two-dimensional hexagonal lattice can be expressed in terms of Pauli matrices associated with the pseudo-spin. The Hamiltonian of these systems comprises the Dirac kinetic energy, a mass term and spin–orbit coupling. The mass term breaks the sub-lattice symmetry of the system's honeycomb structure, and generates a spin-split band-gap with the help of spin–orbit coupling. The buckled lattice generates a staggered sub-lattice potential between silicon atoms at A-sites and B-sites with an electric field, E_z, applied perpendicularly to its plane. The behavior of the system is determined by two Dirac-like cones at K and K′ points. By tuning E_z, a critical value is attained at which the system at the topological insulating phase changes to a semi-metallic state, where the 'spin-down' to 'spin-up' band-gap ratio is much lower than unity for the K valley, and is much greater than unity for the K′ valley. This state is termed a so-called valley-spin-polarized-metal', due to the opposite spin-polarizations of the K and K′ valleys. Upon further increasing E_z, the system turns into a trivial insulator. This is associated with a valley magnetic-moment reversal. Provided that the (non-magnetic) impurity scattering strength, V_0, is moderate (of the same order as the intrinsic spin–orbit coupling of about 4meV), the 'valley-spin-polarized-metal' phase is protected. A substantial increase in V_0 leads however to the disappearance of this phase due to accentuated intra- and inter-valley scattering processes. This disappearance does not occur due to an increase in the Rashba spin–orbit coupling effect.[229] These various states can be plotted on a phase diagram of exchange field, M, versus electric field, E_z. A quantum anomalous Hall insulator, a valley polarized metal, a marginal valley polarized metal, a quantum spin Hall insulator and a band insulator can be delineated. They are characterized by the Chern numbers and/or by the edge modes of a nanoribbon. It is interesting that electrons are moved from a conduction band at the K point, to a valence band at the K' point, for $E_z > 0$ in the valley polarized metal. In the quantum anomalous Hall phase, almost flat gap-less edge modes emerge and the spins form a momentum-space skyrmion to yield the Chern number. It is also remarkable that a topological quantum phase transition can be induced simply by changing the electric field on a single silicene sheet.[230] The interplay between the edge and bulk states, induced by Rashba spin-orbit coupling in a zig-zag nanoribbon in the presence of an external electric field, can be divided into two types. One is the interplay between the edge and bulk states with opposite velocities, and the other is that involving the same velocity direction. The former can open up small direct spin-dependent sub-gaps. A spin-polarized current can be generated in the nanoribbon, as the Fermi energy is in the sub-gaps. The latter can give rise to spin precession in the nanoribbon. Zig-zag silicene nanoribbon can therefore be used as an efficient spin filter and spin modulation device.[231]

One possible way of detecting a topological phase transition is by means of optical absorption monitoring. It is demonstrated that silicene exhibits a strong circular dichroism with respect to optical absorption, and obeys a certain spin-valley selection rule. The selection rule is markedly different for the topological and trivial sectors because a band inversion takes place at the phase transition point. The circular dichroism can thus be used as a signal to distinguish the topological and trivial sectors. The selection rule permits the excitation of electrons, having definite spin and valley indices, by optical absorption. The photo-induced current is spin-polarized and the spin direction is different for the topological and band insulators.[232] The dynamic conductivity should also reveal signatures for these different phases, and thus permit their identification together with the determination of parameters such as the spin-orbit energy gap. An interesting feature is that electric-field tuning of the band-gap could be used to switch the Drude intraband response on and off. In the presence of spin-valley coupling, the response to circularly polarized light as a function of frequency and electric field tuning of the band-gap means that right- and left-handed circular polarization makes it possible to select a particular combination of spin and valley index. The frequency for this effect can be varied by tuning the band gap.[233]

Ab initio studies of Pt(Ni,Pd)-decorated silicene show that all three types of adatom prefer the hollow sites of silicene; thus guaranteeing that the Dirac cones are unbroken. The Pt(Ni,Pd)-decorated systems all exhibit quantum valley Hall states with the gap opened exactly at the Fermi level. The gaps of the quantum valley Hall states can be substantially increased by applying a positive electric field. Interesting phase transitions from quantum valley Hall to quantum spin Hall, and then back to quantum valley Hall again, are found in Pt/Ni-decorated silicene when a negative electric field is applied. The quantum spin Hall state in the palladium case, with a negative electric field, is quenched because of a relatively larger Rashba spin-orbit coupling as compared with the intrinsic spin-orbit coupling of the system.[234]

The Kubo formula has been used to analyze the charge, spin and valley currents which develop intrinsically in armchair nanoribbons. Taking the symmetries of the latter nanoribbons and the structure of the total Hamiltonian into account, it was demonstrated how the matrix elements - together with the summations involved in a suitably-adapted Kubo formalism – can be calculated analytically. The results reveal that all three currents develop in a step-like manner. The heights, indicating a jump in the current, and the plateaux, indicating a constant current, are tunable by adjusting the width and/or the buckling effect. At particular critical widths an inversion in band-gaps can occur, giving rise to quantum phase transitions. The armchair silicene nanoribbons can generate pure spin or valley currents when specific conditions are imposed on the width, buckling effect and Fermi energy.[235]

Electron transport studies of material with both a staggered electric potential and magnetization, the latter arising from the magnetic proximity effect by depositing silicene onto a magnetic insulator, show that the material can be a spin and valley half metal under suitable conditions when spin-orbit interaction is considered. The filtered spin and valley can be controlled by modulating the staggered potential or magnetization. In the spin-valve structure, not only can the anti-parallel magnetization configuration significantly reduce the valve-structure conductance, but the reversing staggered electric potential can cause a high-performance magnetoresistance due to spin and valley blocking effects.[236]

First-principles calculations indicate the possible formation of soliton-like topological domain walls in silicene. The walls between regions of distinct ground states of the buckled geometry should bind electrons when a uniform electric field is applied perpendicularly to the sheet. A topological source of the electron confinement is demonstrated on the basis of numerical calculations of the valley-specific Hall conductivities. Possible experimental signatures of the quantum valley Hall effects are predicted using simulated scanning tunnelling microscopic images.[237]

A new quantum state, the valley-polarized quantum anomalous Hall state, has been identified theoretically. In the presence of Rashba spin-orbit coupling and an exchange field, silicene hosts a quantum anomalous Hall state with a Chern number of 2. Via tuning of the Rashba spin-orbit coupling, a topological phase transition results in a valley-polarized quantum anomalous Hall state. This state exhibits the electronic properties of both the quantum valley Hall state (with a Chern number of 3) and the quantum anomalous Hall state (with a Chern number of -1).[238]

The commensurability oscillations in silicene subjected to a perpendicular electric field, E_z, a weak magnetic field, B, and a weak periodic potential, $V = V_0\cos(2\pi y/a_0)$, are such that the field, E_z, and/or the modulation lift the spin degeneracy of the Landau levels and lead to spin- and valley-resolved Weiss oscillations. The spin resolution is a maximum when E_z is replaced by $E_z = E_0\cos(2\pi y/b_0)$, while the valley one is a maximum for $b_0 = a_0$. For certain ranges of B, the current is fully spin- or valley-polarized. The spin and valley degeneracy of the Landau levels is lifted, due to the modulation, and the interplay between the strong spin-orbit interaction and the potential and electric field modulations can result in spin- and valley-resolved magnetotransport. For very weak magnetic fields, the commensurability oscillations induced by a weak potential modulation can exhibit a beat pattern; depending upon the strength of the homogeneous electric field, E_z. This is not the case when E_z alone is modulated. Additional quantum Hall conductivity plateaux arise due to spin and valley intra-Landau-level transitions.[239] The Hall conductivity plateaus acquire a step structure, due to spin and valley intra-Landau-level transitions, that is absent in unmodulated silicene.[240] Investigation of the electric-field controlled energy gap and Landau levels shows that the energy-gap at various Dirac points has different closing and re-opening conditions, and the two-fold degeneracy induced by K-K′ symmetry is resolved. An externally applied electric field gives rise to two Rashba spin-orbit-couplings between the nearest neighbour and the next-nearest neighbour. Both of these couplings can resolve the spin degeneracy for isolated values of the electric field, where the crossover of successive Landau-levels becomes anti-crossover. Apart from some particular values of the electric field, the 4-fold degeneracy of the energy levels associated with K-K′ symmetry and spin symmetry is completely resolved, with each level having a definite spin polarization; corresponding to quantum Hall plateaux with filling factors of $\nu = 0, \pm 1, \pm 2, \ldots$[241]

Tight-binding and random-phase investigations of intrinsic plasmons indicate that, at finite temperatures, an undamped plasmon is generated from the interplay between intraband and interband-gap transitions. The extent of the plasmon-existence range, in terms of momentum and temperature, depends upon the size of the single-particle-excitation gap and can be further tuned by applying a perpendicular electric field. The plasmon becomes damped in the interband-excitation region. A low-damping zone is created by a field-induced spin split. The field-dependent plasmon spectrum is found to be widely tunable with regard to plasmon intensity and spectral bandwidth.[242] The dielectric properties of two-dimensional Kane-Mele topological insulators in buckled honeycomb lattices have been studied with particular regard to the effect of an on-site potential difference between sub-lattices. The results for real and imaginary parts of the dynamic polarization function show that there are three regimes (topological insulator, valley-spin polarized metal, trivial band insulator), which can be used to extract information on the strength of intrinsic spin-orbit coupling. Study of the inverse dielectric function provides numerical results for the plasmon branch. The behavior of the plasmon as a function of the sub-lattice potential difference shows that the behavior of the plasmon branch, as the on-site potential difference is varied, depends upon the location of the chemical potential with respect to the gaps. A beat phenomenon is detected, in the effective potential, which depends upon the on-site potential difference.[243]

Similar studies of the temperature-dependent Coulomb screening and excitation spectrum of electrons show that, with spin-orbit interaction, monolayer silicene is a narrow-gap semiconductor. As before, at finite temperatures the interplay between intraband and interband transitions can lead to an undamped plasmon mode at low frequencies and the plasmon mode exists only within a limited region of temperature and momentum; corresponding to the constrained gap transition. Beyond this region, another damped plasmon mode dominates the excitation spectrum. The marked change in the plasmon behavior could potentially be observed experimentally and permit the identification of the spin-orbit energy-gap.[244]

Use of the random phase approximation shows how electron-electron interactions are affected by various fields. The combination of a perpendicular electric field, E_z, and an exchange field, M, brings a spin and valley texture to the particle-hole excitation spectrum and permits the formation of spin- and valley-polarized plasmons. When the Fermi level lies in the gap of one spin in one valley, the intraband region of the corresponding spectrum disappears. For zero E_z and finite M, the spin symmetry is broken and spin polarization is possible. The lifetime and oscillator strength of the plasmons depend strongly upon the number of spin and valley type electrons that make up the electron-hole pairs.[245]

In a new model of the graphene/silicene/graphene heterojunction, study of the valley polarization properties in a zig-zag nanoribbon in the presence of an external electric field shows that the energy range associated with the bulk gap of silicene has a valley polarization of more than 95%. Under the protection of the topological edge states of the silicene, the valley polarization remains even if a small non-magnetic disorder is introduced.[246]

When electron transport through ferromagnetic/normal/ferromagnetic silicene junctions with an induced energy-gap is investigated, it is found that the energy-gap can be tuned by applying an electric field or by exchange fields due to the buckled structure. Analysis of the transmission probabilities of four groups and valley conductance shows that they can be turned on or off by adjusting the local electric field and exchange field. A fully valley polarized conductance, in particular, with 80% transmission is found in this junction and is attributed to the interplay of valley-dependent massive Dirac electrons, the exchange potential and the on-site potential difference of sub-lattices.[247]

It is proposed that a highly efficient dual spin-valley filter, consisting of two distinct regions, can be created in silicene. In the first region, angular separation of the two valley spins in momentum-space is induced by a uniaxial strain, with further spin separation being induced by an exchange field. The second region acts as an extractor of the required spin-valley current by means of localized fringe magnetic fields and gate modulation of the electrical potential. Based upon realistic parameter values, controllable and highly efficient filtering can be demonstrated for all four spin-valley combinations.[248]

Due to the valley-opposite gauge field induced by strain, strained silicene having a superlattice structure exhibits an angle-resolved valley and spin-filtering effect when spin-orbit interaction is considered. When an interaction that breaks time-reversal symmetry is introduced, such as spin- or valley-dependent staggered magnetization, the system becomes a perfect spin and valley half-metal in which only one spin and valley species is able to operate.[249]

The conductance of silicene pn junctions, calculated in the bipolar region using the non-equilibrium retarded Green's function method, shows that when an electric field is applied normally the energy gaps of various valleys become different. The conductance of pn junctions is governed mainly by the transport of electrons of the valley with the smaller gap, and the conductance of the other valley can also be observed for short junctions. In spin-orbit coupling dominant cases, edge state bands lie in the bulk gap and contribute a unit tunnelling conductance plateau which is independent of the junction length. Upon continuously changing the electric field, the conductance peaks at a certain position for long junctions but decreases monotonically for short junctions.[250]

Metal/silicene/metal junctions, where the centre silicene sheet is close to a ferromagnet and subjected to a perpendicular electric field, can be analyzed by connecting wave amplitudes which obey the lattice Schrödinger equation for the interfaces, within the tight-binding model, to the tunable conductance of both end-contacted and side-contacted metal/silicene/metal junctions. The current through the metal/silicene/metal junctions is spin- and valley-polarized due to coupling between the valley and spin degrees of freedom. The conductance and polarization exhibit an oscillatory behavior as a function of the length of the silicene sheet. The full spin and valley polarized conductance can be achieved by introducing suitable electric and exchange fields. The conductance is heavily dependent upon the hopping integrals of simple metal, silicene and metal/silicene interfaces for end-contacted junctions. Provided that the hopping integrals satisfy certain condition, for a suitable incident energy, there is no difference between the transport for end-contacted and side-contacted junctions.[251]

The superconducting proximity effect in silicene features massive Dirac fermions with a tunable mass (band-gap). Computation of the conductance across a normal/superconductor silicene junction, of the non-local conductance of a normal/superconductor/normal junction, and the supercurrent flowing in a superconductor/normal/superconductor junction demonstrates that transport processes consisting of local and non-local Andreev reflection can be efficiently controlled by an external electric field, due to the buckled structure. It is possible to obtain a fully spin-valley polarized crossed Andreev reflection process without any contamination of elastic co-tunnelling or local Andreev reflection. This contrasts sharply with ordinary metals. The supercurrent flowing in a superconductor/normal/superconductor junction can be fully spin-valley polarized and is controllable by an external electric field.[252]

In the presence of extrinsic Rashba spin-orbit coupling, silicene can host a quantum anomalous Hall state with spin- and valley-polarized edge states. This can be effectively controlled by the exchange and electric fields. A pair of non-trivial edge states exists in one specific valley and exhibits strong but opposite spin polarization. A distinctive feature is that both the spin and valley indices of the edge states can be switched by reversing the electric field.[253]

Valley-polarized insulating states have been found numerically for zig-zag edged silicene nanoribbons because of strong spin-orbit coupling and spin-polarized electronic structures. The valley polarization splits the degeneracy of the nanoribbons' conductance plateaux and offers valley-polarized transport channels. New optical activation modes appear for the singlet exciton states, due to unequal cancellation of the contributions from different valleys.[254]

In theory, a perfect valley filter can be obtained under the effects of a ferromagnetic stripe, plus an electric field perpendicular to the silicene. A ferromagnetic stripe or an electric field cannot alone generate any valley polarization, but their combination can produce a perfect valley filter. Two configurations, one with the magnetization of the ferromagnetic stripe perpendicular to the silicene plane and one with it parallel to the current direction, can be envisaged. Such a device would be useful for creating a valley beam filter valleytronic device.[255]

An *ab initio* theoretical study of the energetic stability and electronic properties of pristine and hydrogen-adsorbed grain boundaries in silicene shows that grain boundaries in silicene have a lower formation energy as compared with their graphene counterparts. Removal of the inversion symmetry by applying an external electric field perpendicular to the silicene sheet leads to the formation of valley-indexed metallic states lying along the boundaries; characterizing the quantum valley Hall effect. This effect is maintained in the presence of disordered and asymmetrical geometries along the boundaries. The metallic states are suppressed upon the adsorption of hydrogen adatoms along the boundaries. The adatoms promote an imbalance of the electronic occupation of the unsaturated π-electrons beside the hydrogenated grain boundary rows, thus giving rise to a net magnetic moment on the silicon atoms along the edge sites of the hydrogenated grain boundaries and to an electronic band structure characterized by spin-polarized valley states which are protected against back-scattering processes.[256]

Calculation of the intrinsic direct-current spin and valley Hall conductivity of doped ferromagnetic silicene with an electric field perpendicular to the sheet, using the energy spectrum and wave-function and making use of the Kubo formalism, yields a general relationship for the transverse Hall conductivity which can be used to obtain the spin- and valley-Hall conductivity. The results, in the zero limit of the exchange field, reduce to previous results. Necessary conditions can be obtained for observing fully spin- or valley-polarized transport.[257]

Consideration of finite-temperature screened electron-impurity scattering, using a kinetic equation approach to investigate the transport properties of two-dimensional massive fermions, shows that the longitudinal conductivity is always non-vanishing when the chemical potential lies within the energy gap. This residual conductivity arises from interband correlation and depends strongly upon the strength of electron-impurity scattering. The electron-impurity interaction makes a substantial contribution to the spin and valley Hall conductivities, which are almost independent of the impurity density.[258]

Study of spin and valley transport through a ferromagnetic/non-magnetic/ferromagnetic monolayer silicene junction in the presence of an electrostatic gate potential, U, and an on-site potential difference, Δ_z, in the non-magnetic region demonstrates that, away from the band gap (no spin and valley conductances), the spin and valley polarizations of the junction exhibit an oscillatory dependence upon U. Near to it, they exhibit an almost linear dependence upon U. The oscillations are due to the phase difference of the electron wave-functions of the spin and valley-resolved conductances that lead to the chiral nature of the quasi-bound states in silicene. The amplitude and frequency of the spin and valley polarization oscillations of the junction can be tuned by varying U and Δ_z. By increasing U, the second characteristic of the Klein tunnelling is satisfied as a result of the quasi-particle chirality which can penetrate a potential barrier[259]. For special values of U, the junction has a gap in the spin and valley-resolved conductance and the amplitude of this gap is controlled only by Δ_z. Upon increasing the exchange energy the amplitudes of the spin and valley polarization oscillations are suppressed.[260]

The ballistic transport of Dirac fermions in silicene through arrays of barriers in the presence of an exchange field, M, and a tunable potential of height/depth, U, is such that the spin- and valley-resolved conductances as functions of U or M exhibit resonances away from the Dirac point. Close to the latter is a pronounced dip that becomes a gap when a critical electric field, E_z, is applied. This gap widens upon increasing the number of barriers and can be used to permit electric field-controlled switching of the current. The spin, p_s, and valley, p_v, polarizations of the current near to the Dirac point increase with E_z or M and can attain 100% for certain values. These fields widen significantly upon increasing the number of barriers. The p_s and p_v oscillate almost periodically with the separation between barriers or wells, and can be inverted by reversing M.[261] For two barriers, there is no splitting in g_c when the fields, M, are in opposite directions. Most of these phenomena have no equivalent in graphene.[262]

Investigation of the transport properties of electrons in junctions composed of normal and ferromagnetic silicene, with regard to the charge conductance as a function of length of normal region, the net electrostatic potential difference and the barrier potential, shows that this junction is a potential candidate for electric-field controlled switching purposes. Perfect spin- and valley-polarized conductance can be achieved by adjusting the barrier potential. The efficiency of the spin and valley polarization is significantly improved in this junction, potentially leading to 100% spin-polarized spintronics and better than 90% valley-polarized valleytronics.[263]

Theoretical study of valley- and spin-resolved scattering through magnetic barriers in one-layer thick silicene, using the mode-matching method for the Dirac equation, shows that a spin-valley filtering effect can be achieved and also tuned completely by using both a top and bottom gate. Upon reversing the sign of the staggered potential, the direction of the valley polarization is switched while the direction of spin polarization is unchanged.[264]

Due to the valley-spin locking effect, the valley degree of freedom of silicene can be controlled by magnetic means. Valley- or/and spin-selection induced by the exchange field produce a perfect spin-valley filter and a tunnelling magnetoresistance effect in double ferromagnetic barriers on the surface of silicene nanoribbon. There are valley-resolved mini-gaps and mini-bands in the zig-zag silicene nanoribbon, modulated by the magnetic superlattices, which give rise to a periodically modulated spin- or/and valley-polarization and a tunnelling magnetoresistance.[265]

Contrary to the case of graphene, the pseudo Dirac mass in silicene is generated by spin-orbit interaction and is tunable by applying electric or exchange fields, due to the buckled structure. The junction leads to currents which are perfectly split into four groups, spin-up or spin-down in k- or k′-valleys, upon applying various values of the electric field. A perfect spin-valley polarization is due to the interplay of a spin-valley dependent Dirac mass and the chemical potential in the barrier. Perfect spin-valley polarization also occurs only for an energy which is comparable to the spin-orbit energy-gap.[266]

Calculations of the magneto-optical conductivity and electronic density of states show that, in the presence of a perpendicular magnetic field and electric field gating, four spin- and valley-polarized levels are seen in the density of states. Transitions between these levels lead to similarly polarized absorption lines in the longitudinal, transverse Hall and circularly polarized dynamic conductivity. Previous spin- and valley-polarizations, predicted for the conductivity, were present only in response to circularly polarized light. Distinct spin- and valley-polarizations were also predicted here in the longitudinal magneto-optical conductivity at experimentally attainable energies. The frequency of the absorption lines could be tuned, using electric and magnetic fields, so as to appear in ranges from THz to the infra-red.[267]

A silicene-based lateral resonant tunnelling device might be obtained by placing silicene under the modulation of upper non-magnetic/ferromagnetic/non-magnetic sandwich nanogates. Due to the electrically-tunable band-gap of silicene, a lateral double-barrier structure is formed by imposing a flexible electrostatic modulation on the top gates. By aligning the spin- and valley-resolved confined states in a magnetic well, a remarkable spin/valley polarization can be achieved via a spinor-reliant resonant tunnelling mechanism. Under electrostatic-, magnetic- and dimension-manipulation, the confined well state can be engineered and the observed spin- and valley-polarizations can be further tuned.[268]

Charge-, valley- and spin-conductances in a normal/ferromagnetic/normal silicene junction oscillate with the length of ferromagnetic silicene. The current through this junction is valley- and spin-polarized due to coupling between the valley and spin degrees of freedom. The valley- and spin-polarizations can be tuned via the local application of a gate voltage. A fully valley- and spin-polarized current is found upon applying an electric field.[269]

1.9 Magnetic Properties

In contrast to the singlet d+id′ superconductivity in the low dopant-content region, bilayer silicene is an itinerant ferromagnet in the narrow dopant regions around the Van Hove singularities, and is a triplet p+ip′ superconductor in the vicinity of these regions. Addition of a weak Kane-Mele spin-orbit coupling to the system further singles out the time-reversal invariant equal-spin helical p+ip′ pairing as being the leading instability. The present triplet pairing is driven by the ferromagnetic fluctuations, which become strong and markedly increase the superconducting critical temperature near to the phase boundaries between ferromagnetism and superconductivity.[270]

First-principles calculations of the effect of chemical doping on zig-zag nanoribbons, with manganese as a passivating element replacing hydrogen atoms at one edge, show that the system leads to an anti-ferromagnetic state with half-metallic properties. This first-principles approach, based upon the Keldysh non-equilibrium Green's function method, gives the spin-dependent transport properties. The system is quite a good spin-filtering device, with a nearly 80% spin-filtering efficiency over a wide bias-voltage region.[271]

First-principles calculations of the magnetic properties of silicene doped with chromium or iron atoms, under isotropic and uniaxial tensile strains, show that these materials exhibit a strain-tunable magnetism. The magnetism undergoes sharp transitions, from low-spin to high-spin states, under a small isotropic tensile strain. In iron-doped material, an almost non-magnetic state changes to a highly magnetic state under a small isotropic tensile strain. The magnetic moments of iron-doped silicene exhibit a sharp jump to about $2\mu_B$ at a small threshold value of uniaxial strain, and the magnetic moments of chromium-doped silicene increase gradually to about $4\mu_B$ with increasing uniaxial strain. The electronic and magnetic properties of chromium- and iron-doped silicenes are sensitive to the magnitude and direction of the external strain.[272] A similar study of the effect of an external electric field shows that the doped systems exhibit various magneto-electric behaviors. The magnetic moment of iron-doped silicene undergoes a sharp jump at a threshold electric field; indicating a good switching-effect. At low concentrations of iron or chromium, there are two structures for which the changes in magnetism are considerably different under an external electric field. The magnetic moment of the doped systems has an almost linear region as a function of the electric field. The changes in magnetic moment depend strongly upon the direction of the electric field.[273]

Other first-principles calculations of the magnetic properties of samples embedded with transition metals ranging from scandium to zinc, and containing single vacancies and double vacancies, show that the binding energies for the various elements are related to the transition-metal d-shell electrons. Thus, scandium, titanium and cobalt have binding energies as high as 6eV while zinc has the lowest binding energy (about 2eV). The magnetic moment can be modulated by embedding transition-metal atoms ranging from vanadium to cobalt, with this arising mainly from the transition-metal 3d-orbitals plus contributions from neighboring silicon atoms. Iron atoms at single vacancies and manganese atoms at double vacancies have the largest magnetic moment (3μB). The doping of nitrogen or carbon atoms into vacancy sites might greatly enhance the magnetism.[274] First-principles density functional theory calculations of the effect of incorporating transition-metal ions (titanium, niobium, tantalum, chromium, molybdenum, tungsten) to form MSi_2 show that, whereas pure silicene is stable as a distorted honeycomb lattice with opposite out-of-plane displacements of the two silicon sub-lattices, its electronic structure still exhibits a linear dispersion; with a Dirac cone feature similar to that of graphene. The incorporation of transition-metal ions results in a rich set of properties, with a clear dependence upon the resultant structural changes. In particular, $CrSi_2$ forms a two-dimensional magnetic phase which exhibits strong piezomagnetic coupling.[275] In another study, the electronic properties of 3d transition metal decorated silicene were investigated using density functional calculations. Among the ten types of doped material, the titanium-, nickel- and zinc-doped material became semiconducting whereas cobalt- and copper-doping made the substrate half-metallic. In the case of titanium- and copper-doped material, the measured band-gaps were markedly larger than the previously reported band-gap of silicene. The observed band-gap opening at the Fermi level was produced by breaking the sub-lattice symmetry. This was in turn due to structural changes: Jahn-Teller distortion and protrusion of the transition-metal atom.[276]

Density functional theory shows that super-halogen $MnCl_3$ can be used to change the electronic and magnetic properties of silicene from semi-metallic to semiconducting with a wide range of band-gaps, as well as from non-magnetic to ferromagnetic or anti-ferromagnetic, by varying the coverage of super-halogen molecules. The electronic properties can be further modified when a super-halogen and a halogen are both used. Because of their high electron-affinity and rich structural diversity, super-halogen molecules have advantages over conventional halogen atoms for the modification of the properties of silicene.[277]

Investigation of the Ruderman-Kittel-Kasuya-Yosida interaction between two magnetic impurities in monolayer silicene shows that, due to spin-orbit coupling, the interaction can be further divided into various types according to the polarization directions of the magnetic impurities. The spatial behavior of the interaction is closely related to the external electric field. Together with the spin-orbit coupling, this gives rise to various phases in undoped material. By probing the interaction, which contains all of the information on momentum space, differences between the topological phase and the trivial phase can therefore be explicitly identified. The change in chemical potential which is caused by doping silicene can also have a marked influence on the spatial behavior of the interaction.[278]

First-principles calculations of the geometrical, electronic and magnetic properties of silicene nanoflakes and of a corresponding two-dimensional framework assembled from such nanoflakes show that hexagonal silicene nanoflakes exhibit semiconducting behavior while triangular nanoflakes are magnetic. Triangular nanoflakes which are linked directly are anti-ferromagnetic while a system which is linked by an odd number of silicon chains can exhibit ferromagnetic behavior. This is attributed to an anti-parallel spin-rule on silicon atoms; consistent with the Lieb-Mattis criterion. A two-dimensional framework composed of triangular nanoflakes, linked by a silicon atom, exhibits half-metallic behavior and an integer magnetic moment.[279] Similar calculations of the geometrical, electronic and magnetic properties of two-dimensional germanene-silicene hybrid sheet saturated with bromine atoms show that, although the pristine material is semi-metallic, half-saturation with Br of silicon atoms on only one side results in localized and unpaired electrons in unsaturated germanium atoms and long-range ferromagnetic properties with a high Curie temperature. Half-brominated material has a half-metallic character, with 100% spin-polarized currents at the Fermi level.[280] A first-principles study of two-dimensional silicene saturated with hydrogen and bromine atoms shows that the fully-saturated material exhibits a non-magnetic semiconducting behavior, while half-saturation on only one side with hydrogen or bromine results in localized and unpaired electrons of unsaturated silicon atoms, leading to ferromagnetic semiconducting or half-metallic properties respectively. Total energy calculations show that the half-hydrogenated material exhibits ferromagnetic order while half-brominated material exhibits an anti-ferromagnetic behavior.[281]

Density functional theory calculations of edge-hydrogenated and edge-doped zig-zag silicene nanoribbons show that the formation energies of monohydrogenated and dihydrogenated edges depend slightly upon the width of the nanoribbons while the electronic and magnetic properties can be tuned by using differing forms of edge hydrogenation. Dihydrogenated edges can stabilize the anti-ferromagnetic semiconducting ground state of zig-zag silicene nanoribbons. Nitrogen- or phosphorus-doped monohydrogenated nanoribbon is a magnetic semiconductor and becomes a semi-metal under a transverse electric field.[282]

Analysis of the spin Chern number, based upon Dirac theory, shows that the origin is a pseudo-spin meron in momentum space. The pseudo-spin degree of freedom arises from silicene's double sub-lattice structure. The analysis makes clear how a phase transition from a topological insulator to a band insulator occurs under an increasing electric field. The critical electric field can theoretically be determined with the aid of silicene's diamagnetism. The latter is tunable by the external electric field and exhibits a singular behaviour at the critical electric field.[283]

Application of the generalized gradient approximation and the first-principles projector augmented wave potential within density function theory to H-terminated zig-zag edge nanoribbons decorated with a single carbon chain shows that in either a perfect zig-zag nanoribbon or a single carbon-chain decorated zig-zag nanoribbon, the ferromagnetic state is preferred over the anti-ferromagnetic one. A single carbon-chain decorated zig-zag nanoribbon is nevertheless more stable than a perfect one. The electronic and magnetic properties can also be modulated by placing a single carbon chain at various positions.[284] First-principles investigation of zig-zag edge nanoribbons decorated with double carbon chains shows that such materials are always metallic; regardless of the ribbon width. The defect states contributed by double carbon chains comprise two degenerate bands across the Fermi level. Perfect zig-zag nanoribbon has a ferromagnetic ground state, while that decorated with double carbon chains has an anti-ferromagnetic ground state. The carbon chains are always close to straight ones, thus resulting in a transverse contraction near to the carbon chains and a contraction of the ribbon width. The C-Si bond exhibits ionic bonding features and the C-H bond is a typical covalent one because of the electronegativity and the binding-force difference between hydrogen, carbon and silicon atoms.[285]

First-principles calculations of pristine, nitrogen- or boron-doped and nitrogen-plus-B co-doped nanoribbons show that the substitution of nitrogen or boron for silicon occurs preferentially at the ribbon edge-sites. A singly-substituted nitrogen or boron atom at the edge results in a semiconductor-metal transition in armchair silicene nanoribbons because of the appearance of a half-filled impurity band near to the Fermi level. When the N/N or B/B atoms are doped into armchair nanoribbons at two opposite edges, they preserve the metallic character due to negligible impurity-impurity interaction; regardless of the ribbon width. cobalt-doped systems with nitrogen and boron atoms exhibit a semiconducting behavior, with band-gaps smaller than the corresponding pristine case, due to effective charge compensation between the nitrogen or boron atoms. When silicon is replaced by a nitrogen or boron atom in zig-zag silicene nanoribbons, the system has a ferromagnetic character. This is attributed to the perturbation of π- and π*-states localized at the doped edges. A remarkable half-metal and spin gap-less semiconductor with 100% spin-polarized currents around the Fermi level is found in nitrogen-doped zig-zag nanoribbons. Provided that doping with N/N or B/B atoms is performed on made in zig-zag nanoribbons, the spin-polarization at both silicon edges is compressed and they therefore exhibit a non-magnetic behavior. When nitrogen and boron atoms are co-doped into zig-zag nanoribbons at the most stable edge positions, a transition from metallic to semiconducting occurs.[286]

A first-principles simulation of two-dimensional hexagonal silicene shows that weak overlapping of the $3p_z$ orbitals of neighboring silicon atoms leads to a very reactive surface. This results in a more energetically stable semiconducting surface upon fully hydrogenation. Half-hydrogenation breaks the extended π-bonding network of silicene, leaving the electrons in the unsaturated silicon atoms localized and unpaired. It thus exhibits ferromagnetic semiconducting behavior, with a band-gap of 0.95eV. Long-range ferromagnetic coupling between silicon atoms is also predicted, with a Curie temperature of about 300K.[287]

Based upon phonon-mode analysis, one-side semi-hydrogenated silicene is dynamically stable in the ground state. Semi-hydrogenation from only one side produces localized and unpaired 3p electrons in the unhydrogenated silicon sites and then introduces ferromagnetism, with no need for doping, cutting or etching. One-side semi-hydrogenated silicene is a semiconductor with an energy-gap of 1.74eV. In going from pristine to one-side semi-hydrogenated, and then to fully-hydrogenated, the silicene changes from metallic to magnetic semiconducting and then to non-magnetic semiconducting.[288]

Study of the magnetocapacitance of spin- and valley-polarized silicene in an external perpendicular magnetic field shows that the band-gap is electrically tunable and that the magnetocapacitance exhibits beats in low magnetic fields, and splitting of the Shubnikov de Haas oscillations in high magnetic fields.[289] A theoretical study of the magnetotransport coefficients of spin-orbit coupled gated silicene in the presence or absence of periodic spatial modulations shows that the combined effect of spin-orbit coupling and a perpendicular electric field is seen in the formation of a regular beat pattern in Weiss and Shubnikov de Haas oscillations. Analytical results yield a beat condition which can be useful in determining the spin-orbit coupling constant by simply counting the number of oscillations between any two successive nodes.[290]

Ab initio density functional theory investigation of the effects of full- and half-fluorination upon the structure and properties of silicene-graphene hybrid sheets indicates that full- and half-fluorination reduce the gap energy of the pure material. Partial coverage of the sheets reveals large differences between the structural and electronic properties of the two half-fluorinated conformers, and unusual magnetic properties, depending upon how the adsorption of fluorine atoms occurs: on silicon sites or carbon sites. Although F-bonding silicon atoms give rise to an anti-ferromagnetic semiconductor, the hybrid exhibits non-magnetic ordering when the fluorine atoms bond to carbon atoms. A stability analysis shows that the three fluorinated structures are stable and, in terms of energy, that fluorine atoms prefer to be adsorbed on silicon atoms. Chemical modification with fluorine can thus control the electronic and magnetic properties.[291]

Density functional theory calculations of the effect of edge functionalization on nanoribbons show that -H, -F, -Cl, -Br and -I functionalization of armchair, zig-zag, Klein, reconstructed Klein, reconstructed pentagon-heptagon edge types and their combinations show that a trifluorinated Klein edge nanoribbon is the most stable edge structure. The stability of trihydrogenated Klein edge nanoribbon which is periodically replaced by a dihydrogenated zig-zag edge approaches that of dihydrogenated armchair edge nanoribbons as being the most stable hydrogenated ones. Asymmetry in edge functionalization, or the combination of various edge types, can transform symmetrical edge-functionalized zig-zag nanoribbons from anti-ferromagnetic semiconductors to various magnetic states, such as bipolar spin gap-less semiconductors, ferromagnetic metals and semiconductors and anti-ferromagnetic metals. The mono-fluorinated armchair nanoribbons exhibit spin gap-less semiconducting behavior, with 100% spin-polarized currents around the Fermi level, when the silicon edge atom is replaced by a nitrogen or boron atom. Spin gap-less semiconductors and half-metal and ferromagnetic metals are also observed in nitrogen- or boron-doped asymmetrical edge-functionalized zig-zag nanoribbons, fully functionalized Klein edge nanoribbons and combinations of zig-zag nanoribbons with reconstructed Klein edge nanoribbons.[292]

1.10 Optical Properties

The optical conductivity of two-dimensional honeycomb crystals beyond the usual Dirac-cone approximation can be calculated on the basis of the independent quasi-particle approximation to the complex dielectric function for optical inter-band transitions, taking into account the full band structures. In the case of silicene, the influence of excitonic effects can also be studied, with particular care being taken to derive converged spectra with respect to the number of k-points in the Brillouin zone and the number of bands. Both the real and imaginary parts of the optical conductivity are thereby correctly described for low and high frequencies.[293]

The magneto-optical longitudinal, transverse Hall and circularly polarized responses of silicene and other materials described by a Kane-Mele Hamiltonian have been calculated with particular attention being paid to the effects of an external electric field and finite charge doping. The energy of interband transitions can be tuned by varying the electric field. The onset frequency of the absorptive peaks moves differently between the topological insulator and band insulator regimes. This can be exploited experimentally to confirm the existence of the two insulating phases, as well to provide a measure of the spin-orbit band-gap. The zeroth Landau level splits between four spin and valley energies, resulting in valley-spin polarized levels in the density of states. Upon charge doping, transitions between these levels permit a spin- and valley-polarized response of the conductivity, where charge carriers of specific spin and valley index can be isolated by tuning the incident photon frequency. Increasing the chemical potential redistributes the spectral weight from interband transitions to a strong low-energy intraband response. At large chemical potentials, this intraband feature is associated with the semi-classical cyclotron resonance frequency, increases linearly with the magnetic field.[294]

Theoretical investigation, using the inter-sub-band transition theorem at low temperatures, of the infra-red optical properties of a three-terminal silicene-based device irradiated with a circularly polarized terahertz electromagnetic field demonstrates that the electronic structure of bulk silicene can be tuned to the topological insulator or band-insulator state by using the back-gate voltage-induced staggered sub-lattice potential. With enhancement of the staggered sub-lattice potential, the refractive index, extinction coefficient and optical conductivity from the topological insulator state spin-up and spin-down sub-bands are red- and blue-shifted, respectively. Those of the band-insulator state spin-up and spin-down sub-bands are continually blue-shifted. The collective excitations and individual electron-hole pair excitations induce a topological insulator and band-insulator state electron energy loss spectrum with similar red- and/or blue-shifted behaviour to that of the refractive index, extinction coefficient and optical conductivity.[295] Calculation of the complex dielectric function for optical interband transitions shows that silicene possesses the same low-frequency absorbance as does graphene, and is governed by the Sommerfeld fine structure constant. Deviations occur at higher frequencies, when the first interband transitions outside of K or K′ begin to contribute. The low-frequency results are a consequence of the honeycomb geometry, but do not depend upon the type of group-IV atom, the sheet buckling nor the orbital hybridization.[296]

First-principles calculations of two-dimensional multi-layer silicene have revealed the existence of rich dynamically-stable phases whose stability is closely related to the extent of sp^3 hybridization, as evaluated using average bonds and effective bond-angles. Stable Si(111) surface structures are obtained when the silicene thickness attains four layer;

indicating the critical thickness for structural evolution. Multi-layer silicene with π-bonded surfaces offers outstanding opto-electronic properties for exploitation in solar cells and optical fiber communications, due to the incorporation of sp^2-type bonds into the sp^3-type bond dominated system.[297]

Combined experimental and theoretical studies of the optical response of epitaxial silicene on silver show that the silicene/Ag(111) ultraviolet-visible absorption spectra is strongly non-additive. Electronic transitions involving silver states make huge contributions to the optical absorption of silicene; compatible with strong Si-Ag hybridization. The results are independent of the specific silicene configuration, and also apply to thin amorphous silicon. This is taken to reflect a dimensionality-driven dielectric response of the two-dimensional silicon/silver interface, and is confirmed by transient-reflectance spectroscopy data. These indicate a metallic-like relaxation-time, and demonstrate the effect of the strong hybridization, which arises in silicene/Ag(111) systems, upon the charge-carrier dynamic properties.[298]

Density functional theory studies, in the long-wavelength limit, of the optical properties of two-dimensional pure or doped (aluminium, phosphorus) buckled nanosheets show that, unlike the case of graphene, no new electron energy loss spectra peak appears for parallel polarization, regardless of the dopant type. Under perpendicular polarization, two new small but significant peaks appear after phosphorus doping. The appearance of these new peaks is attributed to the buckling effect of stable silicene. Calculations also reveal that the maximum values of the absorption coefficient of the doped system are higher than those of the pristine one. A study of reflectivity modulation as a function of dopant concentration indicates the appearance, for both polarizations of the electromagnetic field, of some strong peaks which possess the robust characteristic of a doped reflective surface. In all of the doped systems, the reflectivity modulation is restricted to low energies (<4eV) and high energies (>8eV) for parallel and perpendicular polarization, respectively. No significant changes are found in the maximum values of the optical conductivity as a function of dopant concentration for perpendicular polarization, but a sudden jump appears in an Al-P co-doped system at a concentration of 18.75%.[299]

Based upon the inter-subband transition theorem for semiconductors, theoretical investigation of the optical properties of a three-terminal silicene-based device under irradiation with a circularly polarized terahertz electromagnetic field shows that spin–orbit coupled electronic structure can be engineered to give topologically-insulated and band-insulated states by the staggered sub-lattice potential from the back-gate voltage. The dielectric functions and optical absorption spectra from the topologically-insulated spin-up and spin-down sub-bands are red-shifted and blue-shifted, respectively, with increasing sub-lattice potential. Those from the band-insulated spin-up and spin-down sub-bands are instead continually blue-shifted by the staggered sub-lattice potential.[300]

First-principles calculation investigation of the electronic and optical properties of silicene nanomeshes indicates that their properties are sensitive to the width of the silicon chains between neighboring holes. The band-gaps of nanomeshes having an odd width remain closed. Band-gaps are opened at the Γ points however in nanomeshes having an even width. In particular, silicene nanomeshes exhibit a broad frequency photo-response, ranging from the far-infrared (0eV) to the ultra-violet (10eV). The optical properties can be tuned by varying the width.[301]

The propagation of short discrete electromagnetic pulses, considered to be optical solitons, in silicene waveguide layers has been studied and a phenomenological equation derived which describes the dynamics of such short light pulses in a system of silicene waveguides.[302]

1.11 Electrical Conductivity

The epitaxial growth and electrical resistance of multilayer silicene on an Ag(111) surface show that the atomic structure of the first silicene layer differs from that of the next layers, and that the adsorption of silicon induces the formation of extended silicene terraces surrounded by step-bunching. Using controlled contact-formation, between the tips of a multiple-probe scanning tunnelling microscope and the extended terraces, a low sheet resistance has been measured. It was however much higher than that of the underlying silver substrate. The results nevertheless indicate the electrical viability of multilayer silicene.[303] *Ab initio* simulations based upon density functional theory, combined with a non-equilibrium Green's function formalism, of variations in the transmission characteristics shows that silicene nanoribbons exhibit voltage-dependent resistances that are due to elastic potential scattering; consistent with the Büttiker formalism. It is noted that the modulation of the transmission spectra of graphene nanoribbons by an applied voltage is lower than that in silicene nanoribbons. This suggests that metallic graphene nanoribbons are better than the silicene analogue for use as metallic interconnects.[304]

The electroresistance in silicene-based normal/ferroelectric-gated/normal junctions is such that the spontaneous electrical polarization in the ferroelectric can be switched by using an external electric field. Due to the combination of these properties, the junction can generate a tunnelling electroresistance in excess of 10^9%, and the conductance ratio of the on to off states is greater than 10^7. This can be further enhanced by increasing the thickness of the barrier or by increasing the degree of electrical polarization in the ferroelectric layer. The giant tunnelling electroresistance effect is directly related to the buckled lattice and to the presence of spin-orbit interactions.[305]

The conductance of charge carriers through silicene-based superlattices consisting of monolayer silicene has been studied by using the transfer matrix method, with consideration of ordered superlattices and the transmission probability of Dirac fermions. The number of resonance peaks increases with increasing number of superlattice barriers. The oscillating conductance as a function of barrier height is suppressed by disorder in the silicene superlattice.[306] The non-linear electrical transport properties of massive Dirac electrons in monolayer silicene, taking account of screened electron-impurity scattering with a Coulombic potential and of all intravalley and intervalley electron-phonon interactions due to deformation potential couplings, can be analyzed by using a balance-equation scheme. This reveals a strong influence of the energy gap on both linear and non-linear drift velocity and resistivity. Tunable values of over 50% for mobility and 1800% for resistivity, by using a vertical electric field, are predicted; a useful prospect for silicene-based devices.[307]

The band inversion in a silicene quantum dot situated in perpendicular magnetic, B, and electric fields depends markedly upon the strength of the magnetic field. Study of the energy spectrum of quantum dots when the electric field provides a tunable band-gap shows that boundary conditions introduce chirality. Negative and positive angular-momentum zero Landau level edge states therefore exhibit quite different behaviors with regard to the band-inversion mechanism underlying the topological insulator transition. Whereas some zero Landau levels suffer band inversion for any $B > 0$, other zero Landau levels suffer only band inversion above critical values of the magnetic field at non-zero values of the gap.[308] Protected edge states are present in antidots and dots; polarized in valley-spin pairs. Studies of the effect of lattice termination upon the properties of single electron energy levels and electron density distribution of antidots and dots, situated in a perpendicular magnetic field, confirm that the topological edge states propagate over the perimeter of the antidot/dot for both ideal and realistic edge terminations which exhibit roughness at the atomic length-scale. The valley polarization, and the slope of the energy lines, as a function of the magnetic field are nevertheless reduced when the antidot or dot has a rough edge.[309] Time-dependent density functional theory investigation of plasmon excitation processes in quantum dots reveals two main plasmon resonance bands, running parallel to the quantum-dot plane, at about 2.0 and 7.0eV. Because delocalized π-electrons can participate in excitation of the two plasmon resonance bands, an increase in the side-length of rectangular quantum dots in the direction of the excitation leads to red-shifting of the two main bands. The plasmon excitation in the dots also depends upon the edge configuration and, because of the relatively high symmetry of hexagonal quantum dots, the plasmon resonance modes of the quantum dots are identical along the various excitation directions running parallel to the quantum-dot plane.[310]

Theoretical study of the electronic transport properties of zig-zag nanoribbon superlattices, subjected to a periodic electric field perpendicular to the surface, shows that the conductivity of the system depends upon the superlattice structural parameters and exhibits effects which are analogous to those found in two-dimensional semiconductor superlattices. For a superlattice with N_b barriers a series of resonant peaks, each split into (N_b-1) sub-peaks, and transmission blockade regions appear in the conductance spectrum, indicating the formation of mini-bands and mini-gaps.[311]

The spin transport features of aluminium-doped zigzag-edged nanoribbons, where the ribbons are Si-H_2 bonded at one edge and Si-H bonded at the other to form an asymmetrical edge hydrogenation, when investigated using non-equilibrium Green's function methods and spin-polarized density functional theory indicate that perfect (100%) spin filtering can be achieved in the calculated bias region. Rectification of spin-polarized currents, at a ratio greater than 10^5, can be achieved by changing the positions of the dopant atoms. A negative differential resistance effect is also observed in the spin-polarized current. Such ribbons are promising materials for use in multifunctional spintronic devices.[312]

The electronic transport properties of zigzag-edged nanoribbons having one or two protrusions at the edges, studied using density functional theory combined with non-equilibrium Green's function methods, are such that the protrusion usually breaks down the edge state along the same edge which carries current in the junction. In ribbons having an even number of zig-zag chains in their width, the protrusions can increase the conductance; except when there are two symmetrical protrusions. In ribbons having an odd number of zig-zag chains in their width, the introduction of edge protrusions can suppress currents. In ribbon-based devices with an antiparallel magnetism configuration, only non-symmetrical or symmetrical protrusion models with a width made up of an even number of zig-zag chains exhibits a perfect spin filter effect.[313]

Density functional theory and non-equilibrium Green's function investigations of the transport properties of three zigzag-armchair-zigzag nanoribbon junctions show that the differential conductances of the systems decrease in the order, 5-ZAZ > 4-ZAZ > 3-ZAZ. A negative differential resistance is observed, within certain bias voltage range, only in 3-ZAZ nanoribbons.[314]

As already noted, the low-energy structure of silicene is described by Dirac electrons having relatively large spin-orbit interactions, due to the buckled structure. A key observation is seen to be that the band structure can be controlled by applying an electric field to a silicene sheet and, in particular, the gap closes at a critical electric field. Examination of the band structure of a nanoribbon shows that a topological phase transition occurs, from a topological insulator to a band insulator, with increasing electric field. It is possible to generate helical zero modes anywhere in a silicene sheet by adjusting the local electric field so as to equal the critical value. The region can then act as a quantum wire or quantum dot, surrounded by topological and/or band insulators. The wave functions for simple geometries can be explicitly constructed on the basis of low-energy effective Dirac theory[315], and these results also happen to be applicable to germanene. When the transport properties through a quantum wire controlled by a gate are studied by using the non-equilibrium Green's function formalism, a pair of gapless and spin-polarized edge states appears only if the gate voltage is strong and the silicene quantum wire has perfect zig-zag or armchair boundaries from which additional silicon atoms are absent. The edge states controlled by the gate are also spin valley-polarized, in that the directions of the spins are opposite in different valleys. This is a different behavior to that of the edge state at the interface between silicene and vacuum.[316]

A theoretical framework, based upon the tight-binding Fisher-Lee relationship, has been used to span non-equilibrium Green's function techniques, scattering methods and semi-classical Boltzmann transport theory. Within this hybrid quantum-classical two-scale framework it is possible to calculate the transmission and reflection coefficients of monolayer and bilayer Ag-silicene-Ag junctions by using the non-equilibrium Green's function method together with density functional theory. Also calculated have been the group velocities and resistance, using the semi-classical Boltzmann equation. The resistances of these junctions are about $0.08f\Omega m^2$ for monolayer silicene junctions and about $0.3f\Omega m^2$ for bilayer ones. These factors, of about 8 and about 2, respectively, are smaller than the Sharvin resistances estimated via the Landauer formalism.[317]

The electron transport properties of the pristine silicon-substituted analogue of pyrene, $Si_{16}H_8$, and of its carbon-doped analogue, $Si_{14}C_2H_8$, between two semi-infinite aluminium nanochain electrodes have been investigated by means of density functional theory and the non-equilibrium Green's function method. The current-bias characteristics have been studied in the bias potential range of 0 to 2.0V, in 0.1V steps, by imposing three gate voltages: -3, 0 and +3V. An important result of the study is the observation of multiple negative differential resistance regions, suggesting that these systems could be used as nano-multiswitches. The current-bias behavior of the systems, together with the observation of a negative differential resistance at each of the above gate voltages can be interpreted in terms of the transmission spectrum. This interpretation can be pursued by integrating the transmission spectrum within the corresponding bias window. The observed negative differential resistance characteristics, including the bias range and the current amplitude, can be changed by varying the applied gate voltage. Controlling the negative differential resistance characteristics of such devices by using variations in the gate voltage will be a major advantage in practical applications.[318]

1.11.1 Superconductivity

The non-coplanar low-buckle structure of silicene immediately inspired a study of the pairing symmetry of the doped system in a perpendicular external electric field. Such a study reveals that the electric field induces a quantum phase transition from the singlet chiral d+id′-wave superconducting phase to the triplet f-wave one. The emergence of f-wave pairing results from sub-lattice symmetry-breaking, caused by the electric field, and ferromagnetic-like intra-sublattice spin correlations at low dopant contents. Due to the increased density of states, the superconducting critical temperature is markedly increased by the electric field.[319] First-principles calculations of undoped bilayer silicene show that the system is intrinsically metallic, with sizable pocket Fermi surfaces. With realistic electron-electron interaction the system is identified as being a chiral d+id′ topological superconductor, mediated by strong spin fluctuation on the border of the antiferromagnetic spin density wave order. Tunability of the Fermi pocket area using strain makes it possible to adjust the spin density wave critical interaction strength near to the real one and leads to a high superconducting critical temperature.[320] First-principles calculations within the rigid-band approximation predict that electron-doped silicene is a good two-dimensional electron-phonon superconductor under biaxial tensile strain. The superconductivity transition temperature of electron-doped silicene can be increased by more than 10K by imposing a 5% tensile strain. The strong interaction between acoustic phonon modes normal to the silicene plane, and the increasing electronic states around the Fermi level induced by tensile strain, is mainly responsible for the increased critical temperature.[321]

A possible superconducting gap, about 35meV, was observed in silicene on Ag(111) substrates by means of scanning tunnelling spectroscopy. The temperature-dependence reveals a superconductor-metal transition, with a critical temperature of 35 to 40K.[322]

Investigation of the emergence of topological superconductivity in honeycomb systems such as silicene shows that photo-irradiation induces the Haldane term into the system. By introducing s-wave superconductivity by proximity coupling, the system becomes a topological superconductor: a photo-induced topological superconductor. The topological phase diagram can be determined by deriving explicitly the four independent 2-band theory and by determining the Chern number. Also explicitly obtained are the wave functions of Majorana zero-energy bound states which appear at the boundary of two distinct topological phases. The position of the Majorana fermion can be arbitrarily controlled by moving the area of photo-irradiation.[323]

Electron-doped silicene, under a biaxial tensile strain, is predicted to be a phonon-mediated superconductor. By using the Eliashberg formalism, an investigation was made of the thermodynamic properties of superconducting silicene in the case when the tension is 5% and the electron-doping is $3.5 \times 10^{14}/cm^2$. Under these conditions, a silicene monolayer is expected to exhibit its highest superconducting transition temperature. Based upon the electron-phonon spectral function, and assuming a wide range of Coulomb pseudopotential values, the superconducting transition temperature decreases from 18.7 to 11.6K. A similar behavior is observed in the case of the zeroth temperature superconducting energy gap at the Fermi level. Other thermodynamic parameters differ from the predictions of Bardeen-Cooper-Schrieffer theory. The ratio of the energy gap to the critical temperature, in particular, changes from 4.14 to 3.87. The ratio of the specific heat jump, to the specific heat in the normal state, ranges from 2.19 to 2.05 and the ratio of the critical temperature and specific heat in the normal state, to the thermodynamic critical field, increases from 0.143 to 0.155. Finally, the maximum value of the electron effective mass is 2.11 times that of the electron band mass.[324]

1.11.2 Hall Effect

Linear response theory and Green's function methods predict an intrinsic contribution to the spin Hall effect in the two-dimensional silicene. When an external voltage is applied normal to the silicene plane, the spin Hall conductivity can exhibit a transition from the spin Hall insulator phase, at a low bias, to the conventional insulator phase at higher voltages. This transition resembles the similar phase transition in bilayer graphene, but the spin-orbit interaction responsible for the transition in silicene is much stronger than that in graphene.[325]

The low-energy behavior of silicene is described by massive Dirac fermions, where the mass is a function of the applied field, E_z, and becomes zero at a critical field, E_{cr}. There are no zero-energy states due to the Dirac mass term, except at E_{cr}. The 4-fold degenerate zero-energy states are completely resolved even without considering Coulomb interactions. These features are very different to those of graphene, and show that silicene has a richer structure. One prominent feature is that, by applying an electric field, the valley degeneracy can be controlled. Depending upon the value of E_z, Hall plateaux appear at the filling factors: $v = 0; \pm1; \pm2;\pm3; \ldots$ except for the points where level crossings occur.[326]

The spin-orbit opened energy gap and band topology in silicene, studied using first-principles calculations, reveals that silicene with topologically nontrivial electronic structures can exhibit the quantum spin Hall effect by exploiting adiabatic continuity and direct calculation of the Z_2 topological invariant. It is predicted that the quantum spin Hall effect can be observed within an experimentally accessible low-temperature regime with a spin-orbit band gap of 1.55meV. This gap can increase to 2.9meV under pressure strain[327]. First-principles calculations of the topological properties of silicene decorated with various functional groups (-H, -F, -Cl, -Br, -I) and subjected to external strain show that normal insulators can transform into topological insulators, having a sizable gap, under a critical level of strain[328].

Analytical models and first-principles Wannier interpolation show that, when silicene is decorated with certain 3d transition metals, such as vanadium, it can sustain a stable quantum anomalous Hall effect. It is also predicted that the quantum valley Hall effect and electrically tunable topological states can be obtained in some transition-metal doped silicene when energy band inversion occurs[329].

Investigation of the transport properties of zig-zag silicene nanoribbon in the presence of an external electric field shows that the staggered sub-lattice potential and two kinds of Rashba spin-orbit coupling can be induced by an external electric field, due to the buckled structure. A bulk gap is opened by the staggered potential, and gap-less edge states appear in the gap upon suitably tuning the two kinds of Rashba spin-orbit coupling. The gap-less edge states are spin-filtered and are insensitive to non-magnetic disorder. Such results prove that the quantum spin Hall effect can be induced by an external electric field in silicene[330]. Further study of the influence of silicene's strong spin-orbit interaction and of an external electric field, E_z, in the presence of a perpendicular magnetic field shows that, for finite E_z, the spin and valley degeneracy of the Landau levels is lifted and leads to additional plateaux in the Hall conductivity at half-integer values of $4e^2/h$, due to spin intra-Landau-level transitions. These are absent from graphene. These plateaux are more sensitive to disorder and thermal broadening than are the main plateaux, occurring at integral values of $4e^2/h$, when the Fermi level passes through the Landau levels.[331]

It is the symmetry of silicene which gives rise to such interesting Rashba spin-orbit effects. A perpendicular electric field is able to couple to the sub-lattice pseudospin, thus making it possible to tune and close the band-gap electrically. An external electric field can provoke a topological phase transition, from a topological insulator to a normal insulator (or semi-metal) and *vice versa*. A group theoretical analysis can systematically construct the Hamiltonian in the vicinity of the K-points of the Brillouin zone, and identify an additional electric-field induced spin-orbit term that is allowed by symmetry. A tight-binding model can then be found which corresponds to the group-theoretically derived Hamiltonian near to the K-points. Starting from this tight-binding model, the topological phase diagram of silicene can be derived by calculating the topological invariant of the band structure. Calculation of the topological invariant of the honeycomb lattice in a gauge-invariant manner permits the incorporation of symmetry-breaking terms, such as Rashba spin-orbit interactions, into the topological analysis. The interplay of Rashba and intrinsic spin-orbit terms can generate a non-trivial quantum spin Hall phase in silicene. In graphene however, Rashba spin-orbit interaction competes detrimentally with the quantum spin Hall effect.[332] The topological phases which arise due to strong spin-orbit interaction in an external perpendicular magnetic field are also relevant: under 3% tensile strain, quantum spin Hall and quantum anomalous Hall phases, respectively, are found below and above a critical field of 10T. At not far above the critical field, and in an experimentally accessible regime, an energy gap in the meV range is observed. This again shows that the quantum anomalous Hall phase can be produced experimentally in silicene. This is not true however of graphene, with its tiny energy gap, nor of germanene … where an enormous field is required.[333]

1.11.3 Semiconductivity and Band Structure

One particular advantage of silicene is its compatibility with current silicon electronics. Both empirical and first-principles techniques have been used to study the electronic properties of silicene.[334]

First-principles total-energy and phonon calculations clarify the structural stability and electronic properties of free-standing bilayer silicene. Extensive structural explorations of previously known stable structures revealed four more dynamically-stable structures; including one possessing the highest cohesive energy. Atomic protrusion from the layer is the main relaxation pattern which stabilizes bilayer silicene and determines the lateral periodicity. Hybrid-functional calculations show that the most stable bilayer silicene is a semiconductor having an energy-gap of 1.3eV.[335] Silicene is expected to be very sensitive to electric fields, due to its buckled structure and, like graphene, can be expected to exhibit a Dirac-Fermion nature.[336]

First-principles calculations, combined with cluster expansion approach studies of chemical functionalization, show that it is generally accompanied by a spontaneous structural transition which originates from the preference of sp^3 hybridization in silicon. In order to produce continuously tunable band-gaps, the chemical functionalization of free-standing silicene at about 900K is recommended.[337]

Bilayer silicene has richer physical properties, than does bilayer graphene, due to its buckled structure together with its trigonal symmetrical structure. The trigonal symmetry involves a particular manner of hopping between two silicenes. It is a topologically trivial insulator since it carries a trivial topological charge, but its properties are closer to those of a topological insulator than those of a band insulator. A bilayer silicene nanoribbon has edge modes which are almost gap-less and helical, and it can be termed a so-called quasi-topological insulator. The band structure is controllable by applying an electric field to bilayer silicene sheet. Just as monolayer silicene undergoes a phase transition from a topological insulator to a band insulator under a certain electric field, bilayer silicene undergoes a transition from a quasi-topological insulator to a band insulator above a critical field. Bilayer silicene is a metal, while monolayer silicene is a semimetal at the critical field. There are several critical electric fields where the gap closes due to a trigonal warping effect.[338] A topological phase transition can be induced by changing the topological class via irradiation with circularly polarized light under a fixed external electric field, E_z. Upon increasing the light intensity at $E_z = 0$, a photo-induced quantum Hall insulator is obtained. Its edge modes are anisotropically chiral, in which the velocities of up- and down-spins are different. At above a critical field, a photo-induced spin-polarized quantum Hall insulator appears. This is a new state of matter: possessing one Chern number and one-half spin-Chern numbers. A single Dirac-cone state also exists along a phase boundary. A characteristic of this state is that one of the two Dirac valleys is closed and the other is open.[339]

First-principles calculation studies of nanoribbons doped with boron-nitrogen bonded pairs at various sites indicate that the boron-nitrogen pair tends to be doped at the edge of ribbons and that boron-nitrogen pair doping of armchair nanoribbons is easier to carry out than unbonded boron-nitrogen pair doping. Unlike the metallic states which are associated with singly-substituted nitrogen or boron atoms at the edge of armchair nanoribbons, the boron-nitrogen bonded pair doping results in semiconducting characteristics where the energy-gap of armchair nanoribbons can be adjusted; depending upon the boron-nitrogen pair doping sites in armchair nanoribbons. In zig-zag nanoribbons, the introduction of boron-nitrogen pairs results in a transition from non-magnetic to spin-polarized states. This is attributed to the perturbation of π and π^* states localized at the doped zig-zag nanoribbon edge. Amazing spin gap-less semiconducting characteristics, with 100% spin polarized currents around the Fermi level have been obtained, when the substitutional boron-nitrogen pair is located at edge or sub-edge sites in zig-zag nanoribbons.[340] Other studies have also confirmed that, when the structures, formation energies and electronic properties of armchair nanoribbons - substitutionally doped with single and double nitrogen or boron atoms at sites ranging from the edge to the center of the ribbon - are investigated using first-principles calculations based upon density functional theory, minimization of the formation energy shows that substitutional doping is favorable at the edge of the ribbon. Single nitrogen or boron atom, and double boron atom, substitution leads to a transition from semiconductor to metal. Double nitrogen atom substitution retains semiconduction. The co-doping of nitrogen and boron at opposite edges also does not affect the semiconducting nature of armchair silicene nanoribbons.[341] Similarly, spin-polarized density-functional theory studies of zig-zag nanoribbons, substitutionally doped with a single nitrogen or boron atom located at sites ranging from the edge to the center of the ribbon further show, by minimalization of the formation energy, that substitutional doping is favorable at the edge of the ribbon. Substitution of a single nitrogen or boron atom for one edge silicon atom can greatly suppress spin-polarization of the impurity atom site and its immediate vicinity, and leads to a transition from an antiferromagnetic to a ferromagnetic state. This is attributed to splitting of the original spin degenerate edge bands. A single nitrogen-atom doped zig-zag nanoribbon will still retain semiconductor properties, but a single boron-atom doped nanoribbon exhibits half-metallic characteristics. Such results reveal that substitutionally doped zig-zag nanoribbons have potential application in silicon-based nano-electronic devices such as field effect transistors, negative differential resistors and spin filter devices.[342]

Unlike planar graphene, the silicon atoms in silicene are not coplanar. By applying a perpendicular field to the two-dimensional silicene sheet, the on-site energy can therefore be modulated and the band-gap at the Dirac point is opened. The buckled structure also creates a height difference between the two edges of silicene zig-zag nanoribbons. External fields can modulate the energies of spin-polarized edge states and their corresponding band-gaps. Due to polarization in the plane, the modulation effect is width-dependent and becomes much more important for narrow ribbons.[343]

It is the strong coupling between unoccupied molecular orbitals and occupied molecular orbitals which leads to pseudo-Jahn-Teller distortion and the characteristic buckling of silicenes. Analysis of σ-π separation reveals that the σ-backbone is stabilized, while the π-backbone is destabilized due to buckling. The stabilization of a puckered σ-backbone overwhelms π-backbone destabilization; contrary to the situation in graphene. Cations such as Li^+ can suppress the pseudo-Jahn-Teller distortions, resulting in a planar structure, and leading to opening of a band-gap (about 1.62eV). Silicon-substituted benzenes bind more strongly with Li^+ than does benzene. Mutual competition/synergy between the orbital interactions of the ring with the cation, and the π-charge density across the surface of the molecule, governs the stability of these complexes.[344] Studies of lithium adsorption predict the stability of completely lithiated silicene sheets in which lithium atoms adsorb on the atom-down sites on both sides of the silicene sheet. Molecular dynamics simulations confirm stability at high temperatures and for phonon frequencies with any k-value. Upon complete lithiation, the band structure is transformed from a zero-gap semiconductor to a 0.368eV band-gap semiconductor.[345]

Bilayer silicene in which 50% of the silicon atoms exhibit sp^3-hybridization and 50% exhibit sp^2/sp^3-hybridization has unusual electronic properties, with two Dirac points formed near to the Fermi level. The two layers are connected by σ-bonding orbitals. Single-gated bilayer silicene can generate a continuously and widely tunable band gap of up to 1.13eV; much wider than that (0.25eV) of dual-gated bilayer graphene. This tuning effect is attributed to dual-polarization of localized π-states from the single-gated electric field, and the polarized σ-bonding orbitals of interlayers.[346] A density functional theory comparison of AA and AB-stacked bilayer silicene shows that there are no excitations of electrons - plasmons at low energies - for out-of-plane polarization. In AB-stacked bilayer silicene, there are 1.48eV plasmons for in-plane polarization; lower than the 2.16eV plasmons of monolayer silicene. Inter-band transitions and plasmons in both bilayer and monolayer silicene occur at relatively lower energies than in graphene. Calculations suggest that the band-gap can be opened-up and varied over a wide range by applying an external electric field to bilayer silicene. In the infra-red region, the imaginary part of the dielectric function for AB-stacked buckled bilayer silicene exhibits a broad structure peak in the range of 75 to 270meV. This contrasts with a short structure peak at 70meV for monolayer silicene, and an absence of such peaks for AA-stacked bilayer silicene. Upon applying an external electric field, the peaks are blue-shifted in the infra-red region.[347]

First-principles calculations show that the band-gap and electron effective mass of various stacks can be effectively modulated by tuning the pressure (interlayer spacing) and stacking arrangement. The electron effective mass is proportional to the band-gap. The band-gap of confined silicene is more sensitive to pressure than is that of confined graphene. A heterogeneous interface structure can be beneficial in effectively regulating the band-gap and carrier effective masses. By using confinement and pressure, the integrity of the honeycomb structure can be preserved so that the small effective mass and high mobility are retained during compression.[348]

The electronic and transport properties of armchair-edge nanoribbons with effective spin-orbit coupling and potential energy, investigated using non-equilibrium Green's function methods, yield the energy gaps. The conductance can be effectively modulated by the effective spin-orbit coupling, λ_{SO}, and the potential energy, V_0. With increasing λ_{SO}, the energy gap of 6-ASiNR and 7-ASiNR decreases, but remains zero for metallic 8-ASiNR. An energy gap appears in 8-ASiNR in the presence of V_0; resulting in the appearance of a conductance gap for 8-ASiNR.[349]

Upon applying uni- and bi-axial strains, the band-gap opens up at small strains. With further increases in strain, to beyond 8%, silicene changes into a metal. The ultimate tensile strength is estimated to be 3.4GPa. The imaginary part of the dielectric function shows that the inter-band transitions are red-shifted for uni- and bi-axial tensile strains and are blue-shifted for uni- and bi-axial compressive strains. The electron energy loss function shows that the π+σ plasmon energies are red-shifted for uni- and bi-axial strains and blue-shifted for compressive strains. The π plasmons disappear for tensile and asymmetrical strains. A bi-axial asymmetrical strain has no effect upon inter-band transitions or π+σ plasmon energies.[350]

The band structure of (3 x 3) silicene monolayers on Ag(111)-(4 x 4), investigated using density functional theory and a k-projection technique that includes the $k_\perp$-dependence of the surface bands, can be separated into contributions arising from the silicene and from the substrate. Such calculations not only reproduce the observed gap and linear dispersion across the K-point of (1 x 1) silicene, but also show that they originate from the $k_\perp$-dependence of Ag(111) substrate states (as modified by interaction with the silicene) and not from a Dirac state.[351] Angle-resolved photo-emission spectroscopy has shed light on the electronic characteristics of (3 x 3) silicene on Ag(111), revealing a silicene-derived band with a clear gap and linear energy-momentum dispersion near to the Fermi level at the Γ symmetry point of the (3 x 3) phase for several distinctive Brillouin zones. The large buckling of ~0.7 of this silicene structure induces the opening of a gap, close to the Fermi level, which is at least greater than 0.3eV. A two-dimensional character of the charge carriers is also revealed by the photon-energy invariance of the gapped silicene band, suggesting limited silicene-silver hybridization.[352] When the electronic band structure of monolayer (4 x 4) silicene on Ag(111) is imaged using angle-resolved photo-electron spectroscopy, a dominant hybrid surface metallic band is found near to the bulk silver sp-band, which is also faintly visible. The two-dimensional character of the hybrid band can be distinguished from the bulk characteristics of the Ag(111) sp-band by means of photon energy-dependence experiments. The surface band exhibits a steep linear dispersion around the $\bar{K}_{Ag}$ point and has a saddle-point near to the $\bar{M}_{Ag}$ point of Ag(111); resembling the π-band dispersion in graphene.[353]

The temporal behavior of the entanglement between electrons and single-mode radiation is such that the corresponding total Hamiltonian and time-evolution operators are block-diagonal. The initial states can be divided into two categories, for which buckling and intrinsic spin-orbit effects are either of the opposite or same sign. Negativity shows that the π-electrons and photons become periodically entangled for both categories. The entanglement exhibits sudden spontaneous variations when the buckling and spin-orbit effects are equal but opposite in sign; leading to quantum phase transitions. As the photonic excitation increases, the entanglement exhibits plateaux of constant duration for these initial states.[354]

Optical resonances in silicene are produced by excitonic effects: a $\pi \rightarrow \pi^*$ excitonic resonance at 1.23eV is contributed by the characteristic dispersion of Dirac fermions, while one at 3.75eV is due to the $\sigma \rightarrow \pi^*$ transition. Hydrogenation or fluorination of silicene removes the conductivity at the Dirac point and produces band-gap opening. The optical absorption of silicane, fluorosilicene and armchair silicene nanoribbons is dominated by strong excitonic effects and the formation of bound excitons with considerable binding energies.[355]

In pn and npn junctions made from silicene, the conductance exhibits the distinct gate-voltage dependences which are peculiar to the topological and non-topological phases, where the topological phase transition is caused by an external electric field. The conductance is suppressed in the np regime when both sides are topological, and in the nn regime when one side is topological and the other side is non-topological. The conductance is not suppressed in the nn regime when both sides are topological, and in the np regime when one side is topological and the other side is non-topological. The conductance is almost quantized to be 0, 1 and 2.[356]

In spite of their similar honeycomb atomic arrangements and linear-band dispersion, the electron-phonon coupling matrix-element squares of the Γ-E_g and K-A_1 modes in silicene are only about 50% of those in graphene. The smaller Fermi velocity in silicene compensates for this reduction by providing a larger joint electronic density-of-states near to the Dirac point, giving rise to comparable phonon linewidths. Kohn anomalies associated with these two optical modes are predicted to be significant in silicene. The electron-phonon coupling-induced frequency shift and linewidth of the Raman-active Γ-E_g mode in silicene, calculated as a function of doping, are comparable to those in graphene; indicating a similar non-adiabatic dynamic origin.[357]

The formation energies and band-gaps of hydrogenated few-layer silicenes increase with decreasing layer thickness. The higher formation energies imply a relatively lower structural stability of the thinner samples, due to their high surface/volume ratio. With decreasing layer thickness, the increased band-gaps lead to an obvious blue-shift of the optical absorption edge. For three differently stacked samples having the same thickness, an ABC stacking has the lowest formation energy and the largest band-gap, due to strong interactions of the silicon layers. A structural transition from ABC stacking to AA stacking causes a relative red-shift of the optical absorption peaks.[358] A systematic investigation of all of the stacking configurations of bilayer silicene shows that strong coupling is found between two silicene layers, destroying the Dirac cones in the band structures of pristine silicene and making bilayer silicene sheets metallic. The intercalation of alkali metals (especially potassium) can decouple the interaction between two silicene layers. In potassium-intercalated bilayer silicene, KSi_4, the Dirac cones are recovered with a small (0.27eV) band-gap located about 0.55eV below the Fermi level. The intercalation of K^+ cations in bilayer silicene, K^+Si_4, results in a semiconductor having a band-gap of 0.43eV.[359]

Many-body effects in silicene can be demonstrated using Green's function perturbation theory: that is, the GW+Bethe-Salpeter equation. Due to confinement, many-body effects play a vital role in quasi-particle excitations and optical absorption spectra; leading to excitonic resonance ($\pi \rightarrow \pi^*$ excitation) in silicene, and strongly bound excitons in silicane - with considerable binding energies.[360] First-principles studies of the optical absorbance of the group-IV honeycomb crystals can account for many-body effects on the optical properties by using a non-local hybrid functional. The optical absorption peaks are blue-shifted due to quasi-particle corrections, while the influence on low-frequency absorbance remains unchanged and reduces to a universal value which is related to the Sommerfeld fine-structure constant. At the Dirac points, spin-orbit interaction opens fundamental band-gaps and parabolic bands with a very small effective mass appear. The low-frequency absorbance is thus modified by a spin-orbit induced transparency region and an increase in the absorbance at the fundamental absorption edge.[361]

The electronic structure and excitonic optical properties of pristine and defected silicene, when investigated within a many-body Green's function and Bethe–Salpeter equation formalism, are such that - compared with pristine one - defects can significantly alter the band structure and form much better metallic characteristics. Low numbers of defects in pristine silicene can considerably alter the optical absorption peaks and change those peaks to lower energies. This has been demonstrated for 5% of defects in pristine silicene and two main peaks in the optical spectrum' are shown, one at low energies and another wider one at the higher energies.[362] These peaks can be used as a tool to define the existence of defects and predict the number of them in experiments.

The structure of the stable configuration has been calculated for group-IV elemental materials and for planar and low (< 1Å) atomic buckling configurations. This revealed a linear band dispersion, giving rise to massless Dirac Fermions having a Fermi velocity which is about two-thirds that of graphene. The monolayer stability is directly attributable to the phonons present, with the instability being driven by the out-of-plane ZA and ZO phonon modes. Long momentum relaxation lengths and high carrier mobilities are predicted for silicene-based devices, as carrier relaxation via phonon scattering is inhibited due to the electron-optical phonon coupling matrix elements being small: a factor of some 25 times smaller than in graphene.[363]

Studies performed using a self-consistent field crystal orbital method, based upon density functional theory, of armchair nanoribbons with a width of 9 to 39 silicon atoms show that the carrier mobilities - obtained from deformation potential theory - oscillate with respect to the width. The values are comparable with those for graphene nanoribbons. A buckled structure, hydrogen saturation, edge reconstruction and edge-roughness all decrease the carrier mobilities; and this can be explained in terms of crystal orbitals.[364] Using first-principles methods incorporating density functional theory, the Boltzmann transport equation and deformation potential theory, the electron mobility is calculated to be 2.57 x $10^5 cm^2/Vs$ while the hole mobility is 2.22 x $10^5 cm^2/Vs$ at room temperature. The latter is smaller than that of graphene, but is still very high.[365]

By using the two-dimensional ionic Hubbard model to describe the electronic structure of silicene in the presence of an electric field induced by the substrate, and by using the coherent-potential approximation to calculate the zero-temperature phase diagram and the associated spectral function at half filling, it is found that any degree of symmetry-breaking induced by the electric field causes the silicene structure to lose its Dirac fermion characteristics. This provides a simple mechanism for the disappearance of the Dirac cone.[366]

The electrical properties are modified by a transverse electric field, and band-gap varies as the absolute value of the field strength. This can be explained in terms of the Stark quadratic effect.[367] In the absence of an external electric field perpendicular to the silicene sheet, the thermopower of pure silicene is of the order of some 80μV/K. When a finite field is applied, a comparatively large band-gap is opened up and the thermopower is then increased by several times as compared to the case without the field.[368]

First-principles density function theory projector-augmented wave potential analyses of silicene nanoribbons with phosphorus doping-sites ranging from edge to center show that the phosphorus atom substitutes preferentially at the edges of both armchair and zig-zag nanoribbons. Although the pristine nanoribbons are non-magnetic semiconductors, the armchair nanoribbons with P-doping at edge sites change to ferromagnetic semiconductors due to spin-splitting induced by the impurity. Armchair nanoribbons with phosphorus doping at inner sites change to non-magnetic metals because of the appearance of a half-filled impurity band across the Fermi level. When phosphorus is doped into zig-zag nanoribbons, an antiferromagnetic to ferromagnetic transition occurs due to spin-suppression at the doped edge.[369]

With regard to the spin-dependent electron transport properties of zig-zag nanoribbons in the presence of an anti-ferromagnetic exchange field, analyzed using a non-equilibrium Green's function method, application of a transverse electric field leads to spin splitting such that the nanoribbon can function as a spin filter. Spin polarization calculations show that the spin filtering is perfect and that the spin states of electrons are fully coherent. The spin direction of transmitted electrons through the filter can be easily controlled by changing the transverse electric field direction. Use of the Hubbard model takes account of the electron-electron interaction and shows that, although this interaction causes some changes in the electron conductance, it has no destructive effect upon the spin filtering properties. The effect of a single vacancy on the electron transport is to decrease the electron conductance, while the spin-dependent properties remain the same. A vacancy near to the edge of the nanoribbon has a less destructive effect upon electron conductance than does one located in the middle.[370]

First-principles calculations suggest that, when heavy-metal (gold, mercury, thallium, lead) atoms are adsorbed on silicene, hollow sites are energetically favored. The heavy-metal adatoms produce substantial energy-gaps and band-splitting in silicene sheet if the binding is strong. This is not always so.[371]

First-principles studies of superlattices, made by alternately stacking silicene and MoS_2 monolayers, show that the distortions of the geometry due to superlattice formation are relatively small and result from relatively weak interactions between the stacking layers. The silicene/MoS_2 superlattice is metallic, with the linear bands around the Dirac points of the pristine silicene apparently being preserved. Small band-gaps are opened up at the Dirac points, due to symmetry-breaking in the silicene layer caused by the introduction of the MoS_2 sheets. Charge transfer occurs mainly within the silicene and MoS_2 layers, as well as some part of the intermediate regions between the silicene and between the stacking sheets in the superlattice.[372]

The properties of silicene paired with a MoS_2 substrate, studied systematically using density functional theory with van der Waals corrections, are such that nearly linear band dispersions can be preserved in the heterobilayers due to the weak interface interactions. The band-gap is opened up because the sub-lattice symmetry is broken by the intrinsic interface dipole. As usual, the band-gap values can be effectively modulated by using an external electric field.[373]

First-principles calculations reveal that a band-gap is opened up in silicene nanomesh when the width, W, of the wall between neighboring holes is even. The size of the band-gap increases with decreasing W, and there is a simple relationship to the ratio of removed silicon atoms to the total number of silicon atoms in silicene. Quantum-transport simulation reveals that sub-10nm single-gated silicene nanomesh field-effect transistors exhibit excellent behavior at zero temperature, but that this is greatly degraded at room temperature.[374]

Density functional calculations show that a sizeable band-gap can be opened up in silicene, without degrading its carrier mobility, by using silicene/substrate hybrid structures with non-covalent interface interactions. More notable charge redistribution in the two sub-lattices of silicene, and thus a larger band-gap, is characterized by a smaller interlayer distance. The band-gap opened up in hybrid structures with SiH/π interaction has attained the quality required for room-temperature use as a field-effect transistor.[375]

Silicene monolayers which are grown onto Ag(111) surfaces have a band-gap that is tunable, using oxygen adatoms, from semi-metallic to semiconducting. Low-temperature scanning tunnelling microscopy shows that the adsorption configurations and numbers of oxygen adatoms on the silicene surface are critically important in its band-gap engineering, which is dominated by various buckled structures in (√13 x √13), (4 x 4) and (2√3 x 2√3) layers. The Si-O-Si bonds are the most energetically-favored species which are formed on (√13 x √13), (4 x 4) and (2√3 x 2√3) structures during oxidation. The monolayers retain their structures when fully covered by oxygen adatoms. It is thus feasible to tune the band-gap of silicene by using oxygen adatoms, unlike the case of graphene oxide.[376]

The band-gap is strongly affected by the substrate. In the case of silicene on $MgBr_2$(00•1), modified by lithium and sodium, charge transfer from silicene to substrate is found for substitutional doping (intercalation). Compared to the band-gap of 0.01eV of the pristine substrate, strongly increased band-gaps of 0.65eV (substitutional doping) and 0.24eV (intercalation) are found, and the band-gap increases with dopant concentration.[377]

Two-dimensional epitaxial silicene forms via surface segregation on zirconium diboride thin films grown onto silicon wafers. A particular buckling of the silicene, imposed by the epitaxial relationship with respect to the diboride surface, leads to a direct π-electronic band-gap at the Γ-point. This again demonstrates that the buckling, and thus the electronic properties of silicene, are modified by epitaxial strain.[378]

First-principles calculations indicate that cobalt-decorated silicene can host an anomalous quantum Hall state. The exchange field induced by the cobalt atoms, combined with the strong spin-orbit coupling of the silicene, opens up a non-trivial band-gap at the K-point. Cobalt-decorated silicene is unique in this respect, as compared to other transition metals, since induced hybridization and spin-polarization usually suppress an anomalous quantum Hall state.[379]

Binding-energy calculations show that boron, carbon and nitrogen adsorbates, apart from the homo N-N dimer, form a stable configuration on silicene. Band-gap opening can be achieved for all of the carbon adsorbates, the homo dimers of boron and nitrogen, the hetero N-B dimer and the B-C-N trimer.[380]

Density functional theory calculations indicate that a semiconducting silicane monolayer is free from dangling bonds and is thus an ideal substrate for silicene. An almost linear band dispersion in silicene, with a band-gap opening of 44 to 61meV, is obtained in all hetero-bilayers. The effective masses of electrons and holes (ranging from 0.033 to $0.045m_0$) near to the Dirac point are very small in hetero-bilayers, and a high carrier mobility ($10^5 cm^2/Vs$) is therefore expected. The characteristics of hetero-bilayers can be modulated by applying a bias voltage or mechanical strain.[381]

First-principles calculations for silicene on hexagonal boron nitride indicate that, because the nitride is semiconducting, the interaction with silicene is weaker than it is for metallic substrates. It is thus possible to open up a 50meV band gap in the silicene. The onsite Hubbard interaction parameter for manganese decoration differs significantly for the top and hollow sites.[382]

First-principles study of the electronic functionalization of monolayers via the chemisorption of hydrogen, lithium, fluorine, scandium, titanium or vanadium adatoms, with consideration of 3 general adatom-monolayer configurations, shows that the induced band-gap is a Γ-gap ranging from 0.2 to 2.3eV. In general, an alternating configuration is the most energetically stable. Boat-like and chair-like conformers are degenerate, with the former having anisotropic effective carrier masses. A top configuration leads to a planar monolayer, and usually to a gapped dispersion. A hollow configuration, with vanadium adatoms, retains the Dirac cone; but with strong orbital planar hybridization at the Fermi level.[383]

When armchair and zig-zag nanoribbons are cut from semi-metallic buckled silicon sheet, the armchair nanoribbons are non-magnetic and exhibit a three-family band-gap decay profile with increasing width; as in graphene nanoribbons. The influence of edge spin polarization is much greater on zig-zag nanoribbons. The antiferromagnetic state is the most stable ground state for zig-zag ribbons, with a monotonic decrease in band-gap as the width increases. Wider ribbons have large transmission values.[384]

Theoretical investigation of Friedel density oscillations, and collective excitation, in the random phase approximation indicates that a tunable spin-valley coupled band structure can lead to exotic properties. An exact analytical and numerical analysis demonstrates that the beating of the screened potential, as well as the undamped plasmon mode, are fingerprints of a topological phase transition in doped silicene. It is concluded that a connection between the topological phase transition, and density oscillations, can be monitored using experimental techniques.[385]

Electron energy-loss spectra, calculated for α-silicene (flat), and β-silicene (buckled), show that - compared to bulk silicon - additional resonances appear which are red-shifted. Buckling gives rise to a richer structure at low energies.[386]

Density-functional calculations demonstrate that bilayer silicene has a stable structure, with AB stacking, and preserves linear energy dispersions near to the K and K′ points in the hexagonal Brillouin zone. The study of four hydrogenated bilayer silicenes using phonon-mode analysis confirms that the configurations of semi-hydrogenated single-layers and semi-hydrogenated double-layers are dynamically stable, with no imaginary phonon frequencies. The semi-hydrogenated single-layers have a ferromagnetic ground-state due to unpaired 3p electrons, while semi-hydrogenated double-layers are predicted to be indirect band-gap semiconductors with a non-magnetic ground state.[387]

By using *ab initio* calculations, a self-consistent scattering potential can be derived for boron, nitrogen, aluminum and phosphorus substitutions in silicene. The quantum-mechanical Landauer-Büttiker approach can be used to evaluate the conductivities of doped silicene ribbons having various impurity concentrations. An individual defect introduces asymmetrical electron-hole conductivities that depend upon both the type of dopant and the position of the foreign atom with respect to the edges. Quantum interference effects at zero temperature show that randomly distributed defects over 1μm long, and realistically wide nanoribbons, widen the intrinsic electronic gap that arises from quantum confinement.[388]

The presence of a perpendicular electric field, E_z, tunes the band gap and level broadening. The interplay of the strong spin-orbit interaction and the field, E_z, gives rise to topological phase transitions. At a critical value of E_z, the direct-current spin-Hall conductivity undergoes a transition from a topological insulator phase to a band insulator phase. The spin- and valley-Hall conductivities exhibit a strong temperature dependence. The longitudinal conductivity scales linearly with n_e/n_i, where n_e is the carrier density for screened Coulomb impurities of density, n_i. It also exhibits an upward jump at a critical value of n_e that is associated with the opening of a new spin sub-band. The contributions of spin-up and spin-down carriers to the power absorption spectrum depend sensitively upon the topological phase and valley index.[389]

A non-equilibrium Green's function technique study, combined with density functional theory, of zig-zag silicene nanoribbons doped with aluminum or phosphorus atoms, shows that the presence of a single aluminium or phosphorus atom introduces quasi-bound states into the nanoribbons which can be observed as new dips in the electron conductance. The aluminium atom acts as an acceptor, whereas the phosphorus atom acts as a donor if it is placed at the center of the ribbon. This behavior is reversed if the dopant is placed on the edges. An acceptor-donor transition is thus observed in zig-zag silicene nanoribbons upon changing the dopant position. Similar results are obtained if two silicon atoms are replaced by two impurities (Al or P) but the conductance is generally modified due to impurity-impurity interaction. If doping breaks the two-fold rotational symmetry about the central line, the transport becomes spin-dependent.[390]

First-principles and tight-binding studies of silicene superlattices show that, according to the band-folding picture for an hexagonal superlattice, (M,N) – where M and N are both integer multiples of 3 - the simultaneous folding of K- and K′-points to the Γ-point results in a four-fold degeneracy. In the orthogonal case, [P,Q], four-fold degeneracy occurs if $Q = 3q$, where q is an integer. By introducing degenerate perturbations to remove the four-fold degeneracy, the band-gap can be opened, whether or not the inversion symmetry remains. In other cases, the band-gaps can be kept closed by inversion-symmetry preservation.[391]

The electronic structure of silicene on Ag(111), studied by means of scanning tunnelling microscopy and angle-resolved photo-emission spectroscopy, contains a flat band at 0.9eV below the Fermi level. This band, as compared to that on flat terraces, is strongly suppressed near to atomic defects, domain boundaries and step edges.[392]

Density functional first-principles calculations, of alkali metal (X = Li, Na, K, Rb) atoms adsorbed on silicene, show that the formation energies of SiX are negative; indicating that the structural stability of these compounds is higher than that of silicene. Electric charge is transferred from the silicon atoms to the hydrogen atoms in SiH but, in SiX compounds, the direction of charge transfer is from the alkali metal atoms to the silicon atoms. The bonds between silicon and alkali metal atoms are mainly ionic, but with some covalent contribution. Band-structure calculations indicate that SiLi is a semiconductor with a direct band-gap of 0.34eV. All of the other compounds exhibit metallic properties.[393]

First-principles investigations of the geometrical structures of porous silicene nanosheets (the silicon analogue of α−graphyne, and known as silicyne) show that elemental silicyne sheets are energetically unfavourable. Following carbon-substitution, hybrid graphyne-like sheets of C-silicyne possess robust energetic and dynamical stabilities. Unlike silicene, C-silicyne is a atomically flat. In C-silicyne, a semi-metallic Dirac-like property is maintained in the non-magnetic state but a spontaneous anti-ferromagnetism produces massive Dirac fermions and opens up a considerable gap between the Dirac cones. Tensile strain can further enhance the anti-ferromagnetism, linearly modulating the gap value without altering the direct-bandgap feature. Strain engineering of C-silicyne can produce type-II band alignment with MoS_2 sheet. A combined C-silicyne/MoS_2

nanostructure offers a power-conversion efficiency of better than 20% for photovoltaic solar cells.[394]

First principles many-body calculations of a two-dimensional silicene-graphene hybrid, based upon phonon-spectrum calculations, indicate an absence of soft modes and a consequent stability of the system. Analysis of optical absorption spectra, and of the correlation in real space between excited electron-hole states, reveals that the excitonic effects in this silicene-graphene hybrid are significant and lead to strong bound excitons. The first active exciton is characterized by a binding energy of 0.81eV, an effective mass of $0.41m_0$ and a Bohr radius of 2.78Å.[395]

Due to the large numbers of degrees of freedom permitted by the buckled structure of silicene, its bilayer structure can have several different stacking configurations. In the lowest-energy conformation, the bilayer material loses its buckled structure and becomes planar. This conformation is possible because there is an energy gain when the system loses its π-cloud to create extra (σ-like) chemical bonds between the two layers. Simulated scanning tunnelling microscopic images give excellent agreement with experimentally obtained images of bilayer silicene. Analysis of the three-dimensional band-structure is deemed to be fundamental for complete understanding of the electronic and transport properties, and differing structures exhibit distinct properties. For some stackings there is an anisotropic behavior of the current as a function of the direction of the applied bias.[396]

Graphene is very effective in cloud computing, but its zero band-gap is problematic. This situation can be improved by preparing a composite with silicene. If graphene is used as an n-type dopant of silicene, a band-gap is then introduced into graphene.[397]

First-principles studies of silicene/silicane hetero-bilayers show that the silicene interacts with silicane, with a binding energy of -50 to -70meV per silicon atom. This suggests a weak van der Waals interaction between the silicene and the substrate. A relative large band-gap with a linear band dispersion of hetero-bilayers is opened up due to the sub-lattice symmetry being broken by the intrinsic interface dipole between the silicene and substrate. The band-gap can also be continually modulated by adjusting the interlayer spacing and strain; independent of the stacking arrangement.[398]

Density functional theory, using both the generalized gradient approximation and hybrid functional calculations, shows that hydrogenated bilayer silicene can have three configurations: AA chair-like, AB chair-like and AA boat-like. Following hydrogenation, the material changes from a zero band-gap semi-metal to an indirect band-gap semiconductor with forbidden band-widths of 1.208, 1.437 and 1.111eV. Hybrid functional calculations yield band-gaps of 1.595, 1.785 and 1.592eV, respectively.

Analysis of hydrogenated bilayer silicene with a strained band-gap indicates that the relationship between strain and band-gap can be continuously adjusted.[399]

Studies of hydrogenation-based surface modifications at experimentally measured coverages of about 33% show that such organic surface modification leads to an increase in the buckling distance of silicene, while the angles of the bonds in the honeycomb structure decrease. Although the initial hydrogenation makes silicene an indirect band-gap semiconductor, subsequent organic surface modification causes further changes in the band structure. Hydrosilylation, phenylation, alkoxylation and amination lead to a reduction in the band-gap of H-silicene. Hydrosilylated and phenylated silicene are indirect band-gap semiconductors while alkoxylated and aminated silicene are direct band-gap semiconductors. Changes in the optical absorption, induced by the organic surface modification, can be closely correlated with the corresponding changes in band structure.[400]

The Rényi-Wehrl entropy of the quantum state in phase space can be used to identify transitions from a topological insulator to a band insulator in silicene in the presence of perpendicular magnetic and electric fields. Electron-hole entropies exhibit an inversion/crossing behavior at the charge neutrality point for any Landau level, and the combined entropy of particles plus holes is a maximum at this critical point. The results are interpreted in terms of a delocalization of the quantum state in phase space. This entropic approach is valid in general two-dimensional gap Dirac materials, having a strong intrinsic spin-orbit interaction, which are isostructural with silicene.[401]

When the Weaire-Thorpe model is applied to silicene with its buckled two-dimensional structure, using the original Weaire-Thorpe model for four hybridized sp^3 orbitals on each atom together with inter-atom hopping, the band structure can be systematically examined in three dimensions. Flat (dispersionless) bands also exist here. In the case of silicene, the Weaire-Thorpe model is re-formulated in terms of the overlapping molecular orbital method, which permits the description of flat bands away from the electron-hole symmetry-point. This overlapping molecular orbital formalism reveals an important difference. That is, in three dimensions the dispersive bands with cones are sandwiched between doubly-degenerate flat bands. In two dimensions the dispersive bands with cones are sandwiched between triply-degenerate and non-degenerate (nearly) flat bands. This is consistent with the original Takeda-Shiraishi band calculations. This helps to explain why the entire band-structure of silicene comprises a pair of dispersive bands with Dirac cones, with each of the bands touching a nearly flat (narrow) band at Γ. For the purposes of band-engineering, the bonds perpendicular to the atomic plane are crucial. Ferromagnetism or structural instabilities are to be expected if the chemical potential can be shifted close to the flat bands.[402]

Effective Hamiltonians for silicene under the influence of strain, electric and magnetic fields have been derived by using the method of invariants. When these Hamiltonians are compared to those for graphene the expression for band-warping, obtained analytically, is of a different order to that for graphene. Uniaxial strain does not open up a gap, contrary to some numerical predictions.[403]

Study of the photo-induced manipulation of charge carriers in monolayer silicene subject to intense electromagnetic terahertz radiation, and consideration of the Dirac cone approximation going beyond the off-resonant condition for large frequencies of the radiation field where only virtual photon processes are allowed, yields the exact zero-momentum pseudo-spin polarization and numerical results for the quasi-energy band structure and time-averaged density of states. Resonant processes, due to real photon emission and absorption processes, induce a band inversion that qualitatively modifies the quasi-energy spectrum. These band-structure changes are detected as an inversion of the averaged pseudo-spin polarization. Analysis of the time-averaged density of states shows that effective photo-induced gap manipulation can be achieved only in the intermediate and strong matter-radiation coupling regime, where the off-resonant approximation fails.[404]

First-principles calculations of the low-energy electronic properties and van Hove singularities under uniaxial strain show that the Dirac points persist when silicene is stretched uniaxially, but are shifted away from the corners (K-points) of the first Brillouin zone. The relative positions of the Dirac points with respect to the K-points, for samples strained along the armchair or zig-zag directions, exhibit opposite tendencies as compared with strained graphene. This is due to the larger deformation of the unit cell of strained silicene as compared with that of strained graphene. For silicene under armchair or zig-zag strain, the Fermi velocities around the Dirac points along the positive and negative directions of the first Brillouin zone exhibit quite significant differences. The character of the van Hove singularities just above the Fermi energy undergoes a transition from $\pi*$-band to $\sigma*$-band under increasing armchair or zig-zag strain, again suggesting that uniaxial strain is effective in tuning the electronic properties of silicene.[405]

First-principles studies of silicene on WSe_2 show that the energy barriers for lateral translation between the two sub-systems are very small, due to weak van der Waals interactions. For this reason, the Dirac phenomena of silicene are preserved. The induced band-gap is sufficient to withstand thermal fluctuations.[406]

Density functional theory investigation of heterostructures consisting of silicene, graphene and BN monolayers under the influence of an electric field shows that, with no electric field, silicene/graphene and silicene/BN have a finite gap of about 50meV although silicene is a zero-gap two-dimensional material. Application of a field perpendicular to the bilayer system facilitates modulation of the band-gap, with an approximately linear relationship to the gap energy. This contrasts with the behaviors of the constituent monolayers. The degree of modulation is determined mainly by the Si-p_z electronic states at the interfaces of the silicene/graphene and silicene/BN bilayers.[407]

Slightly buckled silicene monolayers, being centrosymmetric like graphene and unlike piezoelectric hexagonal boron nitride, are not intrinsically piezoelectric. First-principles calculations show that chemical co-decoration with hydrogen and fluorine on opposite sides of silicene (one side decorated with hydrogen and the other with fluorine) breaks the centrosymmetry. By redistributing the charge density, due to the electronegativity difference between the atoms, non-centrosymmetric co-decoration induces out-of-plane dipolar polarization and a concomitant piezoelectricity into a non-piezoelectric silicene monolayer. The associated piezoelectric coefficients are comparable with those of other two-dimensional piezoelectric materials, such as hydrofluorinated graphene/h-BN, and of some bulk semiconductors, such as wurtzite GaN and wurtzite BN. Because of the lower elastic constants of silicene as compared with those of graphene or h-BN, the piezoelectric strain constants are significantly larger than those of hydrofluorinated graphene/h-BN. It is predicted that a wide range of band-gaps, having an average width of 2.52eV, can be opened up in slightly buckled gap-less semi-metallic silicene monolayers by surface co-decoration with hydrogen and fluorine atoms.[408]

Boltzmann transport-equation investigation of low-field carrier transport in dual-gated silicene field-effect transistors, taking account of screened charged impurity scattering, shows that both the calculated conductivity and band-gap not only depend strongly upon the carrier sheet density, but also depend strongly upon the effective offset density. Screening-induced metal-insulator transition phenomena in buckled silicene can be observed theoretically, and are similar to those found in monolayer graphene.[409]

Application of the tight-binding formalism, taking account of the effect of an exchange field, M_z, to a zig-zag silicene ribbon subjected to an external electric field, E_z, shows that the band structure lacks spin inversion symmetry in the presence of intrinsic spin-orbit interaction in a combination of E_z and M_z fields. Quantum transport calculations indicate that, for certain energy ranges of the incoming electrons, the silicene ribbon can act as a controllable high-efficiency spin polarizer. The polarization maxima occur simultaneously with the van Hove singularities of the local density of states. In this case, the combination of electric and exchange fields is the key to achieving nearly perfect spin polarization, which also leads to the appearance of additional narrow plateaux in the quantum conductance. The output current remains completely spin-polarized for low-energy carriers, even when a few edge vacancies are present.[410]

Density functional theory, with dispersion corrections, shows that for three coverage ratios of nitrophenyl diazonium (1:8, 1:18, 1:32) spin-polarized electronic structures can always be obtained with nitrophenyl diazonium adsorption, although the band-gap decreases upon reducing the nitrophenyl diazonium coverage ratio. The quasi-localized p_z electrons of silicene are responsible for ferrimagnetism in the two-dimensional systems. The system with a nitrophenyl diazonium to silicene ratio of 1:8 is predicted to be a bipolar magnetic semiconductor. Half-metallicity can thus be obtained by applying a gate voltage with reversible spin polarization.[411]

Studies performed using density functional theory, combined with non-equilibrium Green's function methods, of a series of armchair silicene nanoribbons with various torsion angles, show that several key factors, such as the electron transmission coefficient and the band structure, determine the transport properties. The results suggest that the transport properties of armchair nanoribbons are insensitive to the torsional configuration in the scattering region. With increasing torsion angle, the transmission coefficient is well maintained within the limits of the torsion angle. When that angle is increased even to 120°, the current drops by only 22% as compared with initial armchair nanoribbons at a zero torsion angle. All of the configurations behave here as conventional conductors and exhibit an almost linear current-voltage dependence.[412]

Density functional theory calculations in the generalized gradient approximation, of the cooperative effects of uniaxial strain and degenerate perturbations, show that uniaxial strain splits π-bands into π_a- and π_z-bands; resulting in Dirac-cone movement. An hexagonal antidot splits π_a- (π_z-) bands into π_{a1}- and π_{a2}- (π_{z1}- and π_{z2}-) bands. This accounts for band-gap opening in superlattices, with the Dirac cone being folded to the Γ point. This is a different mechanism to sub-lattice equivalence breaking. The energy interval between the split π_a- and π_z-bands can be tuned so as to switch the band-gap on and off. This suggests the occurrence of a reversible switch between the high charge-carrier velocity properties of massless Fermions, attributed to the linear energy dispersion relationship around the Dirac point, and the high on-off properties which are associated with a sizable band-gap. The gap-width can be continuously tuned by manipulating the strain.[413]

First-principles calculations show that silicene, encapsulated between transition-metal dichalcogenide layers, exhibits a two orders of magnitude enhancement of its non-trivial band-gap. This is driven by the strong spin-orbit coupling effect in transition-metal dichalcogenides, via the proximity effect. The cladding layers shield the silicene from environmental gases, which are detrimental to the quantum-spin Hall state in the case of free-standing silicene. Encapsulated silicene is thus a novel two-dimensional topological insulator with a robust non-trivial band-gap.[414]

The dielectric properties of atomically thin silicene, investigated as a function of temperature, shows that - with a zero energy-gap - more carriers are thermally excited as the temperature increases and the intraband transition is strengthened; resulting in a stronger absorption peak.[415]

Density functional theory calculations of the asymmetrical modification of silicene by co-grafting a different group (X, Y) onto each side indicate that chemical functionalization of one side can greatly enhance the chemical reactivity of the opposite side. This suggests that communication occurs between the two adsorbed groups and enhances the stability of the hybrids. When compared with pristine silicene with a zero band-gap, X-silicene-Y sheets exhibit semiconducting properties with a non-zero band-gap. These depend upon the coverage of X and Y.[416]

First-principles calculations, based upon density functional theory, of the electronic properties of armchair silicene nanoribbons perforated with periodic nanoholes considered two different configurations of mono-hydrogenated and di-hydrogenated silicene edges. Pristine armchair silicene nanoribbons can be categorized into three groups with widths, W, of 3P-1, 3P and 3P+1; where P is an integer. The order of the energy gaps changes from (3P-1) < (3P) < (3P+1), for W-mono-hydrogenated armchair silicene nanoribbons, to (3P+1) < (3P-1) < (3P) for di-hydrogenated W-armchair silicene nanoribbons. The band-gaps of mono-hydrogenated W-armchair silicene nanoribbons and di-hydrogenated (W+2)-armchair silicene nanoribbons are slightly different, giving larger band-gaps for wider di-hydrogenated armchair silicene nanoribbons. The band-gaps of armchair silicene nanoribbons perforated with periodic nanoholes changes, depending upon the nanoribbon width, the nanohole periodicity and the position relative to the nanoribbon edge as compared with pristine armchair silicene nanoribbons, because of changes in the quantum confinement strength. Di-hydrogenated armchair silicene nanoribbons perforated with periodic nanoholes are more stable than mono-hydrogenated ones and their band-gaps are noticeably greater than those of mono-hydrogenated armchair silicene nanoribbons perforated with periodic nanoholes. The band-gap for di-hydrogenated 12-armchair silicene nanoribbons perforated with periodic nanoholes, with a periodicity of 2, is 0.923eV. This value is some 2.2 times greater than that for pristine di-hydrogenated armchair silicene nanoribbons and twice that of mono-hydrogenated 12-armchair silicene nanoribbons perforated with periodic nanoholes with a periodicity of 2.[417]

Investigation of the selection rule and optical spectrum for intrinsic armchair-edge silicene nanoribbons with a number of silicon chains in an external electromagnetic field at low temperatures, using the dipole-transition theorem for semiconductors, shows that both semiconducting 9-, 10- and metallic 11- armchair-edge silicene nanoribbons exhibit broad values of optical conductivity and dielectric function, from the infra-red to the ultra-violet. The optical spectra for semiconducting 9- and 10-armchair-edge nanoribbons exhibit transitions between valence and conduction bands having the same and/or different sub-band indices. Those for metallic 11-armchair-edge nanoribbons may originate only from transitions between valence and conduction bands having differing indices.[418]

Gated bilayer silicene exhibits a band insulator phase for small Rashba spin-orbit coupling, and then changes to a strong topological insulator phase with both spin and valley filtering at large Rashba spin-orbit coupling. The strong topological insulator phase is robust in the presence of intrinsic spin-orbit and intrinsic Rashba spin-orbit couplings. When a tilted electric field is introduced, the in-plane component of the electric field gives rise to an interlayer Rashba spin-orbit coupling, and the system turns into a band insulator phase no matter how large the Rashba spin-orbit coupling and bias voltage.[419]

Restructuring of the electronic spectrum in a buckled silicene monolayer can occur when a voltage is applied between its two sub-lattices in the presence of certain impurity atoms. Qualitative restructuring of the quasi-particle spectrum within the initial band-gap, and specific metal-insulator phase transitions, are possible for such a disordered system and can be effectively controlled by varying the electric field bias at a given impurity perturbation potential and concentration.[420]

Systematic examination of the dwell-time for quantum-tunnelling through single and multiple-gate silicene nanostructures shows that, unlike the case of graphene, superluminal tunnelling is observable even at normal incidence. This is due to the sizeable spin-orbit gap of silicene. Together with its field-tunable band-gap, this superluminal tunnelling can be switched on and off electrically. Simulating the dwell-time through symmetrical and asymmetrical double-barrier structures indicates that the dwell-time exhibits a distinct dependence upon the former and latter barrier profiles.[421]

A theoretical realization of quantum capacitance, which has advantages over the scattering problems of traditional transport measurements, can be derived as a function of the Fermi energy and temperature while taking account of electron-hole puddles via a Gaussian broadening distribution.[422]

First-principles calculations of the band-gap opening of silicene nanomesh show that, unlike the mechanism of band-gap opening induced by sub-lattice equivalence breaking, degenerate perturbation through the breaking of bond symmetry can split the π-like bands in the inversion-symmetry preserved nanomesh. This results in π_{a1}-π_{a2} and π_{z1}-π_{z2} band-sets having sizable energy intervals. As well as band-gap opening in nanomeshes with the

Dirac point folded to the Γ-point, the split energy intervals are separate, away from the Fermi level, to leave the semi-metallic nature unchanged for nanomeshes with Dirac points located at opposite sides of the Γ-point, as opposite pseudo-spin wave valleys. A mass band-gap can then be opened, with the aid of uniaxial strain, to cause the nanomesh to be semiconducting with a width that can be continuously enlarged up to a maximum value. The latter could also be tuned by controlling the defect density.[423]

The calculated local density of states around adsorbates clearly reveals that the induced localized states contain band topology information which can be used to determine whether or not the system is a topological insulator. The impact of randomly distributed adsorbates at low concentrations on electron structures and transport properties is such that edge-mode back-scattering is significantly enhanced when the energies of the incoming modes from leads match those of the in-gap localized states.[424]

Investigation of the interactions between 3d transition metals (chromium, manganese, iron, cobalt) and silicene, using DFT+U calculations and the linear response method, shows that in the case of two-dimensional $FeSi_2$ and $CoSi_2$, the metal atoms tend to penetrate into the silicene layer. While $CoSi_2$ is non-magnetic, $FeSi_2$ exhibits a total magnetic moment of $2.21\mu_B$/cell. In two-dimensional MSi_6, a trend towards anti-ferromagnetic favorability in the z-direction is predicted. In the (less stable) ferromagnetic states, each primary unit cell of $CrSi_6$, $MnSi_6$ and $FeSi_6$ has a differing level of total magnetization: 4.01, 5.18 and $2.00\mu_B$/cell, respectively. The absolute magnetization of anti-ferromagnetic MSi_6 structures ranges from 5.33 to $5.84\mu_B$/cell. A direct band-gap of 0.2eV is predicted for anti-ferromagnetic $MnSi_6$, while metastable ferromagnetic $FeSi_6$ structure is predicted to have a band-gap of 0.85eV. Super-exchange interactions between metal atoms in the MSi_6 systems result in anti-ferromagnetic alignments.[425]

Investigation of the infra-red to ultra-violet optical response of perfect armchair-edge silicene nanoribbons with N atomic chains in an external electromagnetic field at low temperatures, using the dipole-transition theorem for semiconductors, shows that the electronic structure can be separated into two distinct families. One is semiconducting and one is metallic. The dipole-transition probabilities, from the valence sub-bands to the smaller indexed conduction ones for semiconducting 9-, 10- and metallic 11-armchair-edge nanoribbons continuously increase, while those to the larger indexed conduction sub-bands first increase and then decrease. Both semiconducting 9-, 10- and metallic 11-armchair-edge nanoribbons have a wide (0.50 to 3.20eV) absorption spectrum, refractive index and extinction coefficient in the vicinity of the K-point. The optical response for semiconducting 9-, 10-armchair-edge nanoribbons arises from transitions between the valence and conduction sub-bands with the same and/or different indices. That for metallic 11-armchair-edge nanoribbons may originate from the different indexed valence and conduction sub-bands.[426]

The electronic structure of a 4 4 silicene monolayer, on a semi-infinite Ag(111) substrate, calculated using density functional theory and the embedded Green's function technique confirms that two-dimensional Dirac bands do not exist on this surface. This is due to symmetry-breaking and to strong orbital hybridization between the silicon π and silver sp states.[427]

The electronic structure of silicene can be simulated by using the tight-binding method, with the basis: $sp^3d^5s^*$. The results are in good agreement with *ab initio* calculations. The effective Hamiltonian in the vicinity of the Dirac point can be constructed by using the method of invariants. The silicon atoms are located on two parallel planes, displaced perpendicularly to each other. The energy spectrum depends essentially upon this displacement.[428]

The effects of thermal and electric fields upon the electronic properties of silicene, studied using a statistical analysis of canonical ensembles combined with the tight binding method, are such that symmetry-breaking caused by the buckling disorder in the thermal field greatly alters the band structures of silicene with small cells. The buckling variation of any atom is compensated by other atoms in the case of large cells. The band-structure features near to the Fermi energy in pristine silicene are thus still protected by the sub-lattice symmetry in a thermal field. The thermal field enhances the effect of the electric field so as to generate a band gap. Randomly buckled silicene requires a much smaller electric field than does pristine silicene. A higher temperature corresponds to a larger gap in a given electric field.[429]

First-principles study of the effects of a small biaxial strain (≤5%) and of a perpendicular electric field on slightly buckled silicene shows that, with increasing biaxial strain, the conduction bands at the high-symmetry Γ- and M-points of the first Brillouin zone shift markedly towards the Fermi level. On the other hand, an electric field changes the band dispersions near to the Γ-point and opens up a small band-gap at the K-point. The field-induced gap-opening can be enhanced by a compressive strain and attenuated by a tensile strain.[430]

Low-energy electron diffraction and scanning tunnelling microscopic studies of hydrogenated silicene with hydrogen on one side (half-silicane), using the so-called 23 x 23 phase of silicene on Ag(111), clearly show that hydrogenation changes the structure of silicene on Ag(111); resulting in a (1 x 1) periodicity with respect to the silicene lattice. The hydrogenated silicene also exhibits a quasi-regular (23 x 23)-like arrangement of vacancies. Angle-resolved photo-electron spectroscopy reveals two dispersive bands which can be unambiguously attributed to half-silicane. The common top of these bands is located at about 0.9eV below the Fermi level. These experimental bands are closely reproduced by the theoretical band-structure of free-standing silicene, with hydrogen adsorbed on the upper hexagonal sub-lattice.[431]

First-principles calculations of the structural stability and electronic properties of penta-silicene nanosheets show that, unlike penta-graphene, pristine penta-silicene sheet has soft modes which originate from the unfavourable highly-buckled configuration of tri-coordinated silicon atoms. The dynamic stability of penta-silicene is restored by surface hydrogen decoration, which also imparts good energetic and mechanical stability. Hydrogenated penta-silicene sheet is a soft material, having good flexibility, which can support biaxial and uniaxial strains of up to 25 and 31%, respectively. Two-dimensional hydrogenated penta-silicene nanosheet is an indirect band-gap semiconductor. In one-dimensional nanoribbons, the edge states induce flat bands at the Fermi level; causing spontaneous spin-polarization. In narrow nanoribbons, the antiferromagnetic state is more favorable than the ferromagnetic one. They become degenerate in wide nanoribbons. Ferromagnetic hydrogenated penta-silicene nanoribbons are bipolar magnetic semiconductors which can be changed into half-metals, with opposite conducting spin channels, by p-type and n-type doping.[432]

First-principles calculations show that the (1 x 1) phase of silicene is energetically more favorable than the (3 x 3) silicene superstructure on a semiconducting Bi/Si(111)-(3 x 3) substrate. The band-gap is strongly influenced by the participation of Bi orbitals, which possess a larger spin-orbit coupling strength as compared to Si. The non-trivial (topological) band-gap of a few meV in free-standing (1 x 1) silicene is increased to 124meV and becomes trivial in the presence of the substrate. An out-of-plane external electric field can be used to tune the band-gap and restore the non-trivial topological phase.[433]

To be able to open up a sizable band-gap in zero-gap silicene, without decreasing the carrier mobility, is a key ambition in nano-electronics. First-principles calculations show that the interaction energies range from -0.09 to 0.3eV per silicon atom; indicating a weak interaction between silicene and a ZnS monolayer. The AB_{Zn} stacking is the most stable one. The band-gap can be tuned from 0.025 to 1.05eV in silicene/ZnS hetero-bilayers. An indirect-direct band-gap cross-over is observed in the hetero-bilayers, depending upon the stacking pattern, the interlayer spacing and the external strain. The characteristics of a Dirac cone with almost linear band dispersion can be preserved in the AB_S pattern, which is a metastable state accompanied by a small electron effective mass. The carrier mobility is therefore not expected to be appreciably degraded.[434]

First-principles plane-wave calculations, within density functional theory, of the adsorption of silver adatoms on silicene surfaces show that the silver adatoms form strong bonds with the silicene; with significant binding energies. Bare silicene, which is a non-magnetic semimetal, becomes either a non-magnetic metal or a semiconductor, depending upon the number of adsorbed silver atoms. Because of charge transfer from the adatoms to silicene, the bonding and anti-bonding π-bands which cross linearly at the Fermi level, shift to 0.35eV below the Fermi level for both single and trimer silver adsorption. Silicene with adsorbed silver dimers becomes a narrow-gap (0.112eV) semiconductor.[435]

First-principles studies of the interfaces between silicene and monolayer MoS_2 substrates show that silicene can be adsorbed on the substrate, via van der Waals interactions, to form silicene/MoS_2 heterostructures with a high-buckle or low-buckle configuration. The lattice mismatch between low-buckle silicene and the substrate leads to the formation of Moiré superstructures. The heterostructures of high-buckle silicene on the substrate are metallic, while those of low-buckle silicene are semiconductors with small band-gaps, due to interface effects. The band-gap depends upon the rotation angle and the stacking pattern, while the formation energy does not. The high carrier mobility of low-buckle silicene is preserved in these heterostructures. The band-gap can be further tuned by applying a vertical external electric field.[436]

Dispersion-corrected density-functional theory studies of the geometric and electronic properties of silicene on ultra-thin silicon nanosheets show that the substrate modifies the electronic properties of silicene, due to a difference in charge population between the two sub-lattices. A sizable band-gap is created at the Dirac-point, which can then be widened by applying a bias voltage.[437]

First-principles calculations of silicene on MgX_2 (X = Cl, Br, I) show that various locations of the silicon atoms are energetically degenerate because of weak van der Waals interaction with the substrate. The silicon buckling height is less than 0.55Å; which is close to the value (0.49Å) for free-standing silicene. The Dirac cone of silicene is well-preserved on MgX_2, being located slightly above the Fermi level, and the band-gaps which are induced by the substrate are less than 0.1eV. Application of an external electric field, and stacking, can increase the band gap.[438]

Density functional theory calculations of various types of periodically repeated junctions made of armchair silicene nanoribbons with various widths, show that the band structures are modulated by the size and strain of the superlattices. The length of narrow and wide nanoribbons can alter the band structure and modify the distribution of localized and dispersive states of the superlattice. With increasing strain, the energy gaps change and are significantly different to those of the constituent nanoribbons.[439]

Calculations of a hybrid bulk system consisting alternating layers of honeycomb silicene and BN predict properties which are similar to those of bulk graphite. There exists a weak van der Waals interaction between the layers of this hybrid system, in contrast to the strong inter-layer covalent bonds which are present in multi-layers of silicene. Results for the electronic band structure and the density of states show that it is a semi-metal. The dispersion around the Fermi level is parabolic in nature, and the charge carriers behave like almost-free particles. This indicates that the electronic properties of the hybrid bulk system closely resemble those of bulk graphite. The electronic band structures exhibit contributions arising only from silicene layers, and the BN layer acts only as a buffer layer. In bi-layers of silicene, with a single BN layer between, a linear dispersion is found which is similar to that of graphene. The characteristic linear dispersion becomes parabola-like when the system is subjected to compression along the transverse direction.[440]

First-principles calculations of bilayer silicene identify several metastable configurations, including a slightly faulted-AA packing structure, termed slide-2AA. Unlike the metallic properties of conventional AA and AB stacking forms, the band structure of slide-2AA bilayers includes an indirect energy gap of some 1.16eV. A metal-semiconductor phase transition along the sliding pathway, with a small energy barrier, is observed; indicating that the electronic properties can be easily tuned by applying a small shear force along the bilayer silicene surface plane.[441]

First-principles calculations of armchair and zig-zag silicane nanoribbons show that they are non-magnetic semiconductors. The band-gap of armchair hybrid silicane-silicene nanoribbons without spin-split, which is determined mainly by the silicene part, can be separated into three different families. The energy-gap of the zig-zag counterparts depends upon the hydrogenation of the zig-zag silicon chain at the silicane/silicene interface. Control of the hydrogenation ratio along the ribbon width can modulate the electronic and magnetic properties of zig-zag hybrid nanoribbons.[442]

Density functional methods show that the cohesive energy of a free-standing monolayer increases with external electric field and decreases with distortion. The electrons behave like Dirac fermions when the bond angle between the silicon atoms is larger than about 102°. Large distortions destroy the electronic structure so that it is no longer a semi-metallic material, and the distorted silicene acts like an n-doped system. An electric field opens up a band-gap around the K-point in the Brillouin zone, and this increases with increasing electric field. The bond-angle between the silicon atoms is a key factor determining the presence or absence of Dirac cones.[443]

Ab initio numerical investigation of silicene nanoribbons with zig-zag edges reveals that, as in the case of graphene, anti-ferromagnetic and ferromagnetic ordering exists, with spin polarization at one edge being anti-parallel or parallel to that at the other edge, respectively. The thermoelectric properties, and especially the Seebeck coefficient, depend markedly upon the electronic band structure and are improved when the Fermi level is in the energy gap. The thermoelectric efficiency is significantly reduced when the phonon contribution to the heat conductance is included. The transition from the anti-ferromagnetic to the ferromagnetic state leads to a large magnetoresistance and to an appreciable magneto-thermopower.[444]

First-principles studies of silicene/graphene bilayers, using various super-cell models, suggest that graphene can be used as a substrate to synthesize monolayer silicene. Multiple phases of monocrystalline silicene, having differing orientations with respect to the substrate, can co-exist at room temperature. Weak interaction between the overlayer and the substrate preserves the low-buckle structure of silicene, as well as its linear energy bands. The gap which is introduced by breaking the sub-lattice symmetry of silicene on graphene can be up to 57meV.[445]

First-principles calculations of silicene under uniaxial strain show that the Poisson ratio and the rigidity exhibit a strong chirality-dependence under large strains. The ultimate strain of silicene under uniaxial strain is smaller than that under biaxial strain. Uniaxial strains introduce Dirac-point deviations from the high-symmetry points in the Brillouin zone, and semimetal-metal transitions; so no band-gap opens up under uniaxial strain. Due to its particular structure, and the variable sp^3/sp^2 ratio of the chemical bond, the deviation directions of Dirac-points from the high-symmetry points of the Brillouin zone and the variations of the Fermi velocities are significantly different to those of graphene. The Fermi velocities exhibit a marked anisotropy with respect to the wave-vector directions, and change slightly before the semimetal-metal transition. The work function increases monotonically with increasing uniaxial strain.[446]

First-principles theory calculations of silicene and silicane under uniaxial tensile strains along various directions show that a chirality effect is more noticeable in the mechanical properties of silicene than in those of silicane. The armchair Young's modulus is 63.51GPa.nm, while the zig-zag Young's modulus is 60.06GPa.nm. Differing failure mechanisms are also found. A band-gap of up to 0.8eV develops from zero under uniaxial tension, and vanishes before the structure attains its in-plane ultimate strength. A pre-existing band-gap of 2.39eV exists in the silicane structure, and decreases to zero with increasing tensile strain, with no chirality effect.[447]

First-principles calculations of silicene nanosheets show that the system exhibits a strain-induced self-doping phenomenon which is closely related to the buckled structure, and thus cannot arise in graphene. Under compressive strain, the Dirac point moves to below the Fermi level, making the nanosheets behave as though n-type doped. Under tensile strain, p-type doping is obtained by shifting the Dirac point to above the Fermi level.[448]

First-principles density functional theory shows that the effect of tensile strain on silicene is such that a semimetal-metal transition occurs when an in-plane strain larger than 7.5% is applied. Downward movement of the lowest conduction band at the Γ-point, which originates from the weakened interaction between neighboring silicon atoms, leads to the transition.[449]

First-principles calculations indicate that semi-hydrogenated silicene is a ferromagnetic semiconductor. *Ab initio* quantum transport theory, applied to silicene with pristine silicene as electrodes, shows that a high on/off current ratio of 10^6 is obtained in single-gated semi-hydrogenated device. A spin-polarized current can also be generated. The spin-filter efficiency increases with gate voltage and reaches 100% at a voltage of 1.9V. The results suggest that a gate voltage can induce half-metallicity in semi-hydrogenated silicene.[450]

First-principles results for planar and simply-buckled silicene confirm its zero-gap nature and reveal a significant renormalization of the Fermi velocity upon including many-electron effects. Other proposed silicene structures exhibit a finite band-gap, indicating that they are gapped semiconductors rather than Dirac-fermion semi-metals. This finite band-gap is preserved when a silver substrate is included. Other calculations reveal increased many-electron effects in these two-dimensional structures.[451]

Zig-zag silicene nanoribbons with asymmetrical sp^2-sp^3 edges are bipolar magnetic semiconductors, due to the incorporation of Klein and zig-zag edge states. These asymmetrical zig-zag nanoribbons can be changed into half-metals, with opposite conductive spin channels, by p-type and n-type doping. The semiconducting properties can be tailored by straining, making the nanoribbons into spin gap-less semiconductors or ferromagnetic metals.[452]

Spin-polarized first-principles studies of zig-zag silicene nanoribbons doped with nitrogen atoms, as well as with nitrogen and silicon vacancy (V_{Si}) complexes, shows that doped nitrogen atom and N-V_{Si} complexes prefer the edge sites of zig-zag nanoribbons. Due to breaking of the degeneracy of the spin-polarization in zig-zag nanoribbons, the substitution of nitrogen for silicon induces a spin gap-less semiconductor property. When a N-V_{Si} complex is introduced, to form so-called pyridine- and pyrrole-like structures in zig-zag nanoribbons, they also exhibit half-metal or spin gap-less semiconductor behaviors with spin-polarized currents in the Fermi level.[453]

First-principles density functional theory calculations of silicene monolayers on various (111) semiconducting surfaces show that the relative stability of the silicene overlayer depends upon whether the interacting top layer of the substrate is metal, or non-metal, terminated. The nature of a silicene monolayer on metal-terminated surfaces can be metallic or even magnetic, depending upon the substrate. The silicene overlayer undergoes n-type doping on metal-terminated surface and undergoes p-type doping on non-metal terminated surfaces of semiconductor substrates.[454]

First-principles investigations of silicene nanoribbons with sawtooth edges show that these nanoribbons are more stable than are zig-zag nanoribbons and have a ferromagnetic ground state, with an intrinsic energy-gap between the majority and minority spin-polarized bands. This shows that sawtooth nanoribbons are spin-semiconductors. Under an external transverse electric field, the energy-gap decreases and can even vanish. The charge densities of the two edge bands near to the Fermi level become spatially separated at different edges. The electric field-induced features can be obtained by applying a suitable uniaxial compressive strain, and this can be understood in terms of the Wilson transition. A small tensile strain makes the sawtooth nanoribbons more sensitive to an electric field.[455]

1.12 Surface Interaction

Ab initio density functional calculations of chemical functionalization by boron, nitrogen, aluminium or phosphorus atoms show that the preferred adsorption sites are valley, bridge, valley and hill sites for boron, nitrogen, aluminium, and phosphorus adatoms, respectively. All of the relaxed systems, with adsorbed/substituted atoms, exhibit a metallic behavior with strongly-bonded boron, nitrogen, aluminium or phosphorus atoms together with appreciable electron transfer from silicene to the boron, nitrogen, or phosphorus adatom/substituent. The aluminium atoms exhibit opposite charge transfer; with n-type doping of the silicene, and weaker bonding. The adatoms/substituents introduce characteristic branches into the phonon spectrum of silicene, which can then be probed via Raman measurements. Molecular dynamics methods show that these systems are stable up to at least 500K.[456]

Study of the interaction of silicene with silicon, carbon, hydrogen, oxygen and titanium atoms and H_2, H_2O and O_2 molecules, using first-principles density functional theory, shows that silicon adatoms initially adsorbed at the top site of silicene push down the silicon atoms beneath so as to form a dumb-bell like structure with 3 + 1 coordination.

This observation is important for silicene research in that it reveals new physical phenomena related to the formation of multilayer Si; deemed to be the precursor state for the missing layered structure of silicon. The dumb-bell structure imparts coverage-dependent electronic and magnetic properties to non-magnetic bare silicene. Silicene with dumb-bells is also energetically more favorable than is pristine silicene and, the more dense the dumb-bell coverage, the stronger is the cohesion. Such structures appear to be intermediate between those of silicene and silicon.

First-principles investigation of the magnetic properties and thermospin effects of zig-zag silicene nanoribbons with a single absorbed silicon atom shows that, following relaxation, a steady dumb-bell like structure is formed. This induces a weaker antiferromagnetic coupling between two zig-zag edges, and a band-gap is thus opened up. The adsorbed nanoribbon meanwhile exhibits a spin semiconducting nature. A large spin thermopower and a weak charge thermopower can be simultaneously achieved in adsorbed zig-zag nanoribbon-based devices.[457] This is attributed to a nearly perfect mirror symmetry between spin-up and spin-down transmission spectra, relative to the Fermi level.

First-principles calculations, based upon self-consistent charge density functional tight-binding, have been used to investigate various band-gap tuning mechanisms in optimized silicene sheet structures, functionalized with H, CH_3 or F, and in nanoribbons having zig-zag or armchair edges. The bond length and buckling value of relaxed silicene agrees with both experimental and other theoretical values. The band-gap is opened by functionalization. The direct band-gap at the K point changes to a direct band-gap at the gamma point.[458] The functionalized silicene band-gap also decreases with increasing strain. The band-gap is near to zero for all sizes of zig-zag nanoribbon. In armchair nanoribbons, an oscillating decay occurs in the band-gap with increasing nanoribbon width. An external electric field can open the band-gap in proportion to the field magnitude.

A carbon adatom, initially adsorbed at the bridge position, replaces one silicon atom if it can surmount a small energy barrier. The oxygen molecule can dissociate on the silicene surface and the constituent oxygen atoms then oxidize the silicene by forming strong bonds. By varying the concentration and degree of decoration with carbon, hydrogen and oxygen atoms, the band gap of silicene can be tuned. Upon adsorbing hydrogen or titanium adatoms, the silicene achieves a spin-polarized state. A half-metallic ferromagnetic behavior is achieved at a specific uniform coverage of titanium adatoms and might function as a spin valve.[459]

Silicene derivatives constructed from periodic dumb-bell units play a crucial role in multilayers of the honeycomb structure. First-principles calculations, based upon density functional theory, clarify dumb-bell formation mechanisms and energetics for adsorbed group-IV atoms on monolayer honeycomb structures. The stabilities of the binding structures can be further confirmed by *ab initio* molecular dynamics calculations for high temperatures. Depending upon the row number of adatoms and substrates, three types of binding structure are found, leading to significant changes in the electronic, magnetic and optical properties of the substrate. In particular, adsorbed silicon, germanium and tin adatoms form dumb-bell structures. Dumb-bell structures not only occur on single-layer monatomic honeycomb structures but also on compounds such as SiC and SiGe. The energy barrier to the migration of a dumb-bell structure is low, due to the concerted action of atoms. This causes dumb-bells to be quite mobile on substrates and to form new single and multilayer silicon and germanium phases.[460]

A mixed sp^2-sp^3 hybridization of silicene imparts an excellent reactive ability, and facilitates chemical modification. Such modification effectively permits tuning of the properties of silicene. A review all types of chemical modification methods for silicene, including hydrogenation, halogenation, organic surface modification, oxidation, doping and the formation of two-dimensional hybrids reveals the effects of these methods upon the geometrical, electronic, optical and magnetic properties of silicene.[461]

First-principles simulations of three conformers of half-fluorinated sheets confirm that sheets with zig-zag, boat-like or chair-like configurations are dynamically stable according to phonon calculations. Upon adsorbing fluorine, energy gaps open in both the zig-zag and boat-like conformations and they are found to be direct-gap semiconductors. Half-fluorinated silicene with a chair-like configuration exhibits an antiferromagnetic behavior which is induced mainly by un-fluorinated silicon atoms. When an isotropic strain is uniformly imposed on chair-like half-fluorinated silicene, the energy difference between the ferromagnetic and antiferromagnetic states decreases, with increasing compressive strain, from 0 to -6% or, with increasing tensile strain, from 0 to 6%.[462] A density functional theory-based screened-exchange local-density approximation of the electronic structures and band-gaps of silicene with adsorbed halogens shows that the band-gaps of silicene, with adsorbed fluorine, chlorine, bromine and iodine exhibit a non-monotonic change with increasing periodic number of the halogen. This is attributed to a transfer of contributions to the band-gaps, arising from silicon-silicon bonding, to silicon-halogen bonding.[463]

Density functional theory studies of nitric oxide adsorption on pristine, aluminium- or P-substituted nanosheets, in terms of adsorption energy, energy gap and Mulliken charge transfer, indicate that the adsorption characteristics are most prominent for pristine and P-substituted nanosheets.[464]

Oxygen adsorption and dissociation on free-standing silicene, studied using first principles methods, shows that the material is not stable in oxygen because O_2 molecules can be easily adsorbed and dissociated into oxygen atoms on the surface without overcoming any energy barrier. Dissociated oxygen atoms migrate with difficulty on the surface, and desorb, leading to the formation of Si-O compounds. In order to enhance the stability of free-standing silicene in oxygen, fully hydrogenated silicene can be as a stabiliser. When compared with the zero-energy barriers on pristine silicene, there are two minor energy barriers to O_2 molecule adsorption and dissociation on fully hydrogenated silicene. This indicates that hydrogenated silicene has a higher stability than does free-standing silicene in oxygen. When the O_2 molecule dissociates into two oxygen atoms on hydrogenated silicene the desorption of the oxygen atoms will be very difficult, due to its high energy barrier.[465]

Density functional calculations of the energetic, structural and electronic properties of graphene and silicene, functionalized with hydrogen, predict that hydrogen atoms bind much more strongly to silicene than to graphene. The adsorbed hydrogen atoms tend to stabilize each other cooperatively, leading to a two-dimensional nucleation and growth mechanism.[466] First-principles calculations indicate that hydrogenated silicenes exhibit peculiar properties, due to the strong H-Si bonds. The various charge distributions, bond lengths, energy bands and densities-of-state depend strongly upon the various possible hydrogen configurations and concentrations. A competition between strong H-Si bonds and weak sp^3 hybridization dominates the electronic properties. Chair configurations are associated with semiconductors, while top configurations have an almost dispersionless energy band at the Fermi level. Both systems have H-related partially flat bands at moderate energies and exhibit recovery of low-lying π-bands upon concentration reduction. Their densities-of-state have prominent peaks at moderate energies, and the top systems have a delta-function like peak at $E = 0$. The intensity of these peaks gradually weakens with decreasing concentration; thus providing an effective method for identifying the hydrogen concentration during scanning tunnelling spectroscopy studies.[467] First-principles calculations based upon density functional theory show that two different atomic configurations are stable, and energetically degenerate. Upon adsorption of hydrogen, an energy-gap opens up. The energy gap has been computed using both the HSE hybrid functional and the G_0W_0 many-body perturbation methods. The hydrogenated material is found to be a wide band-gap semiconductor; the type of gap in silicane (direct or indirect) depending upon its atomic configuration.[468]

Density functional theory investigation, using both the generalized gradient approximation and hybrid functionals, of the structural, electronic and magnetic properties of two-dimensional fully-, and half-hydrogenated, silicene, suggests that silicene is a gap-less semi-metal. The coverage and arrangement of absorbed hydrogen atoms on silicene markedly influence the characteristics of the resultant band structures, such as direct/indirect band-gaps and metallic/semiconducting features. Half-hydrogenated silicene, with a chair-like structure, is a ferromagnetic semiconductor.[469] Density functional theory studies of partially hydrogenated silicene, with hydrogenation ratios of between 3.1 and 100at%, show that the hydrogenation energy increases with hydrogenation ratio: attaining 3.01eV/H at complete hydrogenation. Molecular dynamics simulations demonstrate stability of the adsorption configurations. Partial and patterned hydrogenation, achieved by exposing silicene to hydrogen gas of various densities or by using masking techniques, offers the possibility of obtaining metal/semiconductor/insulator behavior within the same silicon nanosheet.[470]

First-principles calculations, and the cluster-expansion approach, show that hydrogenated bilayer silicene has potential as a new type of opto-electronic material. Hydrogenation converts intrinsic bilayer silicene, which is a strong indirect semiconductor, into a direct-gap semiconductor having a widely tunable band-gap. At low hydrogen concentrations, four ground states of single- and double-sided hydrogenated bilayer silicene are characterized by dipole-allowed direct (or quasi-direct) band-gaps ranging from 1 to 1.5eV; suitable for solar applications. At high hydrogen concentrations, three well-ordered double-sided hydrogenated bilayer silicene structures exhibit direct (or quasi-direct) band-gaps in the red, green and blue color ranges.[471]

The hydrogenation of silicene[472] is an efficient means for opening the band-gap and manipulating its electronic properties, but the energy-barrier to hydrogenation is unfortunately quite high and this limits the occurrence of the chemical reaction. One alternative approach to reducing the energy barrier and facilitating hydrogenation is the use of straining. Theoretical results demonstrate that biaxial strain and uniaxial tensile strain along the armchair direction can significantly reduce the energy barrier to dissociative H_2 adsorption, with the barrier decreasing as the strain increases. Biaxial strain has a more marked effect upon energy-barrier reduction, with the barrier being reduced from 1.71 to 0.24eV when the biaxial strain reaches a critical value of about 12%. Above this value, the silicene structure after hydrogenation would be destroyed. Using this means, the reaction-time for hydrogenation can be astoundingly reduced from 8.06×10^{16} to 1.68×10^{-8}s. The mechanism of the tensile-strain effect is related to the density-of-states of the system and to atomic-charge transfer during hydrogenation.[473] Another alternative approach to the hydrogenation of silicene is to apply an external electric field in order to reduce the reaction energy-barrier. A positive perpendicular electric field can act as a catalyst to reduce the energy barrier to H_2 dissociative adsorption on silicene, thus facilitating the hydrogenation of silicene. The barrier decreases as the perpendicular electric field increases and, when it exceeds 5.14 x 10^{11}V/m, the barrier is low and the hydrogenation of silicene can take place efficiently at room temperature. This catalytic effect of the electric field upon silicene hydrogenation is attributed to the redistribution of atomic charge by the electric field, thus significantly changing the chemical activity of the silicene.[474] When the hydrogenation of monatomic silicene sheet on Ag(111) is investigated using scanning tunnelling microscopy and density functional theory calculations, it can be deduced that the hydrogenation of silicene at room temperature results in a perfectly ordered γ-(3 x 3) superstructure. A theoretical model, involving seven hydrogen atoms and a rearranged buckling of silicon atoms, agrees well with experiment. Upon annealing at about 450K, dehydrogenation occurs and a clean silicene surface can be fully recovered.[475] A monatomic silicene sheet, grown on a substrate, usually reconstructs into various phases and only partially hydrogenated silicene with reconstruction tends to be observed. Half-silicane, in which one silicon sub-lattice is fully H-saturated while the other is intact, forms a perfect (1 x 1) structure. Upon hydrogenating various silicene phases on a Ag(111) substrate, it is found that only the (√2 x √2)R30° phase can produce half-silicane. This phase has sometimes been categorised as being a highly defective or incomplete silicene structure. The structure of the (√2 x √2)R30° phase in fact involves a complete silicene-(1 x 1) lattice rather than defective fragments.[476]

First-principles calculations of structural optimization reveal that buckled silicene is more stable, than its planar counterpart, by about 0.1eV. In comparison with planar graphene, silicene sheet is buckled by 0.537Å but has a similar electronic structure with, like graphene, a zero band-gap at the K-point. Complete hydrogenation studies of various geometrical configurations (chair, boat, table and stirrup) show that the chair-like structure has the highest binding energy per atom when compared with other structures. Hydrogenated silicene (silicane) has a predicted indirect band-gap of 2.23eV.[477]

Ab initio investigation of the spin-polarized transport properties of bare and hydrogenated zig-zag silicene nanoribbons shows that the ribbons with symmetrical edges prefer ferromagnetic ground states with semiconducting properties, while antiferromagnetic states exhibit metallic behavior. Ribbons with asymmetric edges prefer antiferromagnetic ground states with semiconducting properties while ferromagnetic states exhibit metallic behavior. These features result in a giant magnetoresistance behavior between the ferromagnetic and antiferromagnetic states in the low bias-voltage regime. In ferromagnetic ribbons with asymmetrical edges, perfect spin-filtering - with 100%

positive electric current polarization - can be achieved by altering the bias voltage. It is also found that negative differential resistances prefer the metastable states[478]. Theoretical investigation of fluorinated and hydrogenated silicene shows that, unlike the carbon analogue, fluorination prefers a conformation with zig-zag line buckling while hydrogenation prefers a chair conformation as in the case of graphane. A direct band-gap is opened up in silicene fluoride and the gap can be continuously modified by straining. In the case of the hydride, a strain-induced indirect-to-direct band-gap transition occurs.[479]

A study within density functional theory of the effect of hydrogenation upon one-atom thick material made of silicon and carbon atoms, a silicene/graphene hybrid, reveals that the pure hybrid is a non-zero band-gap semiconductor having a stable planar honeycomb structure. The mixing of carbon and silicon in an alternating manner thus offers another means for generating a finite band-gap in one-atom thick materials. Full hydrogenation makes the gap larger, but chemical modification with hydrogen reduces the gap in favor of ferromagnetic order.[480]

Upon comparing the structural stabilities and electronic properties of zig-zag silicene nanoribbons having Klein and pentagon-heptagon reconstructions with unreconstructed zig-zag edges, it is found that deformed bare pentagon-heptagon ones are favored under H-poor conditions while H-rich surroundings stabilize di-hydrogenated Klein edges. These edges exhibit a magnetism which is analogous to that of zig-zag ones; which also possess the electric-field-induced half-metallicity of nanoribbons. Divers magnetic states can be achieved by asymmetrical Klein and zig-zag edges in zig-zag silicene nanoribbons, permitting the transformation of antiferromagnetic semiconductors into bipolar spin-gapless semiconductors and ferromagnetic-metals; depending upon the edge hydrogenation.[481]

Spin-polarized density functional theory studies of the stability and electronic properties of nitrogen and boron impurities in single layers of silicane show that these impurities have lower formation energies when compared to their counterparts in graphane. The adsorption of hydrogen atoms in the vicinity of defects also stabilizes the system. The electronic properties of silicane can moreover be tuned when nitrogen and boron are incorporated, with nitrogen-doping and boron-doping giving rise to n-type and p-type semiconducting properties. The adsorption of hydrogen atoms quenches the doping effects.[482]

When the properties of epitaxial silicene monolayers on Ag(111), at various levels of oxidation, are deduced from complementary density functional theory calculations and soft X-ray spectroscopy, the calculations indicate that moderate levels of oxidation do not cause significant band-gap opening in the epitaxial monolayer. This suggests that oxygen functionalization is not a viable mechanism for band-gap tuning while the silicene monolayer remains on its metallic substrate. Moderate oxidation is predicted to distort strongly the hexagonal silicon lattice, causing it to cluster in the regions of highest oxygen adatom concentration while retaining a two-dimensional sheet structure. The experimental data reveal that beam-induced oxidation is consistent with the formation of islands of bulk-like SiO_2. Complete exposure of the monolayer to ambient conditions results in a fully oxidized material which closely resembles bulk SiO_2, with a considerable portion being completely detached from the substrate.[483]

Density functional studies of the oxides that result from partial or full oxidation of silicene show that the formation of the oxides depends upon the oxidation conditions and oxidizing agents such as oxygen and hydroxyl. The honeycomb lattice can be maintained, distorted or destroyed by oxidation. The silicon charge-state in partially oxidized silicene ranges from +1 to +3, while that in fully-oxidized silicene is +4. Metals, semimetals, semiconductors and insulators can all be found among the oxides, yielding a wide spectrum of electronic structures.[484] The O_2 dissociation and O_2-dissociation-induced oxygen atom adsorption on free-standing silicene, as studied using first-principles calculations, are such that the O_2 molecule dissociates easily, from both the thermodynamic and kinetic points of view, on the free-standing silicene surface; unlike the case of graphene. The dissociation reaction is exothermic, and the dissociated oxygen atoms form strong bonds with silicon atoms, thus substantially lowering the energy of the system. The dissociation reaction occurs spontaneously on free-standing silicene, without surmounting any energy-barrier. The migration and desorption of oxygen atoms are relatively difficult at room temperature, due to the strong Si-O bonds in the oxygen-adsorbed silicene; which favours silicon oxide formation. This is convincing evidence that free-standing silicene is unstable in O_2.[485] First-principles calculations of fully oxidized silicene with a stoichiometric Si:O ratio of 1:1 show that the zig-zag ether-like conformation is the most energetically favorable structure. These nanosheets have notable elastic characteristics, including an unconventional auxetic behavior and negative Poisson ratio. Following oxidation, semi-metallic nanosheets are transformed into semiconductors having narrow direct band-gaps. Due to its anisotropic mechanical and electronic properties, zig-zag ether-like SiO possesses a high axial intrinsic charge mobility of up to $10^4 cm^2/Vs$. This is comparable to that of graphene nanoribbons.[486] The room-temperature oxidation of a one-dimensional grating of silicene nanoribbons grown onto a Ag(110) surface, as studied using high-resolution silicon 2p core level photo-emission spectroscopy and low-energy electron diffraction, is such that oxidation starts at oxygen exposures which are about 10^4 times higher than those for clean Si(111)-(7 x 7) surfaces. This again demonstrates the low reactivity of silicene to molecular oxygen. The use of prior Ar^+ sputtering produces defects, which then enhance oxidation.[487] Thick epitaxial multilayer silicene films, having a (3 x √3)R(30) surface structure, exhibit only mild surface oxidation after 24h in air; as measured by Auger electron spectroscopy. Raman spectroscopy and X-ray diffraction measurements, performed in air without any protective capping or with a thin Al_2O_3 cap, show that the (002) reflection and the G, D and two-dimensional Raman structures are unique fingerprints of thick multilayer silicene.[488]

Epitaxial silicene undergoes strong interactions with a substrate, and this markedly affects its electronic structure. The reconstructions and hybridized electronic structures of epitaxial (4 x 4) silicene on Ag(111), as revealed by scanning tunnelling microscopy and angle-resolved photo-emission spectroscopy, are such that hybridization between silicon and silver results in a metallic surface state which can gradually decay due to oxygen adsorption. The use of X-ray photo-emission spectroscopy confirms the decoupling of Si-Ag bonds following oxygen treatment, as well as the relative oxygen-resistance of the Ag(111) surface as compared with (4 x 4) silicene. First-principles calculations confirm the evolution of the electronic structure of silicene during oxidation. It has been verified, experimentally and theoretically, that the high chemical activity of (4 x 4) silicene is due to the silicon p_z state, while the Ag(111) substrate exhibits a relatively inert chemical behavior.[489]

The chemical stability of buckled silicene results in a low reactivity with O_2 and in progressive oxidation under ambient conditions. This drawback can be countered by aluminium- and Al_2O_3-based encapsulation of the silicene layer. Such encapsulation can usually be applied to any silicene configuration, regardless of the support substrate, and can lead to the fabrication of atomically sharp and chemically intact Al/silicene and Al_2O_3/silicene interfaces that can be used for gated-device fabrication.[490] In order to determine whether aluminum oxide would be a suitable encapsulation material, high-resolution synchrotron photo-electron spectroscopy was used to study the interactions of aluminium atoms and O_2 molecules, as well as combinations of both, with epitaxial silicene on thin ZrB_2(00•1) films grown onto Si(111). The deposition of aluminium atoms onto silicene, up to a coverage of about 0.4 Al per silicon atom, had little effect upon the chemical state of the silicon atoms. The silicene-terminated surface was also hardly affected by exposure to O_2 gas; up to a dose of 4500L but, when aluminium-covered silicene was exposed to the same dose, a large fraction of the silicon atoms became oxidized. This was attributed to the dissociative chemisorption of O_2 molecules by aluminium atoms at the surface, thus producing reactive atomic oxygen species that then led to oxidation. It was concluded that aluminum oxide overlayers, prepared in this manner, were not suitable for encapsulation because they did not prevent the degradation of silicene and could instead enhance it.[491]

First-principles molecular-dynamics simulations demonstrate that the oxidation of silicene can take place easily at either low or high oxygen doses; thus helping to explain earlier inconsistent reports on the oxidation of silicene on Ag(111) substrates. It is shown that, while the energy barrier for an O_2 molecule reacting with a silicon atom depends strongly upon the position and orientation of the molecule, the O_2 molecule immediately dissociates and forms a Si-O-Si configuration when it has found a barrier-less chemisorption pathway around an outer silicon atom of the silicene overlayer. A synergistic effect between molecular dissociation and subsequent structural rearrangement accelerates the oxidation process at high oxygen doses. This also enhances the self-organized formation of sp^3-like tetrahedral configurations (consisting of silicon and oxygen atoms), which results in collapse of the two-dimensional silicene structure and its exfoliation from a substrate. The electronic properties of silicene can be significantly altered by oxidation, and low-flux low-temperature oxygen gas exposure may be keys to control of the oxidation of silicene.[492]

Atomic adsorption of nitrogen, carbon and oxygen, and the molecular adsorption of N_2 and CO, as investigated using density functional theory calculations, is such that the nitrogen atom is the most stably adsorbed; with an energy of 8.207eV. The N_2 molecule undergoes physisorption, while the CO molecule undergoes chemisorption; the corresponding adsorption energies being 0.085 and 0.255eV. Silicene therefore exhibits greater reactivity to CO adsorption than to N_2 adsorption. The differences in charge density, and integrated charge calculations, suggest that the charge transfer for CO adsorption (about 0.015e) is larger than that for N_2 adsorption (about 0.005e)[493]. This further supports the belief that the CO molecule is more active than the N_2 molecule, when adsorbed on silicene.

A first-principles calculation investigation of the possibility of preparing half-metallic silicene nanosheets has shown that silicene could be transformed into a half-metal if it is fully nitrogenated on one silicon sub-lattice. It has a half-metallic gap of about 0.25eV, and the estimated Curie temperature is about 374K. This is very important for the stability of half-metallicity and for practical high-temperature application. The results for two other types of nitrogenated silicene show that both compounds behave as if they had a metallic nature.[494] The conclusion drawn is that nitrogenation, which can be achieved experimentally by preparing silicene on the surface of certain nitrides, may be a route to identifying silicon-based low-dimensional half-metals.

First-principles calculations have been used to study the structural, energetic and electronic properties of 15 different metal atoms adsorbed on silicene (table 3). Among the adatoms, lithium, sodium, potassium, calcium, cobalt, nickel, palladium and platinum exhibit a binding energy which is larger than the cohesive energy of the bulk metal. Whereas the binding of gold and tin to graphene is rather weak, they bind strongly and covalently to silicene. In the case of the alkali metal adatoms on silicene, the bonding is approximately ideal-ionic. When calcium is adsorbed on silicene, hybridization between the Ca 3d states and the silicene states occurs around E_F, together with charge transfer from calcium to silicene. The aluminium, gallium, indium and tin adatoms bind most strongly at the top site above a silicon atom in silicene, and form strong covalent bonds with the nearest silicon atoms. In the case of the titanium, iron, cobalt and gold adatoms on silicene, the adatom d-states are strongly hybridized with the silicene states around E_F, while the hybridization states lie considerably far below E_F, with the latter being located approximately at the Dirac point for nickel, palladium and platinum adatoms on silicene. The strong binding of metal adatoms to silicene and the rich electronic properties of the systems suggest, as usual, further experimental exploration of the functionalization of silicene with metal adatoms.[495]

First-principles density functional theory calculations show that the electronic and magnetic properties of bare and titanium-adatom adsorbed single-layer silicene, when charged or subjected to a perpendicular electric field, can be modified so as to obtain new functionalities. In particular, when subjected to a perpendicular electric field, buckled atoms have the symmetry between their planes broken, thus opening up a gap at the Dirac points. The occupation of 3d orbitals of adsorbed Ti atoms changes with charging, or upon the application of an electric field, thus producing significant changes in the magnetic moment. Neutral silicene, uniformly covered with Ti atoms, becomes a half-metal at a specific value of coverage and this permits the transport of electrons in one spin-direction, but blocks movement in the opposite direction. Such calculated properties exhibit a dependence upon the size of the vacuum spacing between periodically repeating layers if they are treated by using a plane wave basis set with periodic boundary conditions. This spurious dependence can be eliminated by the use of a local orbital basis-set.[496]

Table 3
Binding energies of adatoms to the silicene surface

Adatom	Site	E(eV)
Na	H	1.613
Na	A	1.468
Na	B	1.325
K	H	1.905
K	A	1.799
K	B	1.704
In	H	2.006
In	A	2.138
In	B	1.951
Li	H	2.129
Li	A	1.896
Li	B	1.718
Ca	H	2.188
Ca	A	2.159
Ca	B	2.125
Ga	H	2.305
Ga	A	2.397
Ga	B	2.261
Au	H	2.316
Au	A	2.159
Au	B	1.977
Al	H	2.494
Al	A	2.828
Al	B	2.717
Sn	H	2.458
Sn	A	2.937
Sn	B	2.844
Fe	H	3.430
Fe	A	3.516
Fe	B	3.374
Ti	H	3.899
Ti	A	3.832

Ti	B	3.646
Pd	H	4.200
Pd	A	3.701
Pd	B	3.493
Co	H	4.463
Co	A	4.058
Co	B	3.934
Ni	H	4.776
Ni	A	4.169
Ni	B	3.963
Pt	H	5.871
Pt	A	5.248
Pt	B	5.130

H: above hexagon centre, A: above Si atom of A sub-lattice, B: above Si atom of B sub-lattice,

When silver atoms are deposited onto 3 x 3 and √3 x √3 silicene films via molecular beam epitaxy under ultra-high vacuum, scanning tunnelling microscopy and Raman spectroscopy show that the silver atoms form chemical bonds with neither 3 x 3 nor √3 x √3 silicene films. This is due to the chemically inert surface of silicene. On 3 x 3 silicene films, the silver atoms mainly form stable flat-topped islands. Alternatively, the silver atoms can form nanoclusters and glide on silicene films, thus suggesting a more inert nature. Raman spectroscopy indicates that there is more sp^2 hybridization in √3 x √3 than in √7 x √7/3 x 3 silicene films.[497]

Growth via the controlled adsorption of ferromagnetic elements on two-dimensional silicene is an alternative route to exploiting silicon-based spintronics. Plane-wave density functional theory calculations show that iron, cobalt and nickel adatoms are strongly chemisorbed via strong sigma-bonds, with adsorption energies of 1.55 to 2.29eV. These values are two to six times greater than those for adsorption on graphene. All of the adatoms adsorb more strongly at hole sites than at atom sites, and nickel is the most strongly adsorbed. Hole-hole, b-atom-hole, vertically stacked at hole, vertically stacked at b-atom and bridge site dimer configurations are stable. The cobalt and nickel dimers are most stable when adsorbed in the hole-hole configuration, while the iron dimer is most stable when adsorbed in the atom-hole configuration. The trends which are observed in the adsorption energies and magnetic moments of adatoms and dimers can be attributed to metal-to-silicene and interconfigurational s-to-d electron transfer processes. Adsorption of the metals introduces a small band-gap at the Dirac cone. Cobalt adatom adsorption at the hole site introduces the largest spin-polarized band-gaps: 0.70eV (spin-up) and 0.28eV (spin-down).[498] Another study, using first-principles calculations, of the attachment of two 3d transition metals (iron, chromium) at buckled and planar silicene surfaces, revealed interesting ferromagnetic as well as semi-metallic features of the material. All of the iron adsorption cases are found to be more stable, with the lowest binding energy being 3.39eV, than are chromium adsorption cases. When the metal adsorption rate is high, iron tends to penetrate into both buckled and planar layers. This insertion permits the 3d shells of iron to enhance the bonding interactions with all of the $3p_x$, $3p_y$ and $3p_z$ orbitals of Si; thus producing more stable structures. The adsorption of chromium at a high distribution ratio is more stable than low-chromium distribution structures. The chromium does not penetrate into the layer as iron does. Ferromagnetism generally dominates, with five nanostructures. Two chromium-adsorption cases on planar silicene behave preferentially as antiferromagnets, and one iron adsorption case is non-magnetic. There is an inverse relationship between structural stability and magnetic moments. That is, $FeSi_x$ nanostructures – which are more stable - tend to have lower ferromagnetic moments. Half-metallic characteristics are found in four nanostructures, and are potentially applicable to spin-electronic devices. For those half-metallic nanostructures, the gaps which are derived from spin-down states vary from 0.28 to 0.57eV.[499] When the electronic and topological properties of silicene with adsorbed 4d transition metal atoms are investigated by using *ab initio* methods and tight-binding models, it is found that all six types of studied adatom, ranging from Y to Ru, prefer the hollow sites. The interplay of transition metal-induced exchange interactions, spin-orbit coupling and staggered AB sub-lattice potential, produces many topological states, including quantum anomalous Hall, valley Hall and valley-polarized metallic. In particular, quantum anomalous Hall states having various Chern numbers are obtained: being -2 in Nb/Ru-doped material and 1 in Y-doped material.[500]

First-principles calculations of the adsorption characteristics of alkali, alkaline-earth and transition-metal adatoms on silicene show that, contrary to the case of graphene, the interaction between the metal atoms and the silicene surface is quite strong, due to its highly reactive buckled hexagonal structure. The alkali metals, lithium, sodium and potassium, adsorb at hollow sites without any lattice distortion and, due to the significant charge-transfer from alkalies to silicene, metallization of the silicene occurs. Also revealed are trends which are directly related to the atomic size, adsorption height, work function and dipole moment of the silicene/alkali adatom system. The adsorption of alkaline-earth metals on silicene is entirely different to their adsorption behavior on graphene, in that the adsorption of beryllium, magnesium and calcium turns silicene into a narrow-gap semiconductor. The adsorption of the transition metals, titanium, vanadium, chromium, manganese, iron, cobalt, molybdenum and tungsten, leads – due to their partially-occupied d-orbital - to diverse structural, electronic and magnetic properties. Thus upon adsorption of a transition metal, depending upon the adatom type its atomic radius, the system can exhibit a metallic, semimetallic or semiconducting behavior. With all metal adsorbates, the direction of the charge transfer is from adsorbate to silicene; because of its high surface reactivity. Overall, it is clear that the reactive crystal structure of silicene provides rich opportunities for the useful functionalization at the nanoscale.[501]

When the solidification process of silicon atoms on heterogeneous silicene substrates, with shapes ranging from plane via curved to tubular is studied by means of molecular dynamics simulations, the shape of the nucleus is found to determine the stacking sequence of silicon atoms. Thus silicene plate produces strongly ordered liquid layers, while silicene nanotube makes the silicon follow its cylindrical structure. Growth competition has been observed in the confined nanospace between silicene nanotubes, causing structural changes at the shared interface. The internal potential field around a silicene nanotube is responsible for the formation of spiral structures and competitive growth. The degree of ordering decays with increasing distance from the silicene nanotube, as a result of the decreasing force arising from the nucleus.[502] Classical molecular dynamics simulations, using atomistic many-body potential energy functions, of nanoribbons with finite or infinite lengths at low and ambient temperatures show that finite-length models are more likely to form tubular structures at room temperature.[503]

Because epitaxial silicene is not chemically inert under ambient conditions, its use in practical devices will require the use of an insulating capping layer. An attempt has been made to encapsulate epitaxial silicene, on ZrB_2(00•1) thin films grown onto Si(111) substrates, by using aluminum nitride deposited using trimethylaluminum and ammonia precursors. *In situ* high-resolution core-level photo-electron spectroscopy of the chemical modifications of the surface, due to subsequent exposure to trimethylaluminum and ammonia molecules at 300 and 400C respectively, shows that such an AlN-related layer can be grown. The silicene reacts strongly with both precursor molecules however, resulting in the formation of Si-C and Si-N bonds, so that the use of these precursors does not permit the degree of protective AlN encapsulation that would leave the electronic properties of silicene intact.[504]

The functionalization of silicene using various chemical groups has been suggested as a means of tuning its electronic properties. Density functional theory calculations and *ab initio* molecular dynamics simulations of functionalization by naphthyl or anthracyl groups have been made of the effect of small polycyclic aromatic hydrocarbons, attached at various positions. The carbon attachment position which is farthest from the bond that fuses the aromatic rings gives more stable structures in the case of either functional group. All of the structures have direct band-gaps, and tuning of the band-gap is possible by increasing the length of the polycyclic aromatic hydrocarbon or the thickness of the silicene.[505] The adsorption of H_2Pc molecules on a (3 x 3) silicene/(4 x 4) Ag(111), according to density functional theory, involves electronic redistribution in the central macrocycle of the H_2Pc molecule and the formation of two Si-N covalent bonds between the molecule and the silicene; in accord with a cyclo-addition reaction. On SiC(00•1)(3 x 3) or Si(111) (√3 x √3)R 30°-boron however, the H_2Pc molecule remains planar and takes up a butterfly conformation on the silicene/Ag substrate due to electrostatic or polarization repulsion between the molecule and the silicene.[506] The adsorption of choline benzoate ionic liquid on the surface of various hexagonal nanosheets, studied using density functional theory, is shown to be moderated mainly by van der Waals forces; leading to strong benzoate ion-surface π-stacking. There is also an important degree of charge-transfer from the anion to the sheet. Electronic structure analysis shows that this silicon-based sheet leads to the largest changes in the HOMO and LUMO levels of choline benzoate.[507] Density functional theory calculations of the functionalization of hydrogenated silicene by acetylene, ethylene and styrene have been compared with adsorption on H-Si[111]. In the case of styrene, the binding energies for the intermediate and final states, as well as the energy barrier for hydrogen abstraction, are quite similar for the two systems. The results for acetylene and ethylene however are surprisingly different. In H-silicene, the abstraction barrier is much smaller than in H-Si[111]. These differences are attributed to the differing electrostatic potentials arising from the presence of hydrogen atoms at the bottom of the silicene bilayer, which permit delocalization of the spin density in the intermediate state of the reaction.[508] The adsorption of acetone, acetonitrile, ammonia, benzene, methane, methanol, ethanol and toluene has been investigated using van der Waals density functional theory calculations. The calculated values of the adsorption energy range from -0.11 to -0.95eV; values which, quantitatively, are higher than the corresponding adsorption energies of those molecules on graphene. Electronic structure calculations show that the band-gaps range from 0.006 for acetonitrile to 0.35eV for acetone. The effective mass of the electron is estimated to be relatively small, and this is expected to result in a high electron mobility. When lithium atoms are doped into pristine and acetone-adsorbed silicene, the adsorption energy varies with the number of lithium atoms.[509]

1.13 Defects

Density functional theory and non-equilibrium Green's function method simulations of the structural, electronic and transport properties of various defects in graphene and silicene show that defects are much more easily formed in silicene, as compared with graphene. Although qualitatively similar, the effects of the various defects occur closer to the Dirac point in silicene. Identifying them using scanning tunnelling microscopy is also more difficult due, particularly, to the buckling. This can be overcome by performing direct source/drain measurements. The presence of defects leads to an increase in local current; suggesting that they not only contribute to scattering, but also to heating[510].

1.13.1 Line Defects

The buckled structure of silicene implies the possible existence of new species of line defect which separate regions with reversed buckled phases. It is shown that these new grain boundaries have very low formation energies; an order of magnitude lower than those of grain boundaries in graphene. These defects are stable along various orientations, and can be differentiated in scanning tunnelling microscopic images. All of the defects present a local dimerization between the silicon atoms; with the formation of π-bonds. These defects are thus preferential adsorption sites when compared to pristine regions. The combination of low formation energy and higher reactivity associated with the defects can conceivably be used to design new nanostructures embedded in silicene[511].

The electronic structure of a silicene-like lattice with a line defect is such that, in the bulk energy gap, there are defect-related bands corresponding to spin helical states localized next to the defect line. Spin-up electrons flow forwards on one side near to the line defect and move backwards on the other side; *vice versa* for spin-down electrons. When the system involves a random distribution of spin-flipping scatterers, the electrons suffer much less spin-flipped scattering when they move along the line defect, rather than in the bulk. An electric gate located above the line defect can tune the spin-flipped transmission; turning the line defect into a spin-controllable waveguide[512].

Line defects in two-dimensional materials certainly modulate the various properties of the pristine form. *Ab initio* molecular dynamics simulation investigations of the reconstructions of various types of grain boundary in silicene sheets make it evident that, depending upon the presence of silicon adatoms and the edge-shape of grain boundaries (armchair or zig-zag), stable extended line defects can be introduced in a controlled manner. Further study reveals the stability of these line-defects in silicene grown on Ag(111) surfaces at room temperature. Unlike the case of most two-dimensional sheet materials, 5-5-8 line defects here change the non-magnetic semimetallic pristine silicene sheet into spin-polarized metal. Because ferromagnetically-ordered magnetic moments remain strongly localized at the line defect, a one-dimensional spin channel is created. These spin channels are quite stable since, contrary to the edge of nanoribbons, structural reconstruction or contamination cannot destroy here the ordering of magnetic moments. Zig-zag silicene nanoribbons with a 5-5-8 line defect exhibit interesting electronic and magnetic properties which depend upon their width as well as upon the nature of the magnetic coupling between the edge and defect spin states. Upon incorporating other extended line defects, such as 4-4-4 and 4-8, two-dimensional sheets and nanoribbons of silicene exhibit a non-magnetic metallic or semiconducting ground state[513].

1.13.2 Point Defects

Experiment has provided clear evidence for the presence of larger multi-vacancies in silicene, but their nature and behavior remain largely unexplored. First-principles calculations, together with *ab initio* molecular dynamics simulations, of monolayers containing vacancy clusters, extended line defects and di-adatoms reveal a tendency to coalescence among small defects, and the formation of highly stable vacancy clusters. The 5|8|5 extended line defect, which is the most favored one in both graphene and silicene sheets, is easier to form in the latter material, due to the mixed sp^2/sp^3 hybridization of silicon. Hybrid functional calculations, incorporating part of the Hartree-Fock exchange energy, demonstrate that the introduction of single and double silicon adatoms significantly enhances the stability of the system and provides yet another approach to tuning the magnetic moment and band-gap of silicene[514].

The in-plane strain fields of single-vacancy silicene with various monovacancy concentrations, as well as the corresponding electronic band structures, have been investigated using first-principles calculations. The self-healing monovacancy is the most stable ground structure in silicene, unlike the cases of graphene or h-BN. In the isolated monovacancy center, the bonds along the pentagons are constricted, creating a compression field, while those close to the distorted hexagons are stretched to create a tension field. As the monovacancy concentration increases, the interacting compression fields tend to corrugate the defected silicene, while the interacting tension fields have little effect upon the low-buckle structure. In particular, corrugation is present in super-cells with low monovacancy concentrations, such as (4,5), (4,6), (4,7) and (4,8). The corrugation approaches zero at both low and high monovacancy concentrations, and the (4,6) super-cell - with a monovacancy concentration of about 0.021 - has a peak value of 3.23Å.[515] Electronic calculations show that the linear dispersion at the Γ point in pristine silicene is broken by the lower lattice symmetry of the self-healing monovacancy reconstruction, which also transforms it into a metal.

The morphologies of various point defects in epitaxial silicene on Ag(111) surfaces have been systematically investigated using first-principles calculations combined with experimental scanning tunnelling microscopic observations. The scanning tunnelling microscope signatures for various defects in epitaxial silicene on the Ag(111) surface are identified. In particular, the formation energies of point defects in Ag(111)-supported silicene sheets exhibit an interesting dependence upon the superstructures, which, in turn, may have implications for controlling the defect density during the synthesis of silicene.[516] By estimating the concentrations of various point defects in various silicene superstructures, the mystery of the defective appearance of $\sqrt{13} \times \sqrt{13}$ and $2\sqrt{3} \times 2\sqrt{3}$ silicene in experiments is solved, and 4 x 4 silicene sheet is deduced to be the most suitable structure for future device applications.

The mechanical properties and stability of two layers of defected material, supported by graphene sheets, between which lithium ions pass under the influence of an electrostatic field, have been studied using molecular dynamics methods. The defects consist of mono-, di-, tri- and hexa-vacancies. The graphene and silicene edges are rigidly fixed, and the graphene sheets contacting the silicene take on a convex shape, tending outwards. The mono- and di-vacancies tend to decrease in size, and larger vacancies exhibit a superior stability. Ion-motion control, using an electric field, becomes possible only when using perfect silicene, or silicene with mono- and di-vacancies. The ions penetrate through larger defects, and their motion in the silicene channel becomes uncontrolled.[517] When the ion moves in the channel, the highest stress spikes appear in silicene which contains monovacancies. In the case of fixed edges, perfect silicene intercalated with lithium ions tends to accumulate larger stresses than does silicene which contains defects.

Reactive molecular dynamics simulations of the effect of vacancy defects upon the structural properties and thermal stability of free-standing silicene show that the pristine material is stable up to 1500K; whereupon the system transforms into a three-dimensional amorphous configuration. Vacancy defects produce local structural changes in the system, and considerably reduce the thermal stability. Depending upon the size of the vacancy defect, the critical temperature can decrease by more than 30%. The system is nevertheless still stable well above room temperature. The stability can be increased by saturating the dangling bonds at defect edges with foreign atoms such as hydrogen[518].

First-principles calculations of the structural and electronic properties of armchair silicene nanoribbons containing a monovacancy or a divacancy show that either a monovacancy or a parallel-oriented divacancy changes a direct semiconductor armchair silicene nanoribbon into an indirect one. A slant-oriented divacancy changes its character to metallic. Neither a monovacancy nor a divacancy can change the non-magnetic character of armchair silicene nanoribbons or even metallic armchair silicene nanoribbon with a slant-oriented divacancy. The optimized vacancy structure and the electronic properties are independent of the vacancy positions relative to the edge of the nanoribbon[519].

Defects are conventionally seen as being imperfections which might significantly degrade the performance of materials. At the nanoscale however, defects may be extremely useful in generating novel materials and devices. A generalized gradient approximation, parameterized using the Perdew–Burke–Ernzerhof approach, of a silicene sheet containing 48 silicon atoms with pyridine porphyrin type defects showed that the incorporation of the latter defects resulted in a metallic to semiconducting transition[520].

Density functional theory calculation studies of the structure and properties of perfect and defected zig-zag silicene nanoribbons with a cobalt atom adsorbed on various sites suggest the occurrence of various electronic and magnetic behaviors. The latter depend strongly upon both the vacancy configuration and the cobalt-atom site. More detailed analysis reveals that the vacancy configuration markedly affects the structure and the electronic band dispersion, while the cobalt atom plays a key role in determining the magnetic properties by increasing the magnetic moment of neighboring silicon atoms and suppressing the magnetism of the nearer ribbon-edge[521].

A spin-polarized density functional theory study showed that, upon introducing a monovacancy and a divacancy as well as phosphorus and aluminium atoms into pristine silicene, no magnetism is introduced into the monovacancy system, but there is a significant magnetic moment in the di-vacancy system. Numerical computation also reveals that the magnitude of the magnetic moment is greater when the system is doped with aluminium than with phosphorus[522].

A first-principles study of two-dimensional low-buckled silicene revealed a new type of monovacancy, MV-1, with plane-like sp^3 hybridization at its defect core. This had never been found to exist stably in other two-dimensional single-layer nanostructures such as graphene, h-BN sheet, and single-layer MoS_2. Two other possible monovacancies, MV-2 and MV-3, having higher energies were also found to exist in two-dimensional low-buckled silicene. Here, MV-1 was the most stable ground structure among the three possible monovacancies. High-energy MV-3 is unstable and readily decays into MV-1. Metastable MV-2 can coexist with MV-1 at temperatures below 10K. The predicted diffusion coefficient of MV-1 is 2.3 x $10^{-5}cm^2/s$. This is much higher than that of the monovacancy in graphene. The predicted electronic structures of silicene containing MV-1 and MV-2 show that both of them are metallic[523].

First-principles studies of silicene with monovacancies, on h-BN sheet or a Ag(111) surface, show that due to weak van der Waals interaction with the h-BN sheet, there exist two types of monovacancy in the Si|h-BN composite: the self-healing MV-1 and MV-2 with a closed five- and nine-membered (5-9) pair of rings, with the former being the ground-state monovacancy. This is the same one as in the case of free-standing silicene with monovacancies. The electronic bands around the Fermi level of Si|h-BN composite with silicon monovacancies are all induced by the silicon atoms, leading to a non-magnetic metallic ground state. The Ag(111) substrate has a stronger coupling with the silicon adlayer, leading to the existence of three types of inequivalent silicon atoms, greatly changing the monovacancy structures and the stability of the defective silicene adlayer, together with its corresponding electronic structures. The symmetrical monovacancies are always the stable monovacancy structures in the Si|Ag(111) composite, regardless of which type of inequivalent silicon atom is removed. The silver substrate makes a predominant contribution to the electronic bands of defective Si|Ag(111) composite with silicon monovacancies. Some characteristic bands, introduced by the silicon monovacancy defects can nevertheless be found[524].

Study of the interaction of H_2, O_2, CO, H_2O and OH molecules with the vacancy defects of graphene and silicene shows that atoms around the bare vacancy reconstruct, and that specific chemically active sites are created. The H_2, O_2 and CO remain intact on both pristine graphene and silicene, but they can dissociate when they are placed in close proximity to the chemically active sites, and generate centers for hydrogenation and oxygenation. Saturation of the dangling bonds at the defect sites, by the constituent atoms of dissociated molecules, gives rise to significant modifications of the electronic and magnetic properties. Analysis reveals a concerted action of the surrounding host atoms, together with dissociated molecules, to lower the energy barrier required for dissociation. The dissociation of H_2O or OH is hindered by high energy barriers. It is suggested that graphene and silicene can be functionalized by creating meshes of single vacancies where specific molecules can dissociate, and certain other molecules can be pinned[525].

First-principles calculations of the self-healing of vacancy defects in graphene and silicene show that a silicon adatom, adsorbed on silicene at the top site, forms a dumbbell-like structure by pushing one silicon atom beneath. An asymmetrical reconstruction of the single vacancy in graphene is induced by magnetization via the re-bonding of two dangling bonds and the acquiring of a significant magnetic moment via the remaining unsaturated dangling bond. In silicene, three twofold-coordinated atoms surrounding the single vacancy become fourfold-coordinated and non-magnetic by re-bonding. The energy which is gained by new bond formation becomes the driving force for reconstruction. With an external supply of host atoms, the vacancy defects of graphene heal perfectly but Stone-Wales defects can form in the course of healing of a silicene vacancy. The electronic and magnetic properties of suspended single-layer graphene and silicene are modified by reconstructed vacancy defects[526].

Density functional theory calculations of the chemical activity and sorption, of formaldehyde, on silicene containing point defects show that, compared with the weak adsorption on perfect silicene, formaldehyde molecules tend to be chemisorbed onto the Si-Si bonds of defective silicene; with an appreciable adsorption energy. The electronic conductance changes markedly upon the adsorption of formaldehyde molecules on silicene containing Stone-Wales defects. Meanwhile silicene with double vacancies still exhibits indirect semiconductor characteristics after formaldehyde adsorption. Moreover, the adsorption energy of formaldehyde on silicene with Stone-Wales defects exhibits a continuous increase under tensile strains of up to 10%. This suggests that the chemical reactivity of material containing Stone-Wales defects increases upon applying an external strain[527].

The electronic properties and scanning tunnelling topographical images of X (= F, H, O) functionalized silicene have been investigated by introducing various types of vacancy cluster and chain patterns into monolayers within a density functional theory framework. The relative ease of formation of vacancy clusters and chain patterns is found to be energetically most favorable in hydrogenated silicene. The F- and H-functionalized materials are direct band-gap semiconductors with band-gaps ranging from 0.1 to 1.9eV. The O-functionalized monolayers are metallic in nature. Upon introducing various vacancy clusters and chain patterns, the electronic and magnetic properties are significantly modified. Thus, F- and H-functionalized material with hexagonal and rectangular vacancy clusters are non-magnetic semiconductors with modified band-gap values, while pentagonal and triangular vacancy clusters induce metallicity and a magnetic character into monolayers. Hexagonal vacancy chain patterns introduce a direct-to-indirect gap transition, while zig-zag vacancy chain patterns retain the direct band-gap nature of monolayers.[528] Calculated scanning tunnelling microscopic topographical images exhibit very different characteristics for various types of vacancy cluster and chain patterns. Such characteristics could be used to identify various vacancy patterns which are created during functionalization.

Density functional theory has been used to calculate the properties of hybrid materials in which the pristine structure is modified by substitutional doping or by creating vacancies of the form, Y-X, where one Y atom (silicon or carbon) has been replaced by an X atom (boron, nitrogen, aluminium, phosphorus or a void). The calculations reveal that void systems are unstable, while Si-B and Si-N are the most stable. Some of the doped structures are magnetic in nature, with induced mid-gap states.[529] The silicon-void structure is unstable, but it exhibits the greatest magnetic moment: of the order of $4\mu_B$. The estimated band-gaps of modified silicene/graphene hybrids vary from 1.43 to 2.38eV, and from 1.58 to 2.50eV, for spin-up and spin-down channels, respectively.

1.13.3 Stone-Wales Defect

During the synthesis of ultra-thin materials possessing an hexagonal lattice structure, Stone-Wales defects are quite likely to be formed (figure 4). The existence of such topological defects in a graphene-like structure produces marked changes in their electronic and mechanical properties. The energy barrier for the formation of Stone-Wales defects in free-standing silicene is about 2.4eV, and the barrier for Ag(111)-supported silicene is about 2.8eV; significantly lower than the value (about 9.2eV) for graphene. The buckled nature of silicene presents a high energy barrier to the healing of the Stone-Wales defect, and defective silicene is therefore stable even at high temperatures. Silicene containing Stone-Wales defects is semiconducting, with a direct band-gap of 0.02eV. This value depends upon the concentration of the defects. Nitrogen substitution in Stone-Wales defected silicene indicates that the defect lattice sites are the least-preferred substitution locations for nitrogen atoms. The easy formation of Stone-Wales defects in silicene should provide a guideline for band-gap engineering in silicene-based materials[530].

Investigation of the structural and electronic properties of Stone-Wales defects in silicene sheet and nanoribbons, using periodic boundary conditions and first-principles calculations, leads to a good agreement with available values obtained using earlier first-principles calculations. For all of the concentrations considered, the Stone-Wales defects prefer to locate themselves at the edge, due to the lower formation energy there. The Stone-Wales defects at the center of silicene nanoribbons slightly influence the electronic properties, whereas Stone-Wales defects at the edge of nanoribbons split the degenerate edge states and introduce a sizable gap which depends upon the concentration of the defects.[531] The Stone-Wales defects produce a repulsive perturbation potential, leading to charge on the edge states at the side with the defect transferring to the side without the defect.

First-principles density functional theory calculations of the formation energy and kinetic barrier of the Stone-Wales defect, in infinite and finite sheets of silicene, show that the formation energies and barriers in both cases are significantly lower than those for graphene. Compared with infinite sheets, the energy barriers and formation energies are lower for finite sheets. On the other hand, due to the low barriers the defects are expected to heal out of finite sheets[532].

Density functional theory and non-equilibrium Green's function techniques show that, in zig-zag silicene nanoribbons containing Stone-Wales defects, the predicted formation energy is significantly lower than that of graphene and silicene; implying a high stability of such a defect in nanoribbons. A negative differential resistance is observed, within certain bias-voltage ranges, for both perfect and Stone-Wales defected nanoribbons[533]. Other results, obtained by applying these methods to armchair silicene nanoribbons containing a Stone-Wales defect, an interior/edge vacancy and an edge dangling bond, show that structural imperfections can alter the electronic structure (energy bands, density of states) of nanoribbons and either increase or decrease the ballistic current[534]. *Ab initio* calculations of Stone-Wales defects, single and double vacancies, and adatoms show that Stone-Wales defects can be effectively recovered by thermal annealing. Single vacancies have a much higher mobility than that of double vacancies, and two single vacancies are very likely to coalesce into a single double vacancy in order to lower the overall energy. The existence of Stone-Wales and double vacancies can introduce small gaps into silicene, while that of a single vacancy can transform semi-metallic silicene into the metallic form. Adatoms are unexpectedly stable and can markedly affect the electronic properties. In particular, silicon adatoms as self-dopants in silicene sheets can induce long-range spin polarization as well as an appreciable band-gap; yielding an all-silicon magnetic semiconductor[535].

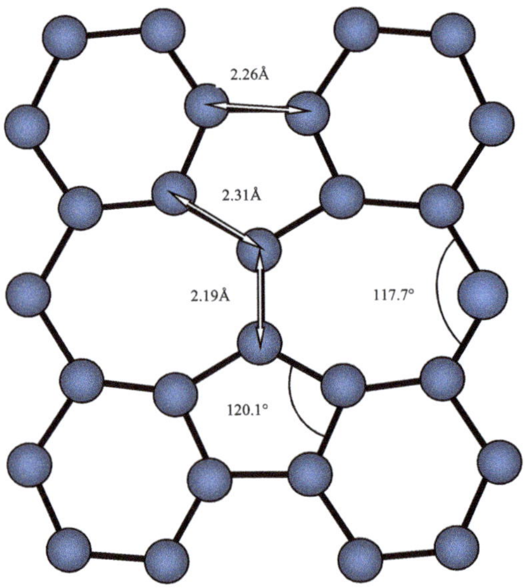

Figure 4. Stone-Wales defect in silicene. Note the changes in bond-angles and atomic spacings as compared to those in the structure at the head of the chapter.

Molecular dynamics finite-element methods, using the Tersoff potential, have been used to predict the Young's modulus, Poisson ratio and uniaxial tensile stress–strain curves for the armchair and zig-zag directions. These show that pristine and slightly defective silicene sheets exhibit almost the same elastic behavior up to fracture. However, the presence of a single defect significantly weakens the sheet and results in a considerable reduction in the fracture strength. One two-atom vacancy at the sheet's center reduces the fracture stress by 18 to 20% and the fracture strain by 33 to 35%. The weakening effect of Stone–Wales defects depends upon the tensile direction and upon the orientation of the defect[536]. First-principles calculations of the properties of zig-zag silicene nanoribbons containing Stone-Wales defects reveal that two types of Stone-Wales defect, SW-I and SW-II, exist in the nanoribbons. The defect is found to be most stable at the edge of the nanoribbon, regardless of orientation. It is even more stable here than it is in an infinite silicene sheet. In addition, the zig-zag silicene nanoribbons can transform from semiconductors to metals or semi-metals upon modifying the Stone-Wales defect location and concentration. For a given defect concentration, the band structures which are affected by the SW-I defect are more distinct than those which are affected by the SW-II defect, when the defect is at the edge. These results suggest the possibility of tuning the electronic properties of zig-zag silicene nanoribbons by using Stone-Wales defects[537].

1.13.4 Miscellaneous

When zig-zag nanoribbons are studied using *ab initio* simulation methods, novel structure-property relationships are found which are influenced by factors such as the width, spin polarization, spin-orbit coupling and extended topological defects. Whereas defect-free nanoribbons exhibit an antiferromagnetic-ferromagnetic transition as a function of the width, all defective nanoribbons are ferromagnets. The role of spin-orbit coupling is also significant because it leads to spin-dependent energy gaps in the electronic structure.[538]

First-principles calculations of the structural and vibrational properties of silicene predict that a structure with the small buckling of 0.44Å and bond-length of 2.28Å is the most favorable energetically, and it does not exhibit an imaginary phonon mode. The calculated non-resonance Raman spectra is characterized by a main peak at about 575/cm (G-like peak). Extensive calculations on armchair nanoribbons, with and without hydrogenation of the edges, reveal the presence of other Raman peaks; distributed mainly at lower frequencies than that of the G-like peak. These are attributed to defects at the edges of the ribbons, and are thus absent from the Raman spectra of non-defective material. In particular, the Raman peak which corresponds to the D mode is located at about 515/cm. The calculated G-like and D peaks are probably the fingerprints of the Raman spectra of the low-buckled structure of silicene.[539]

1.14 Diffusion Processes

Density functional theory, in the generalized gradient approximation, has been used to determine the geometry, electronic structure and diffusion properties of sodium atoms on pristine or single vacancy-defected silicene sheet. The lowest barrier to sodium-atom diffusion is 0.25eV on pristine silicene and 1.44eV on single vacancy-defected silicene sheet. The vacancy exerts a more significant influence, than does adsorption of the sodium atom, on the electronic structure of silicene[540].

The interface stability of hybrid silicene/fluorosilicene nanoribbons, where fluorosilicene is fully fluorinated silicene, has been investigated by using density functional theory calculations. The diffusion of fluorine atoms at the zig-zag and armchair interfaces of the nanoribbons is endothermic, and the corresponding minimum energy barriers are 1.66 and 1.56eV, respectively. These values are marked higher than the minimum diffusion energy barrier of one fluorine atom and two fluorine atoms on pristine silicene: 1.00 and 1.29eV, respectively. The thermal stability of these nanoribbons can thus be significantly enhanced by increasing the fluorine diffusion barriers by interface engineering. The armchair nanoribbons are non-magnetic semiconductors, and their band-gap exhibits an oscillatory behavior when the width of the silicene part changes. In the nanoribbons, the antiferromagnetic semiconducting state is the most stable.[541]

First-principles methods show that the adsorption energy of lithium on silicene is significantly higher than that of lithium on graphene. The adsorption energy also increases with increasing concentration of adsorbed lithium atoms. The diffusion barrier of lithium on silicene is relatively low, and insensitive to the concentration of the adsorbed atoms.[542] First-principles density functional theory calculations of the interaction of lithium with the edges of monolayer and multilayer silicene identify low-energy lithium binding-sites and pathways for lithium diffusion. Both lithium binding and diffusion are significantly affected by the morphology of the silicene edges. Among the known structural forms of silicon, monolayer silicene edges exhibit the strongest binding with Li. The energy barriers for lithium diffusion on monolayer silicene nanoribbons range from 0.14 to 0.26eV, and zig-zag edges can permit up to 80 times faster diffusion than does a silicene monolayer. Monolayer silicene nanoribbons, terminated by zig-zag edges are promising materials for the negative electrodes of lithium-ion batteries.[543] Later studies of the adsorption and diffusion of lithium atoms on silicene sheet and zig-zag nanoribbons have yielded a diffusion energy barrier, for lithium adatoms on silicene sheet, of 0.25eV. This is much lower than the barrier on graphene and bulk Si. The diffusion barriers along the axis of zig-zag silicene nanoribbon range from 0.1 to 0.25eV, due to an edge effect, while the barrier is about 0.5eV for a lithium adatom entering a silicene nanoribbon.[544]

1.15 Applications

Density functional theory-based first-principles calculations of the structural stability and electronic properties of two kinds of silicene domain boundaries which form along armchair edge and zig-zag edges indicate that linkage of tetragonal and octagonal rings (4|8) occurs along the armchair edge, while linkage of paired pentagonal and octagonal rings (5|5|8) occurs along the zig-zag edge. Unlike graphene, the buckled nature of silicene leads to two mirror-symmetrical edge line-defects. The formation energies indicate that the 5|5|8 domain boundary is more stable than the 4|8 domain boundary. Like graphene, the calculated electronic properties show that the 5|5|8 domain boundaries exhibit metallic properties while the 4|8 domain boundaries are half-metallic. Both domain boundaries create perfect one-dimensional metallic wires which, due to their metallic properties, can be used to build silicene-based devices.[545]

First-principles simulations of the interaction of a two-dimensional silicon layer with chalcogenide-layered compounds, MoX_2 and GaX (X = S, Se or Te), reveal a rather weak van der Waals interaction between the silicene and chalcogenide layers. The buckling of the silicene layer can be correlated with the lattice mismatch between the silicene layer and the MoX_2 or GaX template. The electronic properties of silicene on the various templates largely depend upon the degree of buckling of the silicene layer: highly-buckled silicene on MoS_2 is predicted to be metallic, while slightly-buckled silicene on GaS and GaSe is predicted to be semi-metallic, with preserved Dirac cones at the K points. These results suggest new methods for the engineering of silicene nanosheets, permitting the tailoring of the electronic properties of the two-dimensional graphene analogue on non-metallic substrates.[546]

1.15.1 Transistors

Dating from its first successful laboratory synthesis, silicene has been considered to be promising - for post-silicon electronics - as an alternative to graphene and other two-dimensional materials. Offsetting its theoretically high carrier mobility, its zero-bandgap characteristic makes pure silicene, like graphene, impossible to use directly as a field effect transistor at room temperature. There remains the possibility of opening a band-gap in silicene without spoiling its excellent electronic properties[547]. This is because the silicene lattice is slightly buckled and composed of two vertically displaced sub-lattices. By breaking the sub-lattice symmetry a band-gap can be opened[548]. Silicene nanoribbon has a sufficient band-gap to act as field-effect transistor channel material[549]. The two main obstacles to producing a silicene transistor have been its strong interaction with metal templates, and its instability in air[550]. One solution involves the fact that dispersion-corrected density functional theory computations demonstrate that appreciable CH/π interactions exist between graphene, and graphane or its patterned partially hydrogenated (C_4H) derivatives. The equivalence breaking of two sub-lattices of graphene then opens up a 90meV band-gap in the graphene/C_4H bilayer. The band-gap can be further increased to 270meV by sandwiching graphene between two C_4H layers. By taking advantage of analogous SiH/π interactions, a 120meV band-gap can also be opened in silicene.[551] The high carrier mobility of graphene or silicene is preserved, and theoretical results suggest a practical solution for gap-opening in graphene and silicene; allowing them to serve as field effect transistors, etc. Recently prepared silicene field-effect transistors promise electronic applicability, and a multi-terminal spin density matrix approach helps to analyse spin polarizations in the silicene structure of such spin field effect transistors by considering the interaction of intrinsic and Rashba spin-orbit coupling, the gate voltage, Zeeman splitting and disorder. Coexistence of a stagger potential and intrinsic spin-orbit coupling results in spin precession. This renders any in-plane polarization directions reachable by the gate voltage. In particular, intrinsic coupling permits the electrical adjustment of the in-plane components of the polarizations. The Rashba coupling adjusts meanwhile the out-of-plane polarizations. Larger electrically tunable ranges of in-plane polarizations are found in oppositely gated than in uniformly gated silicene. The polarizations in differing phases behave in a distinguishable manner in the weak-disorder regime. Regardless of the phase, stronger disorder leads to saturation.[552]

A ballistic quantum transport model, based upon a non-equilibrium Green's function approach coupled with the three-dimensional Poisson equation, when applied to a double-gate armchair silicene nanoribbon field effect transistor demonstrates that using HfO_2 as the gate insulator can be a good choice. Combined with the proper combination of temperature and strain magnitude, this can give better device performance. Further analysis predicts the electron effective mass as a function of the band-gap of a hydrogen-passivated armchair silicene nanoribbon under strain[553].

Although the carrier concentration of silicene exhibits a similar behavior to that of graphene, there are some differences in the conductance behavior. Modeling predicts an increment in the total carriers and conductance with the gate voltage, as expected for conventional semiconductors which are affected by temperature only at the neutrality point. The minimum conductance increases with temperature, but remains stable in the degenerate regime[554].

Silicene is a quantum spin Hall insulator which undergoes a topological phase transition into other insulators upon applying an external field. The transport properties of nanoribbons can be determined, based on the Landauer formalism, by measuring the density of states and conductance. The latter is quantized, and changes in value when the system transforms into various phases. A silicene nanoribbon, near to zero energy, acts as a field effect transistor. Such a transistor is robust because the zero-energy edge states are topologically protected[555]. According to the bulk-edge correspondence zero-energy edge channels are expected to appear in silicene nanoribbons. The behaviors of the helical edge channels are completely different between armchair and zig-zag edges. Zero-energy states disappear in armchair nanoribbons, in spite of the bulk-edge correspondence, but appear as zig-zag nanoribbons even if the width is quite restricted. This difference is attributed to the penetration depth of the helical edge channel, which is anti-proportional to the spin-orbit gap for the armchair edge but remains as small as the lattice constant for the zig-zag edge. These properties clearly demarcate silicene and graphene nanoribbons; especially for armchair edges. In silicene, edge states emerge as required by its topology, even though the zero-energy states disappear from the energy spectrum. No edge states exist in graphene.[556]

It is rather challenging to grow silicene epitaxially onto conventional semiconductor substrates. First-principles calculations of high-quality van de Waals silicene/Sc_2CF_2 heterojunctions predict that the Dirac cone of silicene exists in the band-gap of the Sc_2CF_2 substrate. A nearly linear band dispersion of silicene with a gap of 36 to 48meV is obtained in all heterojunctions, due to the sub-lattice symmetry broken by the intrinsic interface dipole. The band-gap of all these heterojunctions can be effectively varied by adjusting the interlayer spacing and strain[557]. The structural and electronic properties of two-dimensional silicene, modulated by AsSb van der Waals interaction, when investigated using density functional theory with van der Waals corrections, similarly reveal that there is almost no lattice mismatch introduced into silicene/AsSb hetero-interface and that this is beneficial when compared with the behavior of silicene on other substrates. A sizable band-gap (213 to 563meV) appears in silicene due to breaking of the inversion symmetry, by interface effects, indicating potential applications in room-temperature field effect transistor technology[558]. The almost linear band dispersion of silicene, which guarantees a high carrier mobility, can be preserved in the silicene/AsSb heterostructures. The band-gap can also be effectively tuned by changing the interlayer distance between the silicene and AsSb. An indirect-direct band-gap transition can even occur.

A 2-dimensional silicene and GaP heterobilayer which comprises silicene plus monolayer GaP can be prepared. First-principles calculations show that the interaction energies range from -295.5 to -297.5meV per unit cell, thus indicating a weak interaction between the silicene and the phosphide monolayer. The band-gap changes from 0.06 to 0.44eV in the hybrid heterobilayer. An unexpected indirect-direct band-gap cross-over is also observed in the heterobilayer, depending upon the stacking pattern[559]. These results suggest a possible means for making effective field effect transistors from silicene on a GaP monolayer.

It is found that a silicene field effect transistor, in which a band-gap is introduced by using a perpendicular electric field, exhibits a sub-threshold swing which is smaller than 60mV/decade, plus a switching effect with high on/off current ratio. Such a device also exhibits an output with very good saturation, due to an improved pinch-off of the channel arising from the electrically-induced band-gap[560]. Similarly, by using *ab initio* calculations, it is predicted that a vertical electric field can open a band-gap in semi-metallic single-layer buckled silicene. The size of the band-gap increases linearly with the electric field strength. Quantum transport simulation of a dual-gated silicene field effect transistor confirms that the vertical electric field opens a transport gap. A significant switching effect due to an applied gate voltage is also observed. Thus, biased single-layer material can operate effectively at room temperature as a field effect transistor.[561]

Ab initio quantum transport calculations, for field effect transistors based upon semiconducting armchair silicene nanoribbons, show that these field effect transistors exhibit high performances, with an on/off ratio of over 10^6 and a sub-threshold swing as low as 90mV/decade. The output characteristic exhibits saturation behavior. Drain-current saturation is an advantage with regard to device speed, but is usually absent from graphene field effect transistors[562].

First-principles calculations predict that a sizable band-gap can be opened at the Dirac point of silicene, without degrading its electronic properties, by n-type doping using copper, silver and gold adsorption, p-type doping using iridium adsorption and neutral doping using platinum adsorption. A silicene p-i-n tunnelling field effect transistor can be designed in principle by the adsorption of differing transition metal atoms at various regions. Quantum transport simulations demonstrate that silicene tunnelling field effect transistors have an on/off ratio of 10^3, a small sub-threshold swing of 77mV/decade and a large on-state current of over 1mA/μm under a supply voltage of about 1.7V[563]. Opening an appreciable band-gap without degrading high carrier mobility is, as in the case of graphene, vital for silicene use as a high-performance field effect transistor. Density functional theory calculations predict that a band-gap is opened by the single-side adsorption of alkali atoms; due to sub-lattice or bond-symmetry breaking. The band-gap size can be controlled by changing the adsorption coverage; thus yielding a maximum band-gap of up to 0.50eV. *Ab initio* quantum transport simulation of a bottom-gated field effect transistor, based upon sodium-covered silicene, predicts a transport gap which is consistent with the band-gap[564]. The resultant on/off current ratio is up to 10^8.

Density functional theory calculations of the electronic properties of silicene-based nanostructures, as affected by external factors such as electric fields, strain, etc. predict that, in two-dimensional silicene-based nanostructures, the magnetic moment of iron-doped silicene undergoes a sharp jump at a threshold electric field; offering the possibility of a good switching effect and potential application as a magneto-electric diode. Under an electric field, the good controllability and sharp switching-effect of magnetism suggest potential applications as magneto-electric devices. In the case of one-dimensional nanostructures, silicene nanoribbons having sawtooth edges are more stable than zig-zag silicene nanoribbons, and exhibit spin-semiconducting features. Under an external electric field or under uniaxial compressive strain, gap-less spin-semiconductors are produced; a significant factor in the design of qubits for quantum computing via spintronics. The superlattice structures of silicene-based armchair nanoribbons, another form of one-dimensional silicene nanostructure, can be modulated via the size and strain of the superlattice. With increasing strain, the related energy gaps change; a significantly different behaviour to that of the constituent nanoribbons.[565]

As noted elsewhere, the characteristic buckling permits the opening of a band-gap in silicene by applying a perpendicular electric field. This buckling also permits a highly effective modulation of the conductance by means of an in-plane electric field, applied through side-gates which can be formed concurrently within the same silicene monolayer. This was demonstrated by using silicene to implement self-switching diodes: two-dimensional field-effect nano-rectifiers created within a single silicene monolayer. Quantum simulation results show that such atomically-thin self-switching diodes, with sub-10nm dimensions, achieve a current rectification ratio of more than 200 without any need for doping. This is a thirty-fold improvement over the results for graphene self-switching diodes, and is attributed to band-gap opening by the in-plane electric field; thanks to the silicene buckling.[566]

First-principles calculations suggest that hydrogenation and calcium intercalation can be used to break the bonds between a silicene layer and a silver substrate, leading to the creation of a silicane mono-layer; a wide band-gap semiconductor. The calcium intercalation lifts the silicene sheet up without destroying its sp^2 honeycomb bonding.[567]

First-principles calculations of the effects of doping upon slightly buckled silicene show that, although the lattice is very sensitive to the carrier concentration, it is stable over a wide doping range. The frequencies of the E_{2g}-Γ and A′-K Raman modes can in fact be used to probe the carrier concentration. The phonon dispersion exhibits Kohn anomalies at the Γ and K points, but these are reduced by doping; suggesting that electron-phonon coupling cannot be neglected in field-effect transistor applications.[568]

1.15.2 Spintronics

First-principles calculations predict that an in-plane homogeneous electric field can induce half-metallicity in hydrogen-terminated zig-zag nanoribbons. A dual-gated finite device can exhibit an almost perfect (up to 99%) spin-filter efficiency, while a quadruple-gated finite device can serve as an effective spin field effect transistor with an on/off current ratio of over 100; according to *ab initio* quantum transport simulations.[569]

A mechanism for generating a spin-polarized current in a two-terminal zig-zag silicene nanoribbon is suggested to be that, when a weak local exchange field parallel to the surface is applied to one of the edges of the nanoribbon, a gap is opened up in the corresponding edge states while another pair of gap-less edge states with opposite spin are still protected by time-reversal symmetry. A spin-polarized current can thus be induced in the gap opened up by the local exchange field in a two-terminal system. The important fact is that the spin-polarized current can be obtained even in the absence of Rashba spin-orbit coupling, and in the presence of a very weak exchange field. The mechanism for generating spin-polarized currents can therefore be easily exploited experimentally. The spin-polarized current is also insensitive to weak disorder.[570] Non-equilibrium Green's function studies of the spin-dependent electron transport properties of zig-zag silicene nanoribbon, with ferromagnetic strips deposited onto both edges and with an electric field applied perpendicular to the nanoribbon plane, indicate that complete spin polarization can take place in the presence of the perpendicular electric field for an anti-parallel configuration and that the nanoribbon can function as a perfect spin filter. The spin direction of the transmitted electrons can be easily changed from up to down, and *vice versa*, by reversing the electric field direction. In the case of a parallel configuration, perfect spin filtering can occur even in the absence of an electric field. The spin direction can here be changed by changing the electron energy. The effect of non-magnetic Anderson disorder is that the perfect spin filtering properties of the nanoribbon are destroyed by strong disorder but the nanoribbon retains those properties in the presence of weak disorder.[571]

Spin-polarized electronic transport in buckled bilayer silicene has been studied using a semi-classical Monte Carlo approach. De-phasing of the spin vectors in the buckled bilayer was attributed to Elliott-Yafet and Dyakonov-Perel relaxation mechanisms. The spin de-phasing length is estimated to be of the order of 2μm for buckled bilayer material. Investigation of the ensemble averaged spin vector variation in the buckled bilayer, along the length of the device, as a function of temperature indicated a negligible variation in the spin de-phasing length between 4 and 77K. As the temperature was increased from 77 to 373K, there was a monotonic decrease in the spin de-phasing length. Buckled bilayer silicene is thus a promising candidate for next-generation spintronic devices.[572] Similarly, in N-armchair-edge nanoribbons (where N is the number of atoms in the armchair edge) studied using the semi-classical Monte-Carlo approach, spin vector de-phasing is again attributed to Elliott-Yafet and Dyakonov-Perel relaxation mechanisms and the spin de-phasing length is estimated to be of the order of 3μm. There is again a negligible variation in the spin de-phasing length at 4 to 373K and, from 373 to 1000K, a decrease in the spin de-phasing length upon extrapolating the curve to 1000K. In the higher temperature range, the spin de-phasing length decreases due to hot electron/phonon effects and higher scattering rates. The ensemble averaged spin vector variation along the length of the nanoribbons suggests that there is a negligible effect of the width upon the spin de-phasing length.[573] Finally, in lithiated silicene studied using the same theoretical approach, de-phasing of the spin vectors is again due to Dyakonov-Perel and Elliott-Yafet relaxation mechanisms, the spin de-phasing length is estimated to be about 2μm and there is a negligible variation in the spin de-phasing length between 4 and 373K, with a decrease above 1000K.[574] Finally, the study of spin polarized transport along the length of planar monolayer silicene and buckled monolayer silicene shows that the spin de-phasing length is of the order of 0.5μm.[575]

The topological phase of silicene-nanoribbon holding-edge states in the bulk energy-gap can be easily broken by an external electric field, and low-energy Green's function calculations[576] show that it is possible to localize conducting channels anywhere in a silicene nanoribbon by applying an inhomogeneous electric field. The spin degeneracy of the channels can also be broken in the same way, thus permitting the conduction of spin as well as of charge. This suggests a design for a ternary logic device which could be used in low-power circuits. Silicene, and related group-IV elements having a honeycomb structure, could function as a platform for the efficient manipulation of spin currents using external electric fields; with no requirement to switch magnetic fields for spintronics applications.

Taking dual-gated silicene as an example, theoretical evaluation of the spin-dependent transport in a lateral resonant tunnelling structure with an aligned completely valley-polarized beam and spin-resolved well state shows[577] that a large spin polarization can be predicted, due to a spin-dependent resonant tunnelling mechanism. Under gate electric-field modulation, the forming quantum well state can be externally manipulated so as to trigger further development of an externally controllable spin polarization. By integrating such a structure with a proper valley-filter, constructed in the same manner as in graphene valleytronics, completely-polarized spin beams should be attainable without the assistance of a ferromagnetic component. This therefore permits some profitable strategies for the preparation of non-magnetic spintronic devices on silicene.

The spin-dependent electronic structures of aluminum-doped zig-zag nanoribbons, investigated using first-principles calculations, are such that when they are doped substitutionally with a single aluminium atom on various sites in every three primitive cells they become half-metallic in some cases. This property can be exploited in spintronic devices. Spin-down electrons can moreover be transported at the Fermi energy when the aluminium atom is placed on the sub-edge site. On the other hand, spin-up electrons can be transported at the Fermi energy when the ribbons are doped at sites near to their centre. The magnetic moment on the edge is appreciably suppressed if the aluminium atom is doped at edge or near-edge sites. Analogous results are found for phosphorus- and boron-doped ribbons. When two or more silicon atoms are replaced by aluminium atoms, the half-metallic behaviour is generally replaced by a metallic, spin gap-less semiconducting or semiconducting one. When a line of six silicon atoms along the ribbon's width is replaced by aluminium atoms, the spin resolution of the band structure is suppressed and the system becomes non-magnetic.[578]

First-principles calculations show that field-gated material possesses two gapped Dirac cones which exhibit nearly 100% spin-polarization, situated at the corners of the Brillouin zone. This suggests a possible design for a spin-filter that could allow the spin-polarization of an output current to be switched electrically without switching external magnetic fields. Quantum transport calculations indicate that such a design would be highly efficient, with almost 100% spin-polarization, and would be robust with respect to weak disorder and edge imperfections. A proposed Y-shaped spin/valley separator would also produce a spin-polarized current at two output terminals with opposite spins.[579]

A hybrid spin-filter device has been proposed which is based upon silicene nanoribbon. A ferro-electric polymer, grown on top of the nanoribbon, splits the spin-up and spin-down electron bands and leads to a spin-polarisation of the conductance. The effect of a random distribution of vacancies upon the performance of this spin-filter device is such that disorder induces an Anderson localisation of the electrons. The localisation length depends strongly upon the electron spin. Upon adjusting the Fermi level of the source contact, only electrons with a given spin orientation can reach the drain contact, because its localisation-length is greater than the length of the device. Electrons with an opposite spin are largely back-reflected. The electrical conductance then becomes spin-polarised and the device behaves as a quasi half-metal.[580] A moderate concentration of vacancies has little effect upon the spin-filter capabilities, and opens up the possibility of using it as a tuneable source of polarised electrons. A similar study of the effect of a superlattice defect, representing an abnormal height barrier, upon the spin and valley dependent transport properties shows that the transmission resonance is greatly suppressed because the symmetry of the superlattice structure is destroyed by the defect. The spin-up and spin-down electrons near to the K and K' valleys are dominated by differing effective superlattices and defects. The conductances therefore depend strongly upon the spin and valley.[581] By properly adjusting the defect strength, the spin and valley polarizations can be markedly affected over a wide energy range. These results suggest that the structure could function as a defect-controlled switch.

1.15.3 Environmental Protection

The probability of NO reduction on silicene was studied using density functional theory calculations. One mechanism for NO reduction on silicene is direct dissociation, and the other involves a dimer mechanism. The former mechanism involves direct breaking of the N-O bond. Calculated potential energy surfaces show that the total energy barrier in the favored direct dissociation pathway is 0.466eV. The dimer mechanism involves the formation of $(NO)_2$ on silicene, which then decomposes according to: $N_2O + O_{ad}$ or $N_2 + 2O_{ad}$. The $(NO)_2$ dimer formation on silicene is thermodynamically and kinetically feasible. The energy barriers for $(NO)_2$ dimer formation are lower than 0.231eV. The results indicate that $(NO)_2$ dimers can be readily reduced to N_2O or N_2. The energy barriers, in the favored decomposition pathways, to produce N_2O are lower than 0.032eV. The energy barrier for the release of N_2 is calculated to be 0.156eV. Further results indicate that it is easy to reduce N_2O to N_2 over an energy barrier of 0.445eV. The NO reduction therefore favors N_2 generation via the dimer mechanism rather than by direct dissociation. The conclusion is that silicene can be useful in NO removal and its conversion into environmentally-friendly gases.[582]

Due to its large surface area and sp^3 hybridization, Li-functionalized silicene is a promising CO_2 absorbent for greenhouse gases that is stable up to at least 500K and has the very high storage capacity of 28.6mol/kg (55.7wt%).[583] The adsorption energy of CO_2 on Li-functionalized silicene is enhanced, as compared to pristine silicene, such as to attain an almost ideal value that still facilitates easy release. In addition, the band-gap is found to change sensitively with CO_2 coverage.

1.15.4 Sensors

Quantum conduction is detectably modified by the weak chemisorption of a single CO molecule onto a pristine silicene nanoribbon. This modification is attributed to charge transfer from CO to the silicene nanoribbon and to the deformation induced by the CO chemisorption. The moderate binding energies provide an optimum combination of high detectability and recoverability. When silver contacts are attached to a ~1nm silicene nanoribbon, the interface states mask the conductance modulations caused by CO adsorption, raising length-effect issues for sensor applications.[584]

The adsorption and substitution of platinum on silicene markedly change the electronic structure. The platinum atom prefers to bond to silicene at hollow sites, with an adsorption energy of 5.815eV. To substitute a single platinum atom into the silicene lattice, an energy of 4.401eV is required. Both systems exhibit good catalytic properties and offer some potential as sensors. In particular, the catalytic oxidation of CO on platinum-embedded silicene, with its Eley-Rideal mechanism and two-step route, is suitable for producing CO_2 from CO and activated O_2.[585]

The adsorption of several common gas molecules on silicene, studied using density functional theory, is such that silicene exhibits a significantly high reactivity towards NO_2, O_2 and SO_2; with the adsorption energies being higher than 1.00eV. This suggests its potential value in the development of metal-free catalysts. The NO and NH_3 can be adsorbed onto silicene with energies of 0.35 and 0.60eV, respectively, indicating that silicene could be a good NO or NH_3 sensor. The band-gap is opened upon adsorbing NO, O_2, NH_3 and SO_2 in various ways. Thus NO_2 adsorption renders silicene half-metallic. The presence of Stone-Wales defects and an Ag(111) substrate can enhance the chemical reactivity of silicene.[586] There a therefore a potential application for silicene as a highly sensitive molecule-sensor. Those of NH_3, NO and NO_2 adsorb chemically onto silicene via strong chemical bonds. With distinct charge-transfer from silicene to the molecules, silicene and the chemisorbed molecules form charge-transfer complexes. The adsorption energy and charge transfer in NO_2-adsorbed silicene are larger than those in NH_3- and NO-adsorbed silicene. Depending upon the adsorbate type and concentration, silicene-based charge-transfer complexes exhibit interesting electronic properties, with a tunable band-gap opening at the Dirac point. The calculated charge-carrier concentrations of NO_2-chemisorbed silicene are 3 orders of magnitude larger than the intrinsic charge-carrier concentration of graphene at room temperature.[587] The gas-detection performance of silicene nanosensors for NO, NO_2, NH_3 and CO, predicted using density functional theory and non-equilibrium Green's function methods, is such that pristine silicene exhibits a strong sensitivity for NO and NO_2, but seems to be incapable of sensing CO and NH_3. In order to overcome such sensitivity limitations, due to the weak van der Waals interaction of the latter gas molecules, pristine silicene can be doped with boron or nitrogen atoms. This leads to an increased binding energy and charge transfer, and thus to a significant improvement in sensitivity. Discrimination of the four gases studied, using silicene devices, thus appears to be possible and this promises a new generation of nanosensors for highly sensitive and selective gas detection.[588]

It is possible that silicene could function as an electrical DNA-sequencing device via changes in transmission and conductance caused by the adsorption of each nucleobase. In spite of the relatively weak interaction between nucleobases and silicene, significant changes in the transmittance at zero bias are predicted; particularly for cytosine and guanine. It is suggested that silicene could be developed as an integrated-circuit biosensor as part of a 'lab-on-a-chip' device for DNA sequencing.[589] Alternatively, it has been proposed that a DNA sequencing scheme can be based upon silicene nanopores. First-principles theory has been used to compute the electrical properties of such pores in the absence or presence of nucleobases and, using a 2-terminal geometry, the current-voltage relationship in the presence of nucleobases of various orientations has been analyzed. It is demonstrated that, when nucleobases pass through a pore - even after sampling over many orientations - changes in the electrical properties of the ribbon can be used to discriminate between bases.[590]

1.15.5 Gas Purification

First-principles calculations of the hydrogen permeability and selectivity of a porous silicene membrane show that the sub-nanometer pores have the form of divacancy defects with octagonal and pentagonal rings (585-divacancy). Such porous silicene exhibit a high selectivity which is comparable with that of graphene-based membranes for hydrogen with respect to N_2, CO, CO_2, CH_4 and H_2O. The divacancy defects in silicene are chemically inert towards these gas molecules. The results suggest that porous silicene membranes have great potential usefulness in gas separation and filtering.[591]

The increasing shortage of pure helium has made its purification ever more important. It has been demonstrated, on the basis of first-principles calculations, that porous silicene can be used as an effective ultimate membrane for helium purification. A pristine silicene monolayer is impermeable to helium gas, with a penetration energy barrier of 1.66eV. Porous silicene, with a Stone-Wales defect or divacancy (555, 777 or 585) defect, offers however a surmountable barrier for helium or 0.33 to 0.78eV, but a near-impassable barrier to Ne, Ar and other gas molecules. Porous silicene with divacancy defects in particular exhibits a high selectivity for He/Ne and He/Ar mixtures. In this respect, it is superior to graphene, polyphenylene and traditional membranes.[592]

1.15.6 Hydrogen Storage

Early studies using quantum chemical calculations of the six-membered rings showed chair-like puckering becoming, for large sheet-like clusters, ordered ripples. The binding energies and polarizabilities of silicene clusters revealed monotonic trends analogous to those of polyacenes. The stacking of two silicene layers led to the formation of closed 3-

dimensional clusters with high symmetry and strong Si-Si bonds. The heat of hydrogenation of silicene to silicane was found to be very exothermic[593].

First-principles density functional studies of hydrogen storage on lithium-decorated silicene show that the calculated lithium binding-energy on silicene is much higher than the bulk lithium cohesive energy. This rules out the possibility of cluster formation in lithium-doped silicene, which facilitates reversible hydrogen adsorption and desorption. That is, for every one lithium atom adsorbed on silicene, each lithium atom could adsorb up to five hydrogen molecules. By adsorbing lithium atoms on both sides of the silicene, the hydrogen capacity can attain as much as 6.35wt%. The average binding energy of H_2 molecules falls to between 0.32 and 0.17eV. This is favorable for the development of high-capacity hydrogen storage at room temperature[594]. Similar studies of various metal adatoms (Li, Na, K, Be, Mg, Ca), doped on the most stable chair-like configuration of silicane, show that lithium and sodium adatoms are ideally suitable; not only because of their strong metal-to-substrate binding and uniform distribution over the substrate, but also because of their high capacity for hydrogen storage. The stability of both lithium- and sodium-functionalized SiH was also confirmed by molecular dynamics simulations. Both Li^+ and Na^+ adsorbed four H_2 molecules and attained reasonably high storage capacities of 6.30 and 5.40wt%, respectively, with the average adsorption energies being within the range suitable for practical H_2 storage applications; unlike the case of alkaline-earth metals[595]. The preferential sites of the dopants, stabilities of doped systems, bonding mechanism and hydrogen storage capacities are calculated by using the projector augmented wave method, the Perdew-Burke-Ernzerhof variant of the generalized gradient approximation, the Nosé-Hoover thermostat and Bader charge analysis. In the case of silicene, Li^+ and Na^+, can adsorb five hydrogen molecules, giving storage capacities of 7.75 and 6.9wt%, respectively[596]. The electronic structure, stability and hydrogen-storage capacity of strain-induced magnesium-functionalized SiMg and SiHMg monolayers have been studied by means of van der Waals-induced first-principles calculations. A sharp increase in the binding energy of magnesium adatoms on either type of monolayer, under a biaxial symmetrical strain of 10%, produces a uniform distribution of dopants over the substrate. A significant positive charge on each magnesium atom accumulates a maximum of six H_2 molecules, giving an H_2 storage capacity of 8.10% and 7.95% in the case of SiMg and SiHMg, respectively. The average adsorption energy for H_2 molecules is ideal for the design of practical H_2 storage materials.[597]

Storage and release parameters define the efficiency of a hydrogen storage medium. Density functional theory calculations of the effect of an external electric field on the adsorption and desorption of H_2 on calcium-decorated silicene demonstrate that nine H_2 molecules per calcium atom can be adsorbed and that 6.4wt%H_2 can be adsorbed with an average binding energy of 0.19eV per H_2. An appropriate electric field can be used to enhance hydrogen storage and release in this system. This highly synergetic effect can be attributed to the fact that the field leads to an enhancement of the charge transfer between the H_2 molecules and the calcium-silicene system.[598]

First-principles calculations of potassium-decorated silicene and silicane show that potassium atoms can give a uniform and stable coverage on one side of silicene, and on both sides of silicane. Each potassium atom can absorb a maximum of five H_2 molecules, and the hydrogen storage capacity of potassium-decorated silicane[599] can reach 6.13wt%, with an average adsorption energy of 0.133eV/H_2. Again, similar studies of calcium-decorated silicene sheet and zig-zag silicene nanoribbons predict that calcium atoms prefer to be dispersed on the sheet, or at the edges of zig-zag silicene nanoribbons, without clustering, due to the strong binding between calcium and silicon atoms. By adsorbing calcium atoms on both sides of silicene, the hydrogen storage capacity can attain 6.17wt% (gravimetric density) with an average adsorption energy of 0.265eV/H_2; quite suitable for reversible hydrogen adsorption and desorption under ambient conditions. The hydrogen storage capacity can be further improved to 8.43wt%, with an average adsorption energy of 0.182 to 0.269eV/H_2 in calcium-decorated zig-zag silicene nanoribbons. The adsorption of H_2 on calcium-decorated silicon nanostructures is dominated mainly by polarization and orbital hybridization.[600]

1.15.7 Energy Storage

The lithium-oxygen (Li-O_2) battery is one of the most promising technologies for energy storage, but suffers from problems such as the decomposition of its cathodes, which are normally made of carbon-based materials. A carbon-free two-dimensional silicene cathode has been proposed for Li-O_2 batteries on the basis of density functional theory calculations. In contrast to graphene, oxygen reduction reactions and oxygen evolution reactions can occur on pristine silicene with no defect sites. Reactions on silicene are strongly correlated with strong adsorption of oxygen reduction reaction intermediates, caused not only by ionic bonding between silicene and the oxygen atoms in the oxygen reduction reaction intermediates but also by the structural stabilization of silicene.[601] Density functional theory calculations yield the band structure and densities-of-state of silicene super-cell models. A comparison of the adsorption mechanism of lithium metal atoms and lithium ions onto free-standing silicene (buckled, 101.7°), with results for graphene, shows that the interactions between lithium metal atoms or lithium ions and a silicene surface are quite strong, due to its highly reactive buckled hexagonal structure. The adsorption height, most stable adsorption site and energy barrier to lithium diffusion are propitious to the use of silicene in electronic components such as lithium-ion batteries. Graphene layers are commonly used as anode electrodes in these batteries, but graphene has a very limited lithium-storage capacity and smaller surface area than silicene.[602] Based upon first-principles calculations, it has been proposed that Si/MgX_2 (X = Cl, Br, I) would be suitable hybrid structures for this purpose. It is found that lithium will not cluster and that the lithium capacity is very high: 443mAh/g for silicene on $MgCl_2$.[603] Sandwich structures can be used to enhance the performance further. Low diffusion barriers of less than 0.3eV are predicted for all of the hybrid structures.

Intrinsically metallic boron-substituted silicenes such as H-BSi_3 and R-BSi_3, where H and R denote hexagonal and rectangular symmetry, have been predicted to be global-minimum structures of the BSi_3 monolayer. With their unusual planar geometry and superior electronic conductivity relative to buckled and semi-metallic pristine silicene sheet, boron-substituted silicenes are expected to find application in high-capacity lithium-ion battery anodes. Density functional theory investigations of the adsorption and diffusion of lithium on H-BSi_3 and R-BSi_3 predict a good electronic conductivity, a very high theoretical charge capacity (1410 and 846mAh/g for single and double layers, respectively), rapid lithium diffusion and a low open-circuit voltage. These results all suggest that BSi_3 silicene could serve as a promising high-capacity and fast charge/discharge rate anode material for lithium-ion batteries.[604]

In fact, obtaining two-dimensional aromatic compounds having planar cyclic six-membered silicon rings (c-Si_6) is a great challenge. Density functional theory predicts that c-BSi_3 silicene - the global minimum of the BSi_3 monolayer - contains planar aromatic D_{6h} c-Si_6. In c-BSi_3 silicene, the strong π-p conjugation between c-Si_6 and boron atoms leads to the planar geometry of the aromatic D_{6h} c-Si_6; in sharp contrast to the conventional bonding patterns of (sp^3-bonding) silicon. Two-dimensional c-BSi_3 silicene and its one-dimensional analogues (c-BSi_3 nanotubes and nanoribbons) are metallic regardless of chirality, tube diameter or ribbon width. The metallic behavior of c-BSi_3 silicene is also very robust with regard to mechanical strain and surface chemical functionalization.[605]

Free-standing silicene, a graphene/silicene/graphene heterostructure and a graphene/silicene superlattice have all been investigated, using first-principles calculations, for possible application in Na-ion batteries, The calculated Na capacities of 954mAh/g for free-standing silicene and of 730mAh/g for the graphene/silicene superlattice (under 10% biaxial tensile strain) are highly competitive. The potentials of more than 0.3V against the Na^+/Na potential exceed the corresponding value for graphite.[606] The diffusion barriers are also predicted to be less than 0.3eV.

1.15.8 Catalysis

First-principles studies of gold-decorated silicene as a high-activity catalyst for CO oxidation show that the high binding strength of the gold-decorated silicene system and the high diffusion energy-barrier to gold adsorbates, together with the assisted Coulomb repulsion effect, all jointly prevent the formation of gold clusters. The gold-decorated silicene transfers many more electrons to O_2 than to CO. This facilitates CO oxidation, first via the Langmuir-Hinshelwood mechanism ($CO + O_2 \rightarrow OOCO \rightarrow CO_2 + O$) and then via the Eley-Rideal mechanism ($CO + O \rightarrow CO_2$). These reaction processes have relatively low catalytic energy barriers of 0.34 and 0.32eV, respectively. The underlying mechanism of the high catalytic oxidation of CO is suggested to be electronic-state hybridization among the Au d orbitals and the CO and O_2 $2\pi^*$-antibonding states around the Fermi energy.[607]

Gold clusters supported on solid materials have attracted attention because of their potential applications, and an important objective has been the stabilization of the energetically less stable, but catalytically more active, planar form of finite-sized clusters on a suitable solid support. Theoretical investigation of the morphological transformation of dimer, trimer and Au_{20} clusters on pristine silicene/Ag(111) surfaces shows that, unlike previous schemes for the stabilization of planar structures by appropriate surfaces, the present method does not require doping or the application of an external electric field. The particular characteristics (buckled nature, weak Si-Si π-bonding) of the silicene surface can be exploited so as to enable this surface to form strong Si-Au covalent bonds. This results in the two-dimensional planar Au_{20} isomer being more stable than its three-dimensional tetrahedral structure. By contrast, the planar isomer of the Au_{20} cluster is not energetically favored when it is adsorbed on a pure graphene surface. Bader charge-density analysis indicates that the amount of charge transfer from the silicene surface to the planar Au_{20} cluster is quite high, thus increasing the probability of this composite system acting as a better catalyst.[608]

In closing this chapter, it is interesting to mention what one might term, 'analogues of analogues'. That is, density functional theory calculations can be used to make a systematic investigation of the structures and properties of silicene-like SiX and Si_3X (X = B, C, N, Al, P) hexagonal heterosheets. In the former systems, SiP sheet favors a chair-like buckled structure like that of silicene while SiB, SiN and SiAl prefer washboard-like buckling, and SiC sheet adopts a flat-plane form like that of graphene. Such planarity is also favored by Si_3X sheets involving boron, carbon or aluminium, while those involving nitrogen and phosphorus prefer chair-like buckled structures. All of the heterosheets are found to be stable. Unlike semi-metallic silicene, most of the SiX sheets transform to metals, apart from SiC with its wide band-gap. The Si_3X systems can be metals, semimetals or narrow band-gap semiconductors; depending upon the second element. The Si_3B and Si_3N sheets exhibit metallic behavior, while Si_3Al and Si_3P are semiconductors with narrow indirect band-gaps; dominated by the Si-Si and Si-X bonding/antibonding states. In particular, Si_3C sheet maintains the zero band-gap semi-metallicity of silicene and the p_z orbitals of silicon and carbon atoms contribute to linear Dirac-like bands near to the Fermi level. The dynamic stabilities of Si_3C, Si_3Al and Si_3P sheet, examined using phonon calculations and *ab initio* molecular dynamics simulations, are confirmed by the robust stability of their free-standing states.[609]

Chapter 2

Germanene

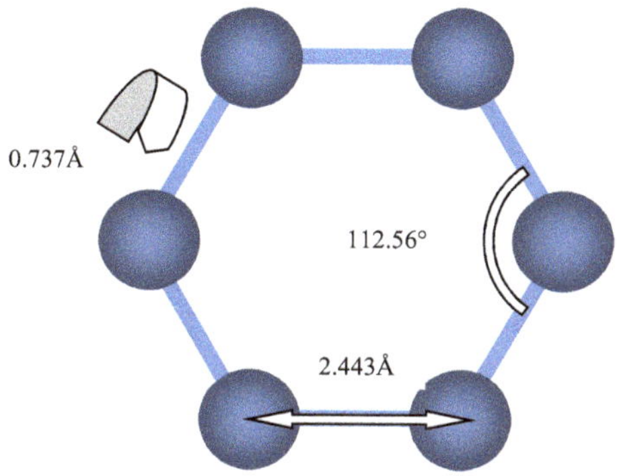

As with silicene, the amazing electronic and opto-electronic properties of free-standing graphene have led to the consideration of other possible two-dimensional materials that might be easily integrated with the present generation of electronic technology. Unlike two-dimensional oxides and dichalcogenides, the elemental two-dimensional analogues of graphene, are promising candidates.[610] Atomically thin two-dimensional materials offer the ultimate thickness-scaling of nano-electronic devices and this could substantially improve electrostatic control and permit a number of low-power applications.[611]

2.1 Preparation

Flat 2-dimensional germanium sheets having a honeycomb lattice which matches that of germanene have been prepared by depositing sub-monolayers of germanium onto graphite at room temperature and then annealing them at 350C. Scanning tunnelling microscopy shows that the germanene islands exhibit small buckling, with no atomic reconstruction, and do not seem to undergo alloy formation or hybridization with the substrate. Density functional theory calculations of the structural properties agree well with experimental results, and indicate that the germanene sheet interacts only weakly with the substrate. Band-structure calculations confirm that the Dirac cone of free-standing germanene is preserved in layers supported by graphite.[612] The germanene islands undergo slight, but characteristic, charge transfer with the graphite substrate. This is predicted by *ab initio* simulations and is again in excellent agreement with scanning tunnelling spectroscopic measurements.

Germanene has been successfully synthesized on molybdenum disulfide, a band-gap material. Pre-existing defects in the disulfide surface act as preferential nucleation sites for the germanene islands. The lattice constant of the germanene layer (3.8±0.2Å) is some 20% greater than the lattice constant (3.16Å) of the disulfide substrate.[613] Scanning tunnelling spectroscopy and density functional theory calculations reveal that, in addition to the linear dispersion bands at the K points, there are two parabolic bands which cross the Fermi level at the Γ point.

Monolayer germanene has been successfully prepared on an Sb(111) surface by means of molecular beam epitaxy. Scanning tunnelling microscopy reveals a dendritic structure at low germanium coverages, and mosaic patterns at high coverages; both having a local 1 x 1 lattice. First-principles calculations confirm the 1 x 1 low-buckle structure of germanene.[614] The dendrite and mosaic patterns are attributed to strain modulation introduced by the large lattice mismatch between germanene and the antimony substrate.

Monolayer germanene can be grown on metallic substrates but, when lying directly on a metal surface, the reconstructed atomically-thin sheets are prone to lose their massless Dirac fermion character and the unique associated physical properties of the free-standing material. Few-layer germanene, created via dry epitaxy on a gold template, possesses Dirac cones thanks to a reduced interaction. This finding establishes, using synchrotron-radiation photo-emission, scanning tunnelling microscopy imaging and surface electron diffraction, that few-layer germanene is one of the rare two-dimensional Dirac materials. Because germanium is already being used in the mainstream silicon-based electronics industry, the possibility of using germanene to scale-down to beyond the 5nm node is very appealing. Another interesting property which might be within reach is a robust quantum spin Hall effect for application in spintronics, together with the engineering exploitation of Floquet-Majorana fermions – using light - for quantum computing.[615]

Atom-thin ordered 2-dimensional multiphase films can be grown *in situ* via germanium molecular beam epitaxy onto a gold (111) substrate; analogous to the formation of silicene layers on silver (111) templates. One phase, which forms large domains as observed using scanning tunnelling microscopy, has a clear almost-flat honeycomb structure. Using synchrotron radiation core-level spectroscopy and density functional theory calculations, this phase can be identified as being a (√3 x √3)R(30°) germanene layer combined with a (√7 x √7)R(19.1°) Au(111) super-cell. The above is therefore a reliable process for the synthesis, on gold, of the germanium-based analogue of graphene.[616] A more detailed examination has been made of the electronic structure of monolayer materials based upon this double hexagonal geometry with (1 x 1) and (√3 x √3)R30° superstructures. Inspired by the two-dimensional root system of an exceptional Lie algebra called G2, this hexagonal atomic configuration is treated as involving two hexagons of unequal side-length set at an angle of 30°. The principal hexagonal unit cell contains 12 atoms instead of the usual configuration which involves only 6, relying only on the (1 x 1) superstructure. *Ab initio* numerical calculations of graphene and germanene with their double hexagonal geometry showed that the usual electronic properties and lattice parameters of such materials are modified: that is, the lattice parameters are increased. In the single hexagonal geometry, germanene behaves like a semimetal. In the double hexagonal geometry it becomes metallic.[617] When fabricated on a Pt(111) surface, germanene exhibits a buckled configuration with a (3 x 3) superlattice that coincides with the (√19 x √19) superstructure of the substrate. Covalent bonds exist throughout the germanene layer.[618] Stable films of germanene can thus be grown on platinum and gold substrates. A germanium adatom, adsorbed on germanene, buries the host germanium atom beneath it to form a dumb-bell structure; an exothermic process which occurs spontaneously. The attractive dumbbell-dumbbell interaction then favors a high coverage of dumb-bells. This presages the existence of stable new germanene phases which are constructed from a periodically repeating coverage of dumb-bell structures which exhibit diverse electronic and magnetic properties.[619] Calculation of the stable configuration for planar, and low (<1Å) buckling configurations, and analysis of the electronic band structure, reveals a linear band dispersion giving rise to massless Dirac fermions with a Fermi velocity that is about two-thirds that of graphene. The monolayer stability is shown to be directly attributable to the phonons present, with the instability being driven by the out-of-plane ZA and ZO phonon modes. Long momentum relaxation-lengths and high carrier mobilities are predicted to exist in germanene-based devices, as carrier relaxation via phonon scattering is inhibited because the electron-optical phonon coupling matrix elements are expected to be small; a factor of about 25 times smaller than in graphene.[620]

First-principles calculations of germanene/graphene bilayers, assuming various super-cell models in order to reduce the strain in the lattice-mismatched bilayer systems, suggest – following energetics analysis and electronic structure deliberations - that graphene can be used as a substrate for the synthesis of monolayer germanene. Multiple phases consisting of single-crystalline germanene having various orientations relative to the substrate can coexist at room temperature. The weak interaction between the overlayer and the substrate preserves the low-buckle structure of germanene, as well as its linear energy bands.[621]

Germanene layers having a Lonsdaleite structure were first synthesized from SiGe thin film by using a N_2 plasma-assisted process. Multi-layered germanene can then be directly observed, and the derived lattice parameters are almost consistent with theoretical predictions. Large-scale multi-layered germanene can be used in various applications.[622]

It remains a challenge to obtain germanene with preserved Dirac electrons. The so-called sandwich-dehydrogenation approach involves fabricating germanene by dehydrogenating germanane within a sandwich structure. Dehydrogenation can spontaneously occur in such a structure, and overcomes the problem of the amorphization which occurs during the heating dehydrogenation approach. The germanene obtained preserves the Dirac electronic properties very well. The Fermi velocity of germanene can moreover be efficiently manipulated by controlling the interlayer spacing between the germanane and the sandwich surfaces. These results offer guidelines to the fabrication of perfect 2-dimensional materials.[623]

2.2 Structure

Computational techniques have been developed in order to construct three-dimensional statistical density maps, and thus to be able to view directly the local atomic structure of amorphous germanium and aid the interpretation of various state-of-the-art experiments and simulations. The continuous random network model has been compared with an experimental model, as refined using reverse Monte Carlo methods. In the refinement, amorphous germanium admits of two dominant structures: a fourfold-coordinated tetrahedron and a buckled threefold-coordinated local structure, similar to that of silicene and germanene. Such structures account for 95.7% of the total number of atoms, in a 5:2 ratio. The present method reveals well-defined structural ordering in the second shell of amorphous germanium. This method of visualization allows the interpretation of complex disordered materials and reveals the bimodal structures of amorphous germanium.[624]

Germanene shares many similarities with graphene but, unlike the planar graphene lattice, the germanene honeycomb lattice is buckled and comprises two vertically-displaced sub-lattices. Indeed, the heavier analogues of graphene, such as silicene and germanene, all tend to be buckled. The buckling leads to interesting properties such as a direct band-gap in hydrogenated germanene. It is interesting that the sequential replacement of carbon by germanium in benzene leads to increasing buckling, with a maximum buckling distance of 0.61Å being found for Ge_6H_6. The origin of a buckling-induced lowering of the symmetry, $D_{6h} \rightarrow D_{3d}$, can be traced to pseudo Jahn-Teller distortion along the b_{2g} normal mode, arising from mixing of the non-degenerate (A_{1g}) ground state with the low-lying (Δ_0 = 4.36eV) excited state of B_{2g} symmetry. Buckling also leads to an increased chemical reactivity of germanene; especially with hydrogen, to yield germanane. The high affinity of germanene for hydrogen explains a reported experimental synthesis of exfoliated layers of germanane. Formation of the latter leads to the opening up of a large band-gap; thus making hydrogenation a chemical route for the control of electronic properties in these two-dimensional materials. The presence of buckling in germanene also leads to greater hole reorganization energies than for polyaromatic hydrocarbons having the same nuclearity.[625]

First-principles calculations of the structural and electronic properties of germanene, and structural optimization, reveal that buckled germanene is more stable than its planar counterpart by about 0.35eV. Compared to planar graphene, with its buckling parameter of 0Å, the germanium sheet is buckled by 0.737Å but has a similar electronic structure with a zero band-gap at the K point. Further investigations of the effects of complete hydrogenation, and consideration of various geometrical configurations (chair, boat, table, stirrup), show that the chair-like structure has the highest binding energy per atom. The hydrogenation of germanene yields germanane, with a direct band-gap of 1.8eV.[626]

Density functional theory calculations reveal that free-standing germanene is a two-dimensional Dirac fermion system. That is, the electrons behave like massless relativistic particles which can be described by the Dirac equation; the relativistic variant of the Schrödinger equation. The spin-orbit gap in germanene is about 24meV, which is much larger than that in graphene: that is, less than 0.05meV. This makes germanene an ideal candidate material for exploiting quantum spin Hall effects at experimentally accessible temperatures. The germanene lattice also offers the possibility of opening-up a band-gap by means of, for example, the imposition of an externally applied electric field, the adsorption of foreign atoms or the use of a suitable substrate. Such opening-up of the band-gap permits the preparation of germanene-based field-effect devices. It is necessary therefore to find means of synthesizing germanene and to predict its properties, using mainly density functional theory calculations.[627]

High-resolution photo-emission and density functional theory calculations of the core-level aluminium 2p and germanium 3d lines, found for germanene grown onto Al(111) by using synchrotron radiation, show that the germanium 3d line is shifted towards low binding energies - with respect to bulk germanium - and has 3 components; reflecting the site multiplicity of the germanene atomic structure. The calculations also reveal appreciable sizable charge localization at the germanene/Al(111) interface, charge transfer from the aluminium surface atoms to the germanene and the existence of 3 non-equivalent germanium sites having 3 different atomic Bader charges; in agreement with the photo-emission measurements.[628]

Structural analysis of germanium layers, deposited by molecular beam epitaxy onto Ag(111) surfaces - with and without an AlN buffer layer - and investigated using X-ray absorption spectroscopy at the Ge-K edge, shows that - for germanium layers deposited onto an h-AlN buffer layer on Ag(111) - an interatomic Ge-Ge distance of 2.38Å exists. This is typical for 2-dimensional germanium layers and is in agreement with theoretical predictions for free-standing slightly-buckled germanene. First-principles calculations, performed within the density functional theory framework, support experimental RHEED and X-ray absorption spectroscopy data, and thus provide evidence for epitaxial 2-dimensional germanium layer formation on h-AlN/Ag(111) templates.[629]

First-principles density functional theory investigation of surface functionalization and strain effects upon full- and half-passivated germanene shows that fully-passivated germanene has a different band structure. That is, the band-gap of GeH is larger than that of GeF or GeCl. When both surface passivation and strain are applied, the germanene undergoes a transformation from semiconducting to semi-metallic. Half-passivated germanene also has a different magnetic character. That is, Ge_2H is a ferromagnetic semiconductor while Ge_2F and Ge_2Cl are anti-ferromagnetic semiconductors. The stability of the magnetic coupling of Ge_2X-type materials can be modulated by applying an external strain. The calculations thus indicate that the electronic and magnetic properties of passivated germanenes depend strongly upon surface functionalization and strain.[630] First-principles calculations of a new type of hybrid phase which comprises buckled germanene, with saturated or half-saturated degrees of alkali metal atom adsorption, similarly suggest that adsorbed lithium, sodium or potassium can be used to synthesize germanene-based new phases in two dimensions. The predicted new phases, Ge_2X_2 or Ge_2X_1 (X = Li, Na, K) on a single germanene sheet could exist at room temperature. The formation energy of the Ge_2Li_2 configuration, resulting from complete lithiation, is even more favorable than that of germanane. Charge transfer is significant between the alkali metal atoms and the germanium; reflecting the ionic interactions between them, and charge-density analysis further indicates that a covalent component exists to some extent in two-dimensional Ge_2X_2 and Ge_2X_1 (X = Li, Na, K) phases. Complete lithiation of germanene even leads to a semiconductor with an energy gap of 0.14eV. The 2-dimensional phases of Ge_2X_1 (X = Li, Na, K) are metallic, with weak polarization on the Fermi level and in unoccupied states. Half-lithiated germanene has a local magnetic moment of $0.48\mu_B$ on germanium atoms with neighboring lithium adatoms.[631]

Density functional theory studies of the stability of germanene under biaxial tensile strain, and of associated modifications of the electronic properties and phonon spectrum, show that the germanene lattice is stable at up to 16% strain; where the Dirac cone shifts towards higher energies and hole-doped Dirac states are created. The latter is due to weakening of the Ge-Ge bonds and to a reduction in the s-p hybridization. The calculated Grüneisen parameter exhibits a similar dependence upon strain.[632]

An hierarchical first-principles investigation of the entangled effects of lattice dimensionality and bond characteristics on the lattice dynamics of germanene shows that bond-bending contributes negatively to the Grüneisen constant, γ, resulting in a negative acoustic γ. Stretching however contributes positively to γ, resulting in a positive optical γ. Layer-thickening caused by chemical functionalization tends to increase the acoustic γ, due to the increased bond-stretching effect. On the other hand, bond-weakening caused by chemical functionalization tends to decrease the optical γ, due to the decreased bond-stretching effect. Excitation of the negative-γ modes results in a negative thermal expansion, while mode excitation and thermal expansion compete with each other in governing the material's thermomechanics. The sensitive structural and electronic responses to functionalization help to deduce a generic physical picture of two-dimensional lattice dynamics.[633]

2.3 Mechanical Properties

First-principles density functional theory calculations have been used to deduce the mechanical properties of single-layer and free-standing germanene. Uniaxial tensile and compressive simulations were carried out to probe and compare stress-strain properties; such as the Young's modulus, Poisson ratio and ultimate strengths (table 4).[634] Others have found a similar value (42.05GPa.nm) for the Young's modulus of the armchair form.[635]

Table 4
Mechanical properties of germanene

Material	**Property**	**Value**
armchair	Young's modulus	44GPa.nm
zig-zag	Young's modulus	43.4GPa.nm
armchair	Poisson ratio	0.29
zig-zag	Poisson ratio	0.35
armchair	UTS	4.7GPa.nm
zig-zag	UTS	4.1GPa.nm
armchair	ε_{UTS}	0.2
zig-zag	ε_{UTS}	0.205

2.4 Semiconduction and Band Structure

First-principles calculations of variations in band structure due to the straining of nanosheets show that the material exhibits a strain-induced self-doping phenomenon which is closely related to the buckled structure and thus cannot occur in graphene. Under a compressive stress, the Dirac point moves to below the Fermi level; making nanosheets behave as if they were n-type doped. Under a tensile stress, p-type doping appears due to a shift of the Dirac point to above the Fermi level. An n-type/zero-band-gap/p-type semiconductor arrangement can be switched by applying a stress.[636]

When a model based upon tight-binding theory is used to calculate the electronic structure of germanene, a 3-dimensional energy-band structure in the first Brillouin zone is obtained which has an open energy-gap of up to 128meV at the Dirac points, K and K′. This open gap shows that germanene is a non-metallic insulator. The first Brillouin zone is almost 3 times larger than that of graphene because germanene has a larger lattice constant (3.95Å).[637] The electron density decreases quickly in the radial direction, while the density is greater in the center of the Brillouin zone in the angular direction.

Like silicene, germanene possesses remarkable electronic properties due to strong spin-orbit coupling and its buckled monolayer structure. Also similar are the band structures of the materials in the presence of perpendicular electric or magnetic fields, taking account of the effects of off-resonant light. Using linear response theory, analytical expressions can be derived and evaluated for the Hall and longitudinal conductivities. Unlike the case of graphene, it can be shown that light leads to a single Dirac-cone state and thence to unusual plateaux and magnetotransport properties which are desirable in electronic applications.[638]

First-principles calculations of the electronic properties of germanene superlattices show that, although a band-gap can be opened-up in all of the superlattices by sub-lattice equivalence breaking, it can be opened up only in some inversion symmetry preserved superlattices with Γ as a four-fold degenerate Dirac point. The D_{6h} antidot, although it does not impair the inversion symmetry, can still open up a band-gap by perturbing the four-fold degeneracy. The band-gap which is opened up by either sub-lattice equivalence breaking or four-fold degeneracy perturbation can moreover be continuously narrowed by reducing the defect density.[639]

First-principles investigation of spin-orbit opened energy gaps and of the band topology demonstrates that spin-orbit coupling opens up a band-gap of 23.9meV in germanene.[640] Density functional theory and density functional perturbation theory studies show that the phonon spectrum is positive for all modes in the first Brillouin zone, and that there is a phonon energy band gap between the acoustic and optical modes at about 50/cm. Spin-orbit coupling opens up a direct energy band-gap at the Dirac point, softens the phonon spectrum and decreases the phonon group velocity of the ZA mode.[641]

First-principles studies of the effects of biaxial strains of up to 5%, and of a perpendicularly applied electric field, upon the electronic and phonon properties of germanene show that – with increasing biaxial strain - the conduction bands at the high-symmetry Γ and M points of the first Brillouin zone shift appreciably towards the Fermi level.[642]

First-principles calculations of the geometrical structures and electronic properties of porous germanene nanosheets, the germanium analogue (germanyne) of α–graphyne, show that elemental germanyne sheets are in fact energetically unfavourable. Following carbon-substitution however, hybrid graphyne-like sheets of c-silicyne/c-germanyne have robust energetic and dynamic stabilities. Unlike germanene, c-germanyne is buckled; with a distinct half-hill conformation. This asymmetrical buckling induces semiconducting behaviour in c-germanyne.[643]

First-principles calculations suggest that 2-dimensional topological insulators can be made from functionalized germanenes, GeX, where X is hydrogen, fluorine, chlorine, bromine or iodine. It is found that GeI is a two-dimensional topological insulator with a bulk gap of about 0.3eV. Meanwhile, GeH, GeF, GeCl and GeBr can be transformed into topological insulators, having sizable gaps, by using reasonable tensile straining. It is noted that an unique mechanism is responsible for the large topologically non-trivial gap which is obtained. Due to the functionalization, the orbitals with stronger spin-orbit coupling dominate the states around the Fermi level, rather than the original orbitals with weaker spin-orbit coupling. Therefore the coupling of the p_{xy} orbitals of germanium and heavy halogens in forming suitable orbitals also plays a key role in further enlarging the gaps in halogenated germanenes. There exists the realistic possibility of exploiting topological effects at room temperature.[644]

The behavior of various regions of a proposed top-gated germanene nanoribbon field effect transistor under compressive strains of 3, 5 or 10% has been predicted. The results indicate that the band-gap and electron effective mass first increase upon increasing the strain, but then decrease. This behavior affects the figure-of-merit. The I_{on}/I_{off} ratio and sub-threshold swing of the unstrained field effect transistor improves when a 3% compressive strain is applied to the whole of the device. Applying a 10% strain to the device also increases the ON current, due to strain-induced self-doping, but the transistor parameters for this structure are unsatisfactory.[645] Application of compressive strain to either the channel, source or sink of the device shows that, due to a high lattice mismatch between the channel and the leads, the performance of the strained transistor is the same, or worse than, as that of the unstrained one.

2.5 Substrate Interaction

A comprehensive first-principles study of two-dimensional Ge/BeO heterostructure shows that there is a weak interaction between the germanene and the BeO monolayer via the van der Waals force, without any site selectivity. Most of the outstanding properties of germanene are therefore conserved; especially a linear band dispersion near to the K point. An energy gap of about 49meV can be opened in the heterostructures. The band-gap can be tuned by varying the interlayer distance, the external strain or the external electric field (14 to 382meV). An electric field is the most effective method.[646] All of the methods of band-gap modulation can maintain the characteristics of a Dirac cone with a linear band dispersion relationship. The improved optical properties are still observed after formation of the heterobilayer system. A BeO monolayer is thus an excellent germanene substrate.

First-principles calculations show that germanene can be attached to the Ag(111) surface to form germanene/Ag superstructures by electrostatic interaction. In optimized superstructures there is a form of epitaxially grown germanene which is similar to isolated low-buckle germanene. The adsorption energy of germanene on the Ag(111) surface is about -464 to -428meV per germanium atom; close to that for silicene on the Ag(111) surface. The germanene on Ag(111) is a continuous layer and p-d hybridization exists between the silver and the germanium. This suggests that the Ag(111) surface is a good substrate for stabilizing germanene. The band structures of germanene are submerged in the electronic states of the metallic silver substrate. In order to preserve the electronic structures of germanene, consideration has also been given to the use of another substrate: hexagonal boron nitride. Germanene can attach stably to an hexagonal boron nitride substrate via Van der Waals interactions and form germanene/BN Moiré superstructures. At equilibrium, a band-gap of about 50meV opens up at the Dirac point of germanene; its value being insensitive to the rotation angle and to displacement between the two lattices. It can however be effectively tuned by changing the interlayer distance. The high carrier mobility of germanene is well preserved in these superstructures and this implies that hexagonal boron nitride can act as an ideal substrate material and permit germanene to fulfil specific applications in nanoscale electronics.[647]

First-principles calculations of a continuous germanene layer, grown onto the Al(111) surface, show that that – in spite of having a different lattice structure to that of its free-standing form – germanene on Al(111) still possesses Dirac points at high-symmetry K and K′ points. There also exist another three pairs of Dirac points on the K(K′)-M high-symmetry lines which have highly anisotropic dispersions, due to the reduced symmetry. These massless Dirac fermions become massive when spin-orbit coupling is included. Hydrogenation of the germanene layer strongly affects its structural and electronic properties: when not fully hydrogenated, ferromagnetism can be induced due to unpaired local orbitals from unsaturated germanium atoms. In particular, one-side semi-hydrogenated germanene is a two-dimensional half-semimetal. This is thus a novel state of matter which makes it both a half-metal and a semimetal.[648] The structure of germanene on an Al(111) surface has been experimentally investigated using the total-reflection high-energy positron diffraction method. The observed spot intensities are asymmetrical, revealing no mirror symmetry in the atomic coordinates of germanene with respect to the <110> direction. Quantitative total-reflection high-energy positron diffraction rocking curve analysis, based on dynamical diffraction theory, has revealed that the germanene layer has a 3×3 structure with asymmetrical buckling due to the protrusion of one of the germanium atoms in the unit cell, which is unlike the structural model proposed in previous studies. The magnitude of the buckling was found to be 0.94Å, and the spacing between the germanene and the Al(111) substrate to be 2.51Å.[649] The new structure proposed in the present investigations, though different from that reported in previous studies, does not contradict the other characteristics which were found experimentally in the previous studies.

Room-temperature scanning tunnelling microscopy and spectroscopy of the structural and electronic properties of germanene-coated Ge_2Pt clusters shows that the interior of the germanene sheet exhibits a buckled honeycomb structure with a lattice constant of 4.3Å and a buckling of 0.2Å. The zig-zag edges of the germanene are reconstructed and have 4x periodicity. The differential conductivity of the interior of the germanene sheet is V-shaped, and recalls the density of states of a two-dimensional Dirac system; the minimum of the differential conductivity being located close to the Fermi level and having a non-zero value. The latter is attributed to the metallic character of the underlying Ge_2Pt substrate. Near to the reconstructed germanene zig-zag edges, the shape of the differential conductivity changes from V-shaped to a more parabola-like shape; revealing that the reconstructed germanene zig-zag edges do not exhibit a marked metallic edge state.[650] A related scanning tunnelling microscopic and spectroscopic investigation of the growth of platinum on Ge(110) showed that the deposition of several monolayers of platinum onto Ge(110), followed by annealing at 1100K, results in the formation of 3-dimensional metallic Pt-Ge nanocrystals. The outermost layer of these crystals has a honeycomb structure which comprises 2 hexagonal sub-lattices that are displaced vertically by 0.2Å with respect to one another. The nearest-neighbor distance of the atoms in the honeycomb lattice is 2.5Å; that is, very close to the predicted nearest-neighbor distance (2.4Å) in germanene. Scanning tunnelling spectroscopy reveals that the atomic layer beneath the honeycomb layer is more metallic than is the honeycomb layer itself. These observations are in accord with a model proposed for metal di-germanides: an hexagonal crystal with metal layers separated by semiconducting layers with a honeycomb lattice. On the basis of such observations, it is proposed that the outermost layer of the Ge_2Pt nanocrystal is germanene.[651]

The stability and electronic structures of germanene on monolayers of GaS, GaSe, GaTe and InSe, investigated using *ab initio* density functional theory, show that germanene preserves its buckled honeycomb structure on all of these substrates and that it is similar to the free-standing case. Germanene remains neutral and preserves its Dirac-cone like band-structure on monolayer GaTe and InSe. In these cases, a band-gap of 0.14 to 0.16eV opens up at the Dirac point of germanene, while the effective masses remain as small as 0.05 to 0.06 times the free-electron mass and the estimated carrier mobility is up to 2.2 x $10^5 cm^2/Vs$. This means that monolayer GaTe and InSe are promising candidate materials for use as substrates for germanene electronic devices.[652]

Use of first-principles methods and density-functional theory has clarified the atomic and electronic structures of germanene on GaAs(111)-(1 x 1). Stable structures exist on both the arsenic-terminated and gallium-terminated GaAs surfaces. The structure of the germanene is similar to that of the free-standing material, with its hexagonal honeycomb geometry. The cohesive energies of germanene on both the arsenic and gallium sides of the GaAs surface are comparable to those for the bulk structures and/or to those on Ag(111) substrates. This suggests the possibility of synthesizing them on both sides of a GaAs surface. The corresponding binding energies range from 0.56 to 1.37eV per germanium atom. This is 10 times larger than the usual van der Waals interaction, reflecting the covalent interaction between the germanene and the GaAs surface. Band-structure calculations show that such a covalent interaction leads to an absence of Dirac electrons for germanene on GaAs surfaces. The Dirac electrons can potentially be recovered by means of hydrogen intercalation. It is found that intercalated hydrogen atoms are chemically bonded to GaAs surfaces, and that the germanene's upward shift-distance from the GaAs surface increases from between 2.50 and 2.58Å to between 3.49 and 3.86Å; where a covalent van-der-Waals interaction transition occurs between the germanene and the GaAs surface. The distances between the germanene and hydrogen atoms are 30% and 8% larger, respectively, than the atomic-radius sum of germanium and hydrogen on arsenic-terminated and gallium-terminated GaAs surfaces. This indicates that the interaction between germanene and hydrogen on the arsenic-terminated GaAs surface is clearly weaker than the typical covalent interaction. On the gallium-terminated GaAs surface meanwhile, it is comparable to the typical covalent interaction.

This difference is attributed to the difference in electronegativity between arsenic and gallium atoms. It is further noted that hydrogen intercalation recovers well the Dirac electrons on the arsenic-terminated GaAs(111) surface, due to the weaker germanium-hydrogen interaction. It does not do so on the gallium-terminated GaAs(111) surface, due to the stronger germanium-hydrogen interaction. These results are confirmed upon performing analogous calculations for germanene on GaAs(111)-(3 x 3) surfaces. There is thus a good theoretical basis for the application of germanene to semiconductor surfaces.[653]

Ab initio calculations of epitaxial germanene on a semiconducting GaAs(00•1) substrate shows that the germanene/substrate interaction is strong for direct contact, but can be substantially reduced by hydrogen intercalation at the interface. It is predicted to be energetically possible to remove the germanene from the GaAs(00•1) substrate. When situated on the substrate, the electronic structure exhibits a distinct Dirac cone shift above the Fermi energy, with a splitting of 175meV. In free-standing sheet there is a band-gap of 24meV which is due to the intrinsic spin-orbit coupling.[654]

Guidelines have been proposed for exploring substrates which stabilize the monolayer honeycomb structure of germanene while simultaneously preserving the Dirac states. In addition to having a strong binding energy with respect to the monolayer, the suitable substrate should be a large-gap semiconductor with a work function such that the Dirac point lies in the gap, and far from the substrate states, when their bands align. The idea has been checked by performing first-principles calculations for germanene on the Al-terminated (00•1) surface of Al_2O_3. The overlaid monolayers on Al-terminated Al_2O_3(00•1) retain the main structural profile of the low-buckle honeycomb structure.[655] An unfolded band structure, derived using the k-projection method, reveals that a gapped Dirac cone is formed at the K point due to the structural distortion and to interaction with the substrate. The gap of 0.3eV for supported germanene suggests that it may have potential application in nano-electronics.

Table 5
Adsorption energies at the germanene surface

Adatom	Site	E(eV)
Zn	H	0.99
Zn	T	0.41
Zn	B	0.42
Mn	H	1.39
Mn	T	0.97
Mn	B	1.06
Cr	H	1.55
Cr	T	0.99
Cr	B	0.97
Cu	H	1.86
Cu	T	1.46
Cu	B	1.20
Fe	H	3.18
Fe	T	2.88
Fe	B	2.61
Co	H	3.22
Co	T	2.92
Co	B	2.67
Ni	H	3.99
Ni	T	3.77
Ni	B	3.45
V	H	4.21
V	T	3.61
V	B	3.20
Ti	H	4.46
Ti	T	4.14
Ti	B	4.07
Sc	H	4.62
Sc	T	4.05
Sc	B	4.10

H: above hexagon centre, T: above Ge atom in upper layer,
B: above Ge atom in lower layer

2.6 Surface Interaction

Spin-polarized density functional theory calculations, of germanene with one of 10 different adsorbed 3d transition metal atoms, show that all but one of the adatoms prefer to bind to the hexagonal hollow site. The exception is zinc, which prefers to bind to the valley site. Strong covalent bonding between the transition metal adatom and the germanene layer is found in most cases. As a result of the adsorption, the germanene can exhibit various electronic and magnetic properties, depending upon the adatom type: non-magnetic metallic (for copper adsorption), non-magnetic semiconducting (for nickel or zinc adsorption), ferromagnetic metallic (for chromium or manganese adsorption), ferromagnetic semiconducting (for vanadium adsorption) and ferromagnetic half-metallic (for scandium, titanium, iron or cobalt adsorption); with 100% spin-polarization at the Fermi level. In particular, chromium adatoms impart the largest magnetic moment, while scandium, titanium, vanadium, manganese, iron and cobalt adatoms all generate almost integer magnetic moments. The effects of the on-site Coulomb interaction, as well as the magnetic interaction between transition metal adatoms, upon the stability of the half-metallic systems are such that the half-metallic states of the scandium/germanene and titanium/germanene systems are all robust. The ferromagnetic transition-metal/germanene systems are expected to have potential application in the field of two-dimensional spintronics devices. Density functional theory studies of scandium, titanium, vanadium, chromium, manganese, iron, cobalt, nickel, copper and zinc, adsorbed on germanene, show that – again based upon the adsorption energy - the transition-metal atoms always prefer to occupy the hollow site (table 5). The resultant values of the total magnetic moment range from 0.97 to 4.95μ_B in going from scandium to manganese. A gap of 74meV, with a strongly enhanced splitting of 67meV, is found in case of scandium adsorption while metallic states are found in the case of titanium, chromium, manganese, iron and cobalt. Non-magnetic states are again found for nickel, copper and zinc adsorption. Semiconduction is found in the non-magnetic cases, with gaps ranging from 26 to 28meV. Vanadium-adsorbed germanene exhibits the quantum anomalous Hall effect. The results demonstrate that the transition-metal atoms and nearest-neighbor germanium atoms are ferromagnetically ordered in the case of vanadium, manganese, iron, cobalt, nickel, copper and zinc, while antiferromagnetic ordering is found for scandium, titanium and chromium. The extent of coverage of all of the transition-metal also has an effect upon the electronic structure. Ferromagnetic and antiferromagnetic coupling is detected in the case of manganese.[656] Analysis of the density of states indicates that the ferromagnetic properties of the transition-metal/germanene systems result mainly from spin-splitting of the transition metal 3d states.[657] When the structural properties of alkali-metal atoms (including lithium, sodium and potassium) adsorbed on germanene at a wide range of coverages are investigated by means of first-principles calculations, all of the atoms are found to prefer to bind to the hexagonal hollow sites of germanene. Unlike the case of graphene, the interaction between the adatom and the germanene surface is quite strong, due to the buckled hexagonal structure. With increasing adatom coverage, the binding between the alkali-metal atom and the germanene sheet weakens due to the increased adatom-adatom repulsion. Due to high charge transfer from the alkali metal to germanene, the adatom-germanium bonds which are formed are of a mainly ionic nature. Following adsorption, the semimetallic germanene becomes metallic, with its Dirac point moving below the Fermi level. The germanene thus behaves as if n-type doped. A small band-gap can open up at the Dirac point, and both the band-gap and the concentration of charge carriers in the alkali-metal/germanene system can be tuned by controlling the adatom coverage.[658] Further first-principles calculations of the adsorption characteristics of 15 different metal atoms again show that, for alkali-metal adatoms (lithium, sodium, potassium), the bonding is approximately ideal-ionic and the semi-metallic germanene finally becomes metallic; with a small band-gap opening at the Dirac point. The bonding of alkaline-earth metal atoms (beryllium, magnesium, calcium) is a mixture of ionic and covalent. Beryllium or magnesium adsorption leads to semiconducting behavior in germanene while, as for lithium, sodium and potassium adsorption, calcium-adsorbed germanene is metallic. For most transition-metal adatoms, strong covalent bonding is found between the adatom and the germanene layer; which causes much larger distortions in the germanene lattice. Due to the partially-occupied d-orbital, the transition metals also exhibit various electronic structures upon interacting with germanene, as previously noted.[659] Other first-principles calculations of the adsorption characteristics of alkali, alkali-earth, group-III and 3d transition-metal adatoms show that the adsorption of alkali or alkali-earth adatoms has a minimal effect upon the geometry of germanene. The significant charge transfer from alkali adatoms to germanene again leads to the metallization of germanene, whereas alkali-earth adatom adsorption, where the interaction is a mixture of ionic and covalence, results in semiconducting behavior with an energy-gap of 17 to 29meV. Group-III adatoms also bind to germanene via mixed, covalent and ionic, bonding. The adsorption characteristics of the transition metals are, as noted previously, rather complicated; though all transition-metal adsorption on germanene involves strong covalent bonding with germanene. The main contributions to this strong bonding arise from hybridization between transition-metal 3d and germanium p_z orbitals. As noted previously, depending upon the transition metal, the adsorbed systems can exhibit metallic, half-metallic or semiconducting behavior. The dipole moment and the work function vary as a function of the adsorption energy.[660] First-principles studies have also been made of the electronic functionalization of germanene monolayers by chemisorption of hydrogen, lithium, fluorine, scandium, titanium or vanadium, with 3 general adatom-monolayer configurations being considered; each having a distinct effect upon the electronic structure and yielding metallic or semiconducting dispersions, depending upon the adatom type and configuration. The induced band-gap is an indirect or direct Γ-gap, ranging from 0.2 to 2.3eV. In general, an alternating configuration is the most energetically stable. The boat-like and chair-like conformers are degenerate, with the former being associated with anisotropic effective carrier masses. The top configuration leads to a planar monolayer and usually to a gapped dispersion. The hollow configuration, combined with vanadium adatoms, retains the Dirac cone; but with strong orbital planar hybridization at the Fermi level.[661]

First-principles density functional theory calculations show that the electronic and magnetic properties of bare and single-layer germanene with adsorbed titanium adatoms, which is charged or subjected to a perpendicular electric field, can be modified so as to impart new functionalities. When subjected to a perpendicular electric field, buckled atoms have the symmetry between their planes broken; thus opening up a gap at the Dirac points. The occupation of 3d orbitals of the adsorbed titanium atom changes due to charging or to an applied electric field; thus causing significant changes in the magnetic moment. The calculated properties exhibit however a dependence upon the size of the vacuum spacing between the periodically repeating germanene layers if they are treated using a plane-wave basis set with periodic boundary conditions. This spurious dependence can be eliminated by using a local orbital basis set.[662]

The germanene derivatives which are constructed from periodic dumb-bell units play an important role in multilayers of the honeycomb structure. First-principles calculations, based upon density functional theory, reveal the dumb-bell formation mechanisms and energetics of group-IV atoms adsorbed on graphene, silicene, germanene or stanene monolayer honeycomb structures. The stabilities of the binding structures are further confirmed, apart from stanene, by *ab initio* molecular dynamics calculations for high temperatures. Stanene instead undergoes structural instability upon adsorbing atoms. Depending upon the row-number of the adatoms and substrates, three types of binding structures are found; leading to significant changes in the electronic, magnetic and optical properties of the substrates. In particular, silicon, germanium and tin adatoms - when adsorbed on silicene or germanene - form dumb-bell structures. Such dumb-bell structures occur not only on single-layer monatomic honeycomb structures but also on compounds, including SiC and SiGe. The energy barrier to the migration of a dumb-bell structure is low, due to concerted action of the atoms. This makes dumb-bells quite mobile on substrates when preparing new single or multilayer silicon and germanium phases.[663]

Organic functionalization using molecules which contain unsaturated C-C bonds, by means of radical-initiated reactions on hydrogenated surfaces, is a favorable route for the attachment of organic layers. Such organic functionalization of hydrogenated germanene, with acetylene, ethylene and styrene, when compared with respect to adsorption on hydrogenated graphene and hydrogenated silicene surfaces, shows that the results arising from organic functionalization of hydrogenated germanene and hydrogenated graphene are very different. In the case of hydrogenated germanene, the adsorption of acetylene and ethylene is energetically favorable while the adsorption of styrene – in spite of being energetically favorable - leads to a final state whose structural configuration does not favor a chain reaction. The adsorption of these molecules on hydrogenated graphene is less likely to occur in the case of acetylene and ethylene and, for styrene, it is not energetically favorable. The differences in the surface reactivity of H-graphene and H-germanene are attributed to the greater lattice constant of hydrogenated germanene and to differences in the electronegativity of carbon, germanium and hydrogen atoms.[664]

First-principles calculations, based upon density functional theory, of the electronic properties of hydrogenated germanene show that two different atomic configurations are stable, and energetically degenerate. Upon adsorbing hydrogen, an energy-gap opens up. The energy-gap can be predicted by using the Herd-Scuseria-Emzerhof hybrid functional as well as a many-body perturbation method. The germanene is a wide band-gap semiconductor, while germanane is predicted to be a direct-gap material, regardless of its atomic configuration, with an average energy-gap of about 3.2eV. This makes the material potentially interesting for use in opto-electronic applications in the blue/violet spectral range.[665]

First-principles calculations of the electronic and magnetic properties of hydrogenated germanene show that, like silicene, buckled pristine germanene is also more stable than is pristine germanene with a planar configuration. Metallic properties can be detected in the planar structures, and semi-metallic (with zero band-gap) properties can be found in the buckled structure. As the most stable structure, buckled pristine germanene is found with hydrogen atoms in several configurations. Of the 5 configurations which have been investigated, stable structures are found only for the chair-like obtuse and boat-like obtuse configurations; both being semiconductors. Only in the chair-like obtuse configuration however does the system have a non-zero magnetic moment. Overall, as compared with 100% hydrogenated silicene and graphene, 100% hydrogenated germanene has the smallest band-gap energy.[666]

The electronic properties and scanning tunnelling microscopic topographic images of X (= F, H, O) functionalized germanene have been investigated, using density functional theory, by introducing various types of vacancy cluster and chain patterns into monolayers. The formation of vacancy clusters and chain patterns is energetically most favorable in hydrogenated germanene. The F- and H-functionalized materials are direct band-gap semiconductors, with band-gaps ranging from 0.1 to 1.9eV. The O-functionalized monolayers are metallic in nature. Upon introducing the various vacancy clusters and chain patterns, the electronic and magnetic properties are significantly modified. The F- and H-functionalized materials, with hexagonal and rectangular vacancy clusters, are non-magnetic semiconductors with modified band-gap values, while pentagonal and triangular vacancy clusters introduce metallicity and magnetic characteristics.[667] Hexagonal vacancy chain patterns induce a direct-to-indirect gap transition while zig-zag vacancy chain patterns retain the original direct band-gap nature. Calculated scanning tunnelling microscopic topographic images reveal distinct characteristics for the various types of vacancy cluster and chain patterns. This can be used to identify various vacancy patterns during functionalization.

First-principles calculations of the electronic properties of chlorine- and hydroxyl-substituted half-hydrogenated or fully-hydrogenated material show that both the concentration and the distribution of the substituent greatly affect the electronic properties. The half-hydrogenated germanene evolves, from a magnetic semiconductor to a half metal, with increasing chlorine or hydroxyl concentration. The chlorine or hydroxyl decoration of fully-hydrogenated germanene does not change the intrinsic electronic structure of the direct band-gap, but can reduce the band-gap with increasing concentration. Fully-hydroxylated and fully-chlorinated germanene are semiconductors, like fully-hydrogenated germanene. These results demonstrate that chlorine and hydroxyl treatment can tune differently the electronic properties of half- and fully-hydrogenated germanene.[668]

First-principles calculations of the intriguing electronic and magnetic properties of one-side semi-hydrogenated germanene show that, based upon phonon-mode analysis, the material is dynamically stable in the ground state. Semi-hydrogenation from one side only leads to localized and unpaired 4p electrons existing in the unhydrogenated germanium sites. This then introduces ferromagnetism into the germanene sheet, with no need for doping, cutting or etching. One-side semi-hydrogenated germanene is a semiconductor with a direct energy-gap. The gap value, as deduced using the Herd-Scuseria-Emzerhof functional, is estimated to be 1.32eV. In going from pristine to one-side semi-hydrogenated, and then to fully hydrogenated, the germanene changes from metallic to magnetic semiconducting and then to non-magnetic semiconducting. Hydrogenation is a novel means of tuning the properties, and has enormous potential for future nano-electronics use.[669]

Investigation of the stability and electronic properties of nitrogen and boron impurities in single layers of germanane, using spin polarized density functional theory within the GGA-PBE and Herd-Scuseria-Emzerhof approaches for the exchange correlation term, reveals that these impurities have lower formation energies in germanane than do their counterparts in graphane. The adsorption of hydrogen atoms in the vicinity of the defects stabilizes the system. The electronic properties of germanane can be tuned when nitrogen and boron are incorporated into the germanium network. The use of nitrogen-doping and boron-doping gives rise to n-type and p-type semiconducting properties, but the adsorption of hydrogen atoms quenches the doping effects.[670]

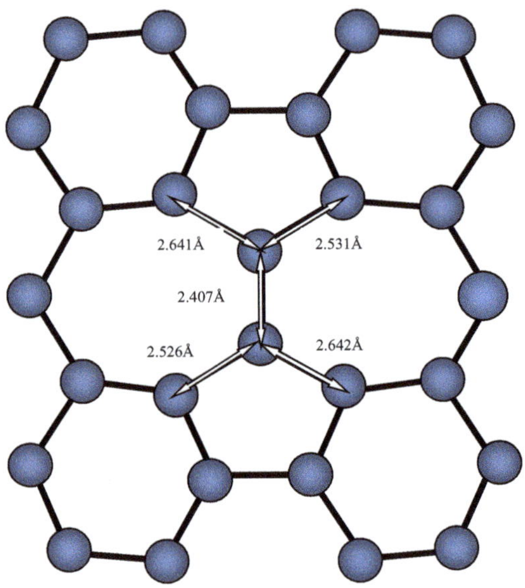

Figure 5. Stone-Wales defect in germanene. Note the changes in atomic spacings as compared to that in the structure at the head of the chapter.

First-principles studies of O_2 dissociation and oxygen-atom adsorption on free-standing germanene show that, compared with the extremely active silicene in an oxygen atmosphere, germanene is less active; due to a dissociation-energy barrier of about 0.57eV. A dissociated oxygen atom then follows two opposite migration pathways on the germanene surface; quite distinct to the case of silicene. The migration and desorption of oxygen atoms at room temperature is relatively difficult, due to strong Ge-O bonding;

resulting in the formation of germanium oxides. The results reveal an interplay between germanene and O_2, and suggest an enhanced stability of germanene in an oxygen atmosphere, as compared with silicene.[671]

Density functional theory studies of the adsorption of the usual gas molecules (N_2, CO, CO_2, H_2O, NH_3, NO, NO_2, O_2) on germanene show that N_2, CO, CO_2 and H_2O are physisorbed on germanene via van der Waals interactions while NH_3, NO, NO_2 and O_2 are chemisorbed on germanene via strong covalent (Ge-N or Ge-O) bonds. The chemisorption of gas molecules opens a band-gap at the Dirac point of germanene. The NO_2 chemisorption on germanene reveals strong hole-doping. Meanwhile O_2 is easily dissociated on germanene at room temperature. The differing adsorption behaviors of common gas molecules on germanene suggest a feasible means for exploiting chemically modified germanene.[672]

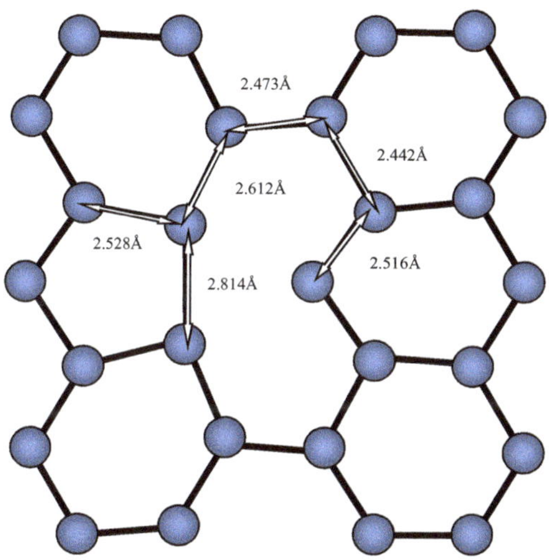

Figure 6. 5-9-6 vacancy in germanene.

First-principles calculations of germanene with small adsorbed nitrogen-based molecules (N_2, NH_3, NO_2, NO) reveal that all of the molecules bind much more strongly to germanene than to graphene. This is due to the hybridized sp^2-sp^3 bonding of germanium atoms. The N_2, NO and NO_2 molecules all act as acceptors, while the NH_3 molecule donates electrons. Sizeable band-gaps (2 to 158meV) are opened up at the Dirac point of germanene by N_2, NH_3 or NO_2 adsorption, with only slight spoiling of its Dirac-cone shape. The NO_2 molecule also exhibits a marked p-type doping behavior, and renders germanene metallic. With adsorbed NO molecules, germanene can change into a ferromagnetic half-metal with 100% spin-polarization at the Fermi level.[673] The various adsorption behaviors of small nitrogen-based gas molecules on germanene offer a feasible means for exploiting chemically modified germanene in a wide range of applications, such as field-effect transistors, gas sensors and spintronic devices.

Density functional theory studies of the adsorption of the ionic liquid, choline benzoate, at the surfaces of hexagonal nanosheets such as graphene and germanene show that the adsorption involves mainly van der Waals forces; leading to strong benzoate ion-surface π-stacking. There is also substantial charge-transfer from the anion to the sheet. Electronic structure analysis shows that germanium-based sheets lead to the largest changes in the HOMO and LUMO levels of choline benzoate. This provides new insights into the ability of density functional theory methods to furnish useful information concerning the adsorption of ionic liquids on nanosheets, and intimates just how ionic-liquid features might be tuned by adsorption on a suitable nanosheet material.[674]

2.7 Defects

Ab initio electronic structure and transport calculations of two-dimensional hexagonal germanium containing Stone-Wales (figure 5), single vacancy (5-9-6) (figure 6) and two types of divacancy (5-8-5) (figure 7), (555-777), show that these defects constitute a local reconstruction that can be clearly identified in scanning tunnelling microscopic images. Among these defects, the Stone-Wales defect has the lowest formation energy. In the presence of structural defects, the two-dimensional hexagonal germanene maintains its Dirac-cone feature only in the case of the single vacancy. Divacancies and Stone-Wales defects destroy the linear dispersion relationship of the electrons near to the Fermi level in this system. These defects also create scattering centers, which diminish the current by some 42% in the case of the Stone-Wales or single vacancy, by 55% in the case of the 5-8-5 divacancy and by 68% in the case of the 555-777 divacancy.[675]

First-principles calculations of the structural and vibrational properties of germanene predict that a structure with a small buckling of 0.7Å and a bond length of 2.44Å is the most favored one energetically and does not exhibit an imaginary phonon mode. In calculated non-resonance Raman spectra, the highest peak is at about 290/cm. Extensive calculations of armchair germanene nanoribbons, with and without hydrogenation of the edges, reveal other Raman peaks which are distributed mainly at lower frequencies than the G-like peak that can be attributed to defects at the edges of the ribbons and are thus not present in the Raman spectra of non-defective germanene. The Raman peak which corresponds to the D mode is located at about 270/cm. The calculated G-like and D peaks are expected to be the fingerprints of the Raman spectra of low-buckle germanene.[676]

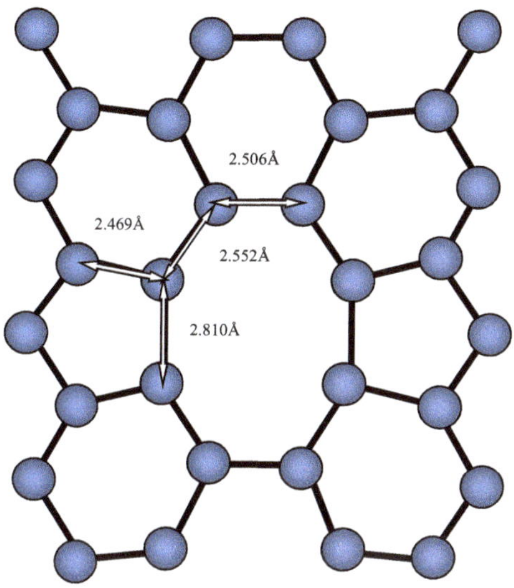

Figure 7. 5-8-5 divacancy in germanene.

2.8 Thermal Conductivity

First-principles calculations and self-consistent solutions of the linearized Boltzmann-Peierls equation for phonon transport within a three-phonon scattering framework, have been used to characterize the lattice thermal conductivities of free-standing germanene under various isotropic tensile strains and temperatures. It is found that the single-mode relaxation time approximation, which works reasonably well for bulk silicon, fails to describe properly the phonon thermal transport in germanene; within the temperature range considered. An increase in strain hardens and stabilizes long-wavelength out-of-plane acoustic phonons, and leads to a similar conductivity behavior to that of silicene.

Such results advance the understanding of phonon dynamics in all group-IV buckled monolayers and should guide the choice of transfer and fabrication techniques for free-standing samples, as well as the modification of the thermal conductivity using suitable choices of size and strain.[677]

2.9 Thermo-Electricity

Non-equilibrium Green's function methods combined with tight-binding Hamiltonian studies, of the spin Seebeck effect in a germanene p-n junction, show that the thermal bias can generate spin thermopower when a local exchange field is applied to one edge of a germanene nanoribbon. The magnitude of the spin thermopower can be modulated by using the potential drop across the two terminals of the p-n junction. When the value of the potential drop is smaller than the spin-orbit interaction strength, the spin thermopower is increased by 2 orders of magnitude, as compared with the case of zero p-n voltage. The optimum temperature corresponding to the maximum spin thermopower is insensitive to the potential drop. In the p-n region, the maximum spin thermopower is attained at relatively higher temperatures. When the potential drop value is larger than that of the spin-orbit interaction, the spin Seebeck effect then decays rapidly with increasing potential drop or temperature. By optimizing the structural parameters, the magnitude of the spin thermopower can be markedly increased, due to the coexistence of the exchange field and the potential drop.[678]

Non-equilibrium Green's function studies of the spin thermo-electric performance of a germanene nanoribbon demonstrate that a temperature bias can generate spin thermopower when an exchange field breaks the edge states of germanene leads. The spin thermo-electric efficiency is however quite low, with a maximum of 0.01. Upon applying a strong electric field to the central region, a relatively large spin-dependent band-gap can be opened up, and the spin figure-of-merit is then predicted to be more than 100 times larger than the case without an external field. The marked enhancement of the spin thermo-electric efficiency (greater than unity) is attributed to suppression of the thermal conductance and to an improvement of the spin Seebeck effect. The striking properties render ferromagnetic germanene nanoribbon a promising pure spin thermo-electric nanogenerator.[679]

First-principles density functional and linear response studies have been made of the thermo-electric properties of 1-dimensional and 2-dimensional germanium. Two-dimensional germanene nanoribbons of various widths have also been investigated with regard to the possibility of nano-structuring systems by mixing silicon and germanium[680]. The room-temperature figure-of-merit of the systems is particularly high and approaches 2.5.

2.10 Optical Properties

First-principles studies, using the non-local hybrid Herd-Scuseria-Emzerhof functional, of the optical absorbance of group-IV honeycomb crystals can account for many-body effects on the optical properties. The optical absorption peaks are blue-shifted, due to quasi-particle corrections, while the influence on the low-frequency absorbance remains unchanged and reduces to a universal value which is related to the Sommerfeld fine-structure constant. At the Dirac points, spin-orbit interaction opens up fundamental band-gaps; parabolic bands with a very small effective mass emerge. The low-frequency absorbance is thus modified, with a spin-orbit induced transparency region and an increase in the absorbance at the fundamental absorption edge.[681] Green's function perturbation theory studies, using the GW+Bethe-Salpeter equation, of many-body effects in germanane show that - due to confinement - many-body effects play an essential role in quasi-particle excitation and optical absorption spectra. This leads to excitonic resonance ($\pi \rightarrow \pi^*$ excitation) and to strongly bound excitons having considerable binding energies.[682] Upon calculating the complex dielectric function for optical interband transitions, it is found that germanene has the same low-frequency absorbance as does graphene and is governed by the Sommerfeld fine-structure constant. Deviations occur at higher frequencies when the first interband transitions outside of K or K′ make a contribution. The low-frequency results are a result of the honeycomb geometry, and do not depend upon the nature of the group-IV atom which is involved, nor sheet buckling nor orbital hybridization. It is expected that the 2-dimensional crystals will be useful as absorption normals in silicon technology.[683]

By using calculations which are based mainly upon the independent-quasiparticle approximation of the complex dielectric function for optical interband transitions, the optical conductivity of two-dimensional honeycomb crystals beyond the usual Dirac-cone approximation can be estimated, taking the full band structures into account. Particular care is taken to derive convergent spectra with respect to the number of k points in the Brillouin zone and the number of bands. In this way both the real and imaginary parts of the optical conductivity are correctly described for low and high frequencies. The results can be applied to predict the reflection, transmission and absorption over a wide range of photon energies.[684]

First-principles density functional theory study of arsenic- or gallium-doped free-standing germanene monolayers reveals that the electronic, magnetic and optical properties can be tailored by choosing the type, concentration and site location of substitutional doping elements. It is possible to tune the splitting of the Dirac cones at, or near to, the Dirac K point by means of proper doping. An induced magnetism which results from double-doping of free-standing germanene can be traced to the 4p level electrons in the spin-polarized projected density of states.[685] Analysis of the optical properties predicts that the static dielectric constant and the refractive index are significantly enhanced by proper doping in the case of parallel polarization. Optical conductivity studies indicate that the height of the maximum peak is amplified by 5.50% by double gallium doping at non-equivalent sites.

2.11 Composites

First-principles studies of silicene/germanene heterobilayers show that the silicene interacts overall with germanene with a binding energy of -50 to about -70meV per germanium atom; suggesting the existence of a weakly van der Waals interaction between silicene and the substrate. A relative large band-gap, with a linear band dispersion of heterobilayers is opened up due to the sub-lattice symmetry broken by the intrinsic interface dipole between silicene and the substrate. The band-gap of the heterobilayer can also be continuously tuned by adjusting the interlayer spacing and strain; regardless of the stacking arrangement.[686] First-principles study of the electron-phonon coupling in germanene shows that it is much reduced in comparison with that in silicene or graphene.[687] First-principles calculations of the geometrical, electronic and magnetic properties of 2-dimensional germanene/silicene hybrid sheet saturated with brominate atoms shows that, although the pristine material is semi-metallic, half-saturation with brominate atoms on only one side of the silicon results in localized and unpaired electrons in unsaturated germanium atoms, exhibiting long-range ferromagnetic properties with a high Curie temperature. Half-brominated material has a half-metallic nature, with 100% spin-polarized currents at the Fermi level. This suggests the use of germanene/silicene hybrid sheet as a component of spintronic devices.[688]

The reduced silicene/germanene single-particle Hamiltonian in the buckled two-dimensional hexagonal lattice can be expressed in terms of the Pauli matrices associated with the pseudo-spin. The Hamiltonian for these systems comprises the Dirac kinetic energy, a mass term and spin–orbit coupling. The mass term breaks the sub-lattice symmetry of the honeycomb structure and generates a spin-split band-gap, with the help of the spin–orbit coupling. The buckled lattice generates a staggered sub-lattice potential between silicon atoms at A-sites and B-sites when an electric field, E_z, is applied at right-angles to its plane. The physics of the system is then determined by two Dirac-like cones at the K and K′ points. By tuning E_z, a critical value is attained at which the topological insulating phase goes into a semi-metallic state, where the ‘spin-down’/‘spin-up’ band-gap ratio is much lower than unity for the K valley and is much greater than unity for the K′ valley. This state is deemed a ‘valley-spin-polarized-metal’, due to the opposite spin-polarizations of the K and K′ valleys. Upon further increasing E_z, the system changes into a trivial insulator, and this is associated with reversal of the valley magnetic moment. Preliminary investigation has shown that, provided that the (non-magnetic) impurity scattering strength, V_0, is moderate (of the same order as the intrinsic spin–orbit coupling of about 4meV) then the 'valley-spin-polarized-metal' phase is protected. A substantial enhancement of V_0 leads however to disappearance of this phase, due to accentuated intra- and inter-valley scattering processes. This disappearance does not occur due to an increase in the Rashba spin–orbit coupling effect. [689]

First-principles studies have been made of the structural and electronic properties of superlattices prepared by the alternate stacking of two-dimensional hexagonal germanene and a MoS_2 monolayer, with the results being compared with those for graphene/MoS_2 superlattices. The distortions of the geometry of the germanene and MoS_2 layers, due to the formation of the superlattice, are relatively small and result from the relatively weak interactions between the stacked layers. The germanene/MoS_2 superlattice is clearly metallic, with the linear bands around the Dirac points of the pristine germanene apparently being preserved. A small band-gap is opened up at the Dirac point of the superlattice, due to symmetry breaking in the germanene layer caused by the introduction of the MoS_2 sheets. Charge transfer occurs mainly within the germanene and MoS_2 layers (intralayer transfer), as well as in some part of the intermediate region between the germanene and MoS_2 (interlayer transfer), thus suggesting the existence of more than just van der Waals interactions between the stacked sheets.[690]

Germanene nanorods, embedded within completely hydrogenated germanane exhibit a quantum spin Hall effect. The nanorods can be created experimentally by local hydrogen dissociation from germanane. First-principles calculations predict that germanene nanorods with zig-zag interfaces will possess dissipation-less conducting channels having in-plane and out-of-plane spin textures.[691]

2.12 Applications

2.12.1 Solar Energy

Computation of the intrinsic electronic and optical properties of two-dimensional methyl-terminated germanene, $GeCH_3$, is possible on the basis of the Herd-Scuseria-Emzerhof hybrid functional within the framework of density functional theory. The $GeCH_3$ monolayer exhibits a direct band-gap of 1.76eV which can be tuned by applying elastic strains: it decreases with increasing tensile strain and increases under small compressive strains. Anisotropic carrier mobility is observed in the armchair and zig-zag directions of the $GeCH_3$ monolayer. The latter also exhibits significant light absorption in the visible and ultra-violet ranges of the solar spectrum; thus making it an attractive material for light-harvesting purposes.[692]

2.12.2 Energy Storage

Based upon first-principles calculations, it has been proposed that Ge/MgX_2 (X = Cl, Br, I) would be suitable hybrid structures for this purpose. It is found that lithium will not cluster and that the lithium capacity is very high: 279mAh/g for germanene on $MgCl_2$.[693] Sandwich structures can be used to enhance the performance further. Low diffusion barriers of less than 0.3eV are predicted for all of the hybrid structures.

2.12.3 Catalysis

First-principles electronic structure calculations of the chemical functionalization of two-dimensional hydrogenated and germanene, known as germanane, can be a powerful means for increasing photocatalytic water-splitting. Spin-polarized density functional theory, within the GGA-PBE and Herd-Scuseria-Emzerhof types of exchange correlation, predicts the structural, electronic and optical properties of germanane functionalized with non-metals (nitrogen, phosphorus, sulfur), alkali metals (lithium, sodium, potassium) or alkaline-earth metals (magnesium, calcium). The surface-adsorbate interaction between the functionalized systems, and H_2 or O_2 molecules, leads to hydrogen and oxygen evolution.[694]

2.12.4 Transistors

The opening up of a sizable band-gap in naturally zero-gap germanene, without incurring a heavy loss of carrier mobility, is a key issue in its application in nano-electronic devices such as high-performance room-temperature field effect transistors. First-principles calculations indicate that a band-gap is opened up at the Dirac point by the single-side adsorption of alkali-metal atoms. This band-gap is tunable, from 0.02 to 0.31eV, by varying the coverage and the type of alkali-metal atom, with a maximum global band-gap of 0.26eV. Because the effective masses (ranging from 0.005 to 0.106me) of the electrons and holes near to the Dirac point following surface adsorption are small, the carrier mobility is not expected to degrade much. Thus, germanene is a potential candidate for the achievement of an effective room-temperature field-effect transistor channel following surface adsorption.[695]

By combining the Boltzmann transport equation with a relaxation-time approximation at the first-principles level, it can be calculated that the intrinsic carrier mobility of germanene sheet can attain about 6 x 10^5cm^2/Vs; larger than that of graphene. The high intrinsic carrier mobility of germanene is attributed to the large buckling distance and to the small effective mass. Since germanium enjoys good compatibility with silicon in the existing semiconductor industry, these results indicate that germanene will play an important role in the future nano-electronics industry.[696] It is nevertheless challenging to grow germanene epitaxially onto conventional semiconductor substrates. First-principles calculation investigations of the structural and electronic properties of germanene/germanane heterostructures indicate that a Dirac cone with an almost linear band dispersion of germanene is maintained in the band-gap of the substrate. The band-gaps which are opened in these heterostructures can be effectively modulated by using external electric fields, or straining, and also feature very low effective masses and high carrier mobilities. These results suggest routes to the design of high-performance field-effect transistors operating at room temperature in nanodevices.[697]

Ab initio calculations predict that a vertical electric field is able to open a band-gap in semimetallic single-layer buckled germanene. The size of the band-gap increases linearly with the electric field strength. Biased single-layer germanene can therefore function effectively at room temperature as a field effect transistor.[698]

Assessment of the performance potentials of germanene and graphene nanoribbons, which both have a sufficient band-gap to act as field-effect transistor channel materials, demonstrates that, when comparing them at the same band-gap of ~0.5eV, the graphene nanoribbon field-effect transistor maintains an advantage over the germanene nanoribbon field-effect transistor under ideal transport condition but that germanene nanoribbon still remains an attractive channel material for high-performance field-effect transistors.[699]

A spin polarized density-functional theory study of the electronic and magnetic properties of pristine or doped germanene nanoribbons of various widths shows that a germanium atom at the ribbon edge always prefers to be replaced by an impurity atom. Single nitrogen- or boron-atom substitution induces a semiconductor-to-metal transition in armchair-oriented germanene nanoribbons, as reflected by the appearance of a half-filled band with lower dispersion. Nitrogen and boron co-doping at the ribbon edges instead only modifies the band-gaps, due to the occurrence of effective charge compensation. Single nitrogen or boron atom substitution generally converts antiferromagnetic semiconducting zig-zag germanene nanoribbons into ferromagnetic semiconductors. This transition is attributed mainly to the perturbation of π and π^* states, localized at the doped edge. Double atom substitution (regardless of the N-N, B-B or N-B configuration) at the edges of the nanoribbons removes the spin-polarization from both edges and transforms them into non-magnetic semiconductors. Some single-atom doped nanoribbons can exhibit a ferromagnetic half-metallic nature, with 100% spin-polarization at the Fermi level. It is suggested that doped armchair and zig-zag nanoribbons have potential application in germanium-based nano-electronic products such as field effect transistors, negative differential resistances and spin filter devices.[700]

2.12.5 Optical Devices

First-principles calculations predict that germanene, decorated with ethynyl-derivatives (GeC_2X; X = H, F, Cl, Br, I) can be a topological insulator with a large band-gap; suitable for room-temperature applications. The GeC_2I and GeC_2Br films are intrinsic topological insulators with a gap ranging up to 180meV over a wide range, while GeC_2H, GeC_2F and GeC_2Cl transform from trivial to non-trivial phases under tensile strain. This topological characteristic is confirmed by s-p_{xy} band inversion, by the topological invariant and by time-reversal symmetry protected helical edge states. In particular, the characteristic properties of edge states - such as the Fermi velocity and edge shape - can be tuned via edge modification. Hexagonal BN sheet is an ideal substrate for the experimental preparation of GeC_2X, and maintains the non-trivial topology. In view of their higher thermo-stability, such GeC_2X films should be good quantum spin Hall effect platforms for topological electronic device design and fabrication in the field of spintronics.[701] When the geometric and electronic properties of germanene, with small adsorbed organic molecules, are similarly studied using first-principles calculations, the adsorption energies are found to range from -0.459 to -0.231eV; higher than those for small adsorbed organic molecules on graphene or silicene. This indicates the presence of strong interactions between the organic molecules and germanene, and leads to a large enhancement of the band-gap to between 3.9 and -81.9meV; due to sub-lattice symmetry breaking[702]. The characteristics of a Dirac cone with a low effective mass and a high electron mobility are preserved, and this offers the possibility of designing germanene-based opto-electronic devices.

2.12.6 Spintronics

Germanene is a 2-dimensional topological insulator which exhibits helical edge states. Global and local manipulation at the edges, by exposing them to a charge-density-wave order, to a superconductor, to an out-of-plane antiferromagnetic field or to an in-plane antiferromagnetic field shows that such perturbations affect the helical edge states in different manners. As a result, it is possible to obtain a quantum spin Hall effect without edge states. Edge manipulation also leads to promising applications, such as giant magnetoresistances and perfect spin filters. The effect of the manipulation of a very few edge sites of a topological insulator nanodisk can be envisaged.[703]

First-principles calculation investigations of the spin-dependent electronic transport properties of zig-zag nanoribbons with odd (5) and even (6) widths reveal symmetry-dependent transport properties, although the σ mirror plane is absent. Even and odd zig-zag nanoribbons have very different current-voltage relationships. The even (6) nanoribbon exhibits a dual spin-filter effect in an antiparallel magnetism configuration, but the odd (5) nanoribbon behaves like a conventional conductor, with a linear current-voltage dependence. When the two electrodes are arranged in a parallel configuration, the width-6 zig-zag nanoribbon system is in a low-resistance state, but can switch to a much higher resistance state when the electrodes are arranged in an antiparallel configuration; where a magnetoresistance of 270% can then be observed.[704]

Ab initio investigation of the edge states of group-IV honeycomb crystals, while fully taking account of spin-orbit coupling into account, reveals topological properties which depend upon the nature of non-ferromagnetic, antiferromagnetic and ferromagnetic edges of the ribbon. Unlike model studies, the dispersion of the edge states is strongly non-linear and destroys gap-less Dirac bands. Applying an electric field permits the further study of a topological phase transition to a trivial insulator above a critical field strength.[705] Theoretical investigation of two types of density oscillation (Friedel, collective), within the random-phase approximation, shows that the tunable spin-valley coupled band structure can lead to exotic behaviors of these phenomena. Using exact analytical and numerical analyses, it can be demonstrated that the beating of the screened potential, and the undamped plasmon mode, are fingerprints of a topological phase transition in doped germanene. There is a connection, between the topological phase transition and the density oscillations, which can be accessed by using various experimental techniques.[706]

Semi-classical Monte Carlo simulations, taking account of Dyakonov-Perel and Elliott-Yafet spin relaxation mechanisms, of the effect of magnon scattering on spin relaxation in germanene at 4K show that the spin relaxation length attains a peak value at a magnon energy of 1.5meV, and then saturates at a value of 520nm for higher values of magnon energy. Germanene thus has a lower peak spin relaxation length than that of graphene.[707] Dephasing of the spin vectors in germanene is due mainly to the Dyakonov-Perel and Elliott-Yafet relaxation mechanisms but, because germanene has a high carrier mobility, the latter is not the predominant relaxation mechanism. Study of spin-polarized transport along the length of the germanene monolayer structure shows that the spin de-phasing length is about 1.5μm. Investigations have also been made of the ensemble averaged spin vector variation along the length of planar and buckled monolayer germanene as a function of temperature.[708]

Chapter 3

Stanene (Tinene)

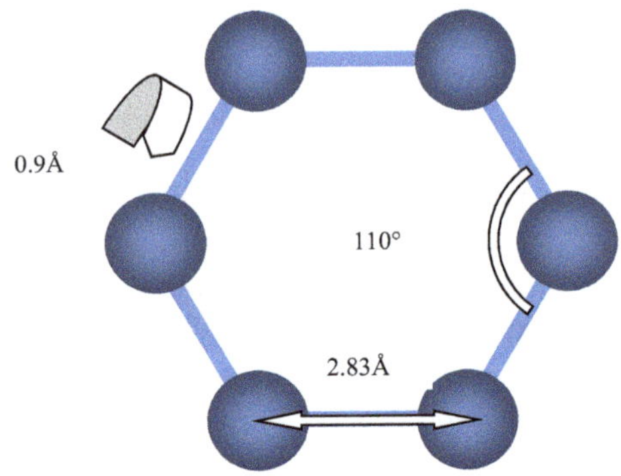

Stanene is one of most important 2-dimensional materials because of its potential to exhibit room-temperature topological effects due to opening of a spin-orbit gap. Synthesis of a few layers of stanene, and atomic-scale morphological and elemental characterization confirms the presence of an hexagonal lattice of Sn atoms. The position of the Raman peak, together with the interplanar spacing, are in good agreement with those deduced from first-principles calculations, and confirm that the sheets are not merely (111) α-Sn sheets. The optical signatures at about 191nm and about 233nm, calculated using density functional theory, for low-buckle stanene are in qualitative agreement with the measured UV-Vis absorption spectrum. Atomic force microscopic measurements indicate an interlayer spacing of about 0.33nm; in good agreement with that reported for epitaxial stanene sheets. No traces of oxygen were detected, thus suggesting the absence of any oxidized phases.[709] This was also confirmed by comparative Raman measurements of oxidized stanene sheets. Stanene has also been successfully fabricated by means of molecular beam epitaxy, as confirmed by scanning tunnelling microscopy and angle-resolved photo-emission spectroscopy in conjunction with first-principles calculations.[710] Monolayer hexagonal tin, stanene, is an ideal candidate for the exploitation of graphene-type physics in nano-electronic applications. Its higher out-of-plane buckling and appreciable spin-orbit coupling lend themselves to improved Dirac-cone engineering. Investigation of the ballistic charge-transport regime of armchair tin nanoribbons with widths of 3m-1, 3m and 3m+1, where m is an integer, shows that tin ribbons exhibit earlier current onsets and carry currents which are 20% larger than those carried by C/Si/Ge-nanoribbons.[711]

3.1 Structure

Ab initio density functional studies of various two-dimensional analogues show that four of them are based upon dumb-bell geometries, one on a Kagome lattice and one on the hexagonal structure. Calculations of the cohesive energies clearly show that the hexagonal structure (graphene) is the most stable one for carbon, as expected. In the case of silicon and germanium, dumb-bell structures and particularly large honeycomb dumb-bell geometries are energetically favored as compared with sp^2/sp^3-bonded hexagonal lattices such as silicene and germanene. The main reason for this is the opening-up of a band-gap in the honeycomb dumb-bell arrangements. The large honeycomb dumb-bell sheet crystals are indirect semiconductors with a K→Γ gap of about 0.5eV. In the present case, tin, the MoS_2-like symmetry is predicted to be more stable, in contrast to the stanene or large honeycomb dumb-bell geometries.[712] First-principles density functional theory predicts tinene to have a honeycomb lattice with a lattice parameter of 4.62Å and a buckling of 0.92. The electronic dispersion indicates a Dirac cone with zero gap at the Fermi energy and a Fermi velocity of 0.97 x 10^6m/s. The Fermi velocity and the strength of spin-orbit interaction in stanene can be calculated within the framework of density functional theory.[713] Other considerations yield a band-gap of 0.10eV. The monolayer is thermally stable at up to 700K, according to first-principles molecular dynamics simulation, and exhibits phonon dispersion without imaginary frequencies. Control of the external electric field and the applied strain produces an induced band-gap of up to 0.21eV, whilst retaining linear dispersion but with degraded electronic transport.[714] Monolayer gray tin, tinene, is a two-dimensional material having both Dirac characteristics and a 72meV band-gap, according to density functional calculations. As compared with silicene and germanene, tinene has a similar hexagonal honeycomb monolayer structure, but the larger buckling height of about 0.70Å. This buckling leads to phonon dispersion, without appreciable imaginary modes, indicating strong dynamic stability. It is deduced from the band structure that six Dirac cones would appear at high-symmetry K-points in the first Brillouin zone when gray tin is thinned from bulk to monolayer thickness, but a band-gap as large as 72meV could be preserved.[715]

3.2 Preparation

Using first-principles calculations, the potential of the Ag(111) surface as a substrate for the epitaxial growth of monolayer stanene has been explored. This reveals that the hexagonal structure of the stanene monolayer is well preserved on the Ag(111) surface. The height of the epitaxial monolayer is comparable to the step height of the substrate, thus permitting growth to cross the surface step and form large-area stanene. The perfect lattice structure of free-standing stanene can be obtained when the epitaxial stanene monolayer is detached from the Ag(111) surface. The barrier to tin adatom diffusion on the Ag(111) surface is only 0.041eV, thus permitting the epitaxial growth of stanene monolayers even at low temperatures.[716] This all suggests that the Ag(111) surface is an ideal candidate for growing large-area high-quality monolayer stanene.

3.3 Mechanical Properties

As usual, it is interesting to consider firstly the mechanical properties. Molecular dynamics simulations, using a modified embedded-atom method potential, show that both the fracture strength and strain of α-stanene decrease with increasing temperature. Uniaxial loading in the zig-zag direction yields a higher fracture strength and strain as compared with the armchair direction. No noticeable variation in the mechanical properties is observed for biaxial loading. At a higher loading-rate, the material exhibits a higher fracture strength and strain.[717] Theoretical study of the properties of graphene and its analogues is most complete for planar structures and for structures having small deviations from planarity. The limit of large deformations is more relevant because it governs material properties near to the point of failure. Discrete differential geometry concepts appear to be promising for linking material properties to shape, regardless of how great is the given deformation.[718] Density functional theory and simple molecular mechanics models for Young's modulus estimate the in-plane stiffness to be 40N/m.[719] Group theory analysis can be used to study the effect of strain, to understand structural phase transitions and to characterize the number of layers, crystallographic orientations and non-linear phenomena.[720]

First-principles density functional theory calculations have been used to investigate the mechanical properties of single-layer and free-standing material. Uniaxial tensile and compressive simulations probed and compared the Young's modulus, Poisson ratio and ultimate strength, and evaluated the effect of chirality upon the mechanical response and bond structure of the 2-dimensional substrates. The armchair and zig-zag Young's moduli are estimated to be 25.2GPa.nm (table 6).[721] Others have calculated the armchair Young's modulus to be 24.46GPa.nm.[722] Such first-principles simulations suggest that the application of uniaxial loading can change the electronic nature of the buckled structures to metallic. The elastic properties of stanene under equiaxial or uniaxial tensions along the armchair and zig-zag directions have similarly been investigated by means of first-principles calculations. The stress-strain relationship is calculated, and the relaxation of the internal atomic positions is analyzed. High-order elastic constants are calculated by fitting polynomial expressions. The Young's modulus and Poisson ratio are thus calculated to be 24.14N/m and 0.39, respectively.[723] Stanene has a lower Young's modulus than that of the preceding group-IV elements. This is attributed to the smaller sp^2-sp^3 bond energy in stanene, as compared to that in silicene and germanene. The calculated values of ultimate stresses and strains, second-order elastic constants and in-plane Young's modulus are all positive; thus proving that stanene is mechanically stable.

Table 6
Mechanical properties of stanene

Material	**Property**	**Value**
armchair	Young's modulus	25.2GPa.nm
zig-zag	Young's modulus	23.5GPa.nm
armchair	Poisson ratio	0.36
zig-zag	Poisson ratio	0.42
armchair	UTS	2.6GPa.nm
zig-zag	UTS	2.2GPa.nm
armchair	ε_{UTS}	0.17
zig-zag	ε_{UTS}	0.18

3.4 Substrate Interaction

Ever since the quantum anomalous Hall effect was observed in magnetic topological insulators, research on the topic has intensified. The very difficult experimental conditions limit however its practical application. *Ab initio* methods predict that non-magnetic PbI_2 film is an ideal substrate for the two-dimensional stanene honeycomb structure. A quantum anomalous Hall effect, with a band-gap of up to 90meV, can be achieved in functionalized stanene/PbI_2 heterostructures. In spite of van der Waals interactions within the heterostructure, band inversion occurs between the tin (s and $p_{x,y}$) and lead ($p_{x,y}$) orbitals. This plays an essential part in determining the non-trivial topology and large band-gap of the system. An absence of magnetic atoms is essential for the triggering of the quantum anomalous Hall effect. A very stable proof-of-principle device, based upon a Sn/PbI_2 heterostructure can be proposed which exhibits quantum anomalous Hall effects.[724]

Free-standing stanene is predicted to have a sizable band-gap of 100meV at the Dirac point, due to spin-orbit coupling, leading to quantum spin Hall, quantum anomalous Hall and quantum valley Hall effects. A systematic study of the interfacial properties of stanene/metal interfaces, where the metal is silver, gold, copper, aluminium, palladium, platinum, iridium or nickel, using *ab initio* electronic structure calculations and considering spin-orbit coupling effects, shows that the honeycomb structure of stanene is preserved on the metal supports, although the buckling height is changed. The buckling of stanene on gold, aluminium, silver or copper supports is greater than that of free-standing stanene. A planar graphene-like structure is instead stabilized in the case of stanene on iridium, palladium, platinum or nickel supports. The band structure of stanene is destroyed on all of the supports, and is accompanied by the metallization of stanene because covalent bonds are formed between the stanene and the support and the stanene structure is distorted.[725] No tunnelling barrier exists between stanene and the metal support, so that the stanene and the metal maintain a good vertical ohmic contact.

3.5 Surface Interaction

The electronic and magnetic effects of transition metal (vanadium, chromium, manganese, iron, cobalt, nickel) atom adsorption on stanene have been investigated by means of first-principles calculations (table 7). The results indicate that most atoms prefer to be relaxed at a hydrogen site on stanene, apart from vanadium, which prefers a valley site. Iron-absorbed stanene is a spin gap-less semiconductor with up-spin electrons and down-spin hole carriers which permit the coexistence of a charge current and a pure spin current. Cobalt-absorbed stanene exists as a half-metal phase.[726] Vanadium-, chromium-, manganese- or copper-absorption turns stanene into a metal, while nickel- or zinc-absorption opens a narrow band-gap. In vanadium-, chromium-, manganese-, iron- or cobalt-absorbed stanene, the magnetic moment of the transition metal survives, while nickel-, copper- or zinc-absorbed stanene is non-magnetic.

First-principles plane-wave calculations within spin-polarized density functional theory of the adsorption of alkali and alkaline-earth metal atoms, at various degrees of coverage, on single-layer stanene show that the most favorable adsorption sites for alkali atoms (lithium, sodium, potassium) are the hollow sites. This is similar to the behavior of other group-IV monolayers, but the case of alkaline-earths on stanene is different to that of silicene or germanene. While magnesium and calcium are bound to stanene at hollow sites, the bridge site is energetically preferable to beryllium adatoms.[727] All of the adsorbed atoms are positively charged due to charge transfer from the adatom to the stanene monolayer. The semi-metallic bare stanene then becomes metallic, except in the case of beryllium adsorption. The latter gives rise to a non-magnetic semiconducting ground-state. These results show that stanene has a reactive and functionalizable surface similar to that of graphene or silicene. First-principles density functional theory calculations of the interaction of sodium or lithium ions on stanene films have identified the most stable binding sites and their corresponding binding energies. Bader charge analysis confirmed complete charge transfer between lithium or sodium ions and the 2-dimensional sheets.[728]

Table
Adsorption energies at stanene surface

Adatom	Site	E(eV)
Zn	H	0.38
Cr	H	2.31
Mn	H	2.45
Cu	H	2.53
V	V	2.80
Fe	H	3.09
Co	H	3.53
Ni	H	3.89

H: above hexagon centre. V: above lower-layer Sn atom

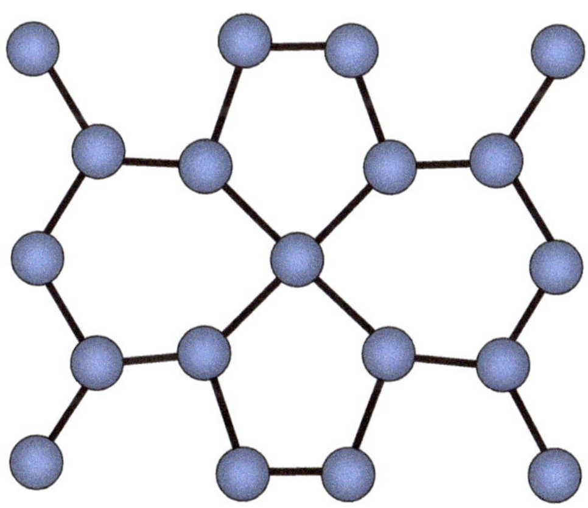

Figure 8. 1(55|66) monovacancy in stanene

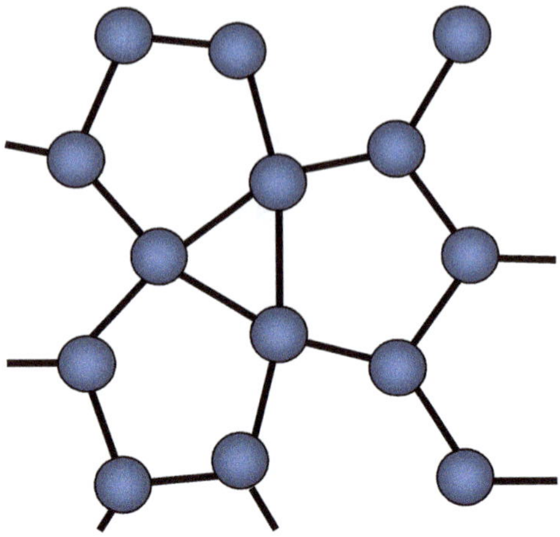

Figure 9. 2(3|555) monovacancy in stanene

3.6 Superconductivity

An energy band-gap of about 0.1eV has been obtained for the buckled honeycomb structure of stanene and phonon-mediated superconductivity in this material is thereby ruled out. First-principles calculations and density functional theory suggest the possibility of producing phonon-mediated superconductivity by doping. Doping with calcium leads to superconductivity with a transition temperature of about 1.3K. Doping with lithium leads to a transition temperature of about 1.4K. Both transition temperatures are lower than that (3.7K) of bulk β-tin.[729] With regard to the emergence of topological superconductivity in a honeycomb system such as stanene, photo-irradiation induces the Haldane term into the system. Introducing s-wave superconductivity by proximity coupling makes the system become a topological superconductor. The topological phase diagram can be determined by deriving the four independent 2-band theory and determining the Chern number. Majorana zero-energy bound states appear at the boundary of two distinct topological phases. The position of the Majorana fermion can be arbitrarily controlled by moving the area of photo-irradiation.[730]

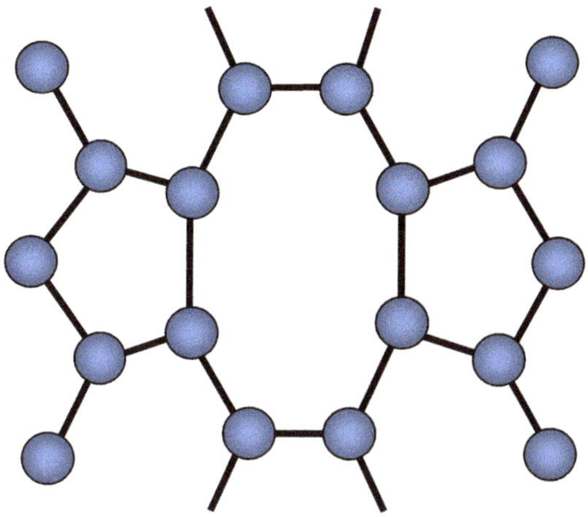

Figure 10. 2(5|8|5) divacancy in stanene

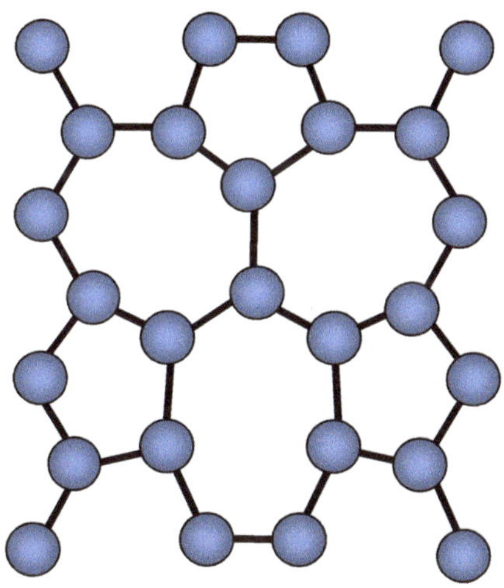

Figure 11. 2(555|777) divacancy in stanene

3.7 Defects

During the synthesis of stanene monolayers, defects are inevitably created, and always affect the properties. *Ab initio* predictions have been made of the structures, diffusion behavior and related properties of typical point defects (figures 8 to 11, table 8), such as the Stone-Wales defect, single vacancies (SV-1[55|66], SV-2[3|555]) and double vacancies (DV-1[5|8|5], DV-2[555|777]). Scanning tunnelling microscopic images have also been simulated. The results reveal that the Stone-Wales defect can be easily recovered by annealing, due to its low reverse barrier. The SV-1[55|66] and SV-2[3|555] vacancies are the most stable. The energetically favored DV-1[5|8|5] di-vacancy can be formed by the coalescence of 2 monovacancies. The DV-2[555|777] can arise from DV-1[5|8|5] via a bond rotation which has to overcome a diffusion barrier of 0.89eV. The point defects have a non-trivial influence upon the electronic properties. The Stone-Wales defect can open up a direct gap in the energy band, without harming the high-velocity carriers.[731] Meanwhile, the SV-1[55|66] makes the stanene metallic while SV-2[3|555], DV-1[5|8|5] and DV-2[555|777] can change stanene into an indirect or direct band-gap semiconductor. Spin orbit coupling effects influence the electronic bands; especially the band-gaps.

Table
Binding and formation energies for stanene

Process	**Energy (eV)**
binding	3.40
formation, Stone-Wales defect	1.21
formation, monovacancy-1	1.58
formation, monovacancy-2	1.58
formation, divacancy-1	1.75
formation, divacancy-2	1.80

First-principles calculations of grain boundaries in stanene reveal four typical structures having particularly low formation energies. These extended defects act as quasi one-dimensional semiconducting or metallic wires, depending upon their geometrical structure. They are also reactive, and the adsorption of hydrogen atoms in the grain-boundary region is more stable than in bulk stanene. Adsorption of a single hydrogen atom has a drastic effect upon the electronic behavior of the grain boundaries, and the band structure can be tuned by varying the coverage of hydrogen adsorption.[732]

3.8 Diffusion

First-principles calculations of dumb-bell formation mechanisms and of the energetics of group-IV atoms adsorbed on stanene monolayer honeycomb structures show that, depending upon the row-number of the adatoms and substrates, three types of binding structures are found. Dumb-bell structures occur not only on single-layer monatomic honeycomb structures but also on compounds such as SiC and SiGe. The energy barrier to the migration of a dumb-bell structure is low, due to the concerted action of the atoms. This causes dumb-bells to be rather mobile on substrates.[733]

3.9 Magnetism

Theoretical study of the indirect spin coupling strength between two magnetic impurities located on honeycomb Kane-Mele zig-zag ribbons with periodic and open boundaries shows that the spin interaction in periodic boundaries exhibits an anti-ferromagnetic-ferromagnetic oscillation behavior with increasing staggered potential and electron density. It approaches a maximum value at the edges. The spin coupling in open-boundary Kane-Mele zig-zag ribbons exhibits a trivial smooth anti-ferromagnetic coupling with varying staggered potential. This novel spin interaction behavior is the combined effect of finite size, topological edge states and inversion-symmetry breaking induced by the staggered potential.[734] The interplay of ferromagnetic order and spin-orbit interactions at the zig-zag edges of graphene-like systems reveals an in-plane magnetic anisotropy which opens up a gap in the otherwise conducting edge channels and should result in large changes in the electronic properties upon rotation of the magnetization.[735] Manipulation of the edges by exposing them to out-of-plane antiferromagnetic and in-plane antiferromagnetic fields affects helical edge states in various fashions. Quantum spin-Hall effects without edge states can then be obtained. Edge manipulation can lead to giant magnetoresistance and a perfect spin filter.[736]

3.10 Band Structure

In general, the band-gap of stanene can be tuned by applying a perpendicular electric field, which induces a topological phase transition. Phase diagrams of topological insulators and superconductors in honeycomb systems can be explicitly determined.[737] Universal properties of the magneto-electric Hall effect can be extracted by projecting experimental transport data directly onto the phase diagram. When data which depend on dominant parameters - such as temperature - are available, flow lines that include the location of quantum critical points and phase boundaries can be traced.[738]

Two-dimensional dumb-bell stanene is proposed to be a promising material for exploiting the quantum spin Hall effect, but the small bulk-gap limits its application at room temperature. First-principles calculations predict that its band-gap can be enhanced to 148meV by methyl functionalization, and further tuned using lattice strain. The quantum spin Hall effect is confirmed by s-$p_{x,y}$ band inversion, a topological invariant, Z_2, of unity and a helical gap-less edge within the bulk band-gap. The characteristic properties of edge states, such as the large Fermi velocity and Dirac cone, can be modulated by edge modification.[739]

Quantum spin Hall insulators feature edge states that are topologically protected from back-scattering, but the main obstacle to the application of the effect is the lack of suitable quantum spin Hall insulators having a large bulk gap. First-principles calculations predict a class of large-gap quantum spin Hall insulators in ethynyl-derivative functionalized stanene (SnC_2X; X = H, F, Cl, Br, I). The topological electronic properties of ethynyl (SnC_2H) are highly tunable by external strain.[740] In particular, SnC_2Cl, SnC_2Br and SnC_2I are quantum spin Hall insulators with a bulk gap of about 0.2eV, while SnC_2H and SnC_2F can be transformed into quantum spin Hall insulators by tensile straining. A single pair of topologically protected helical edge states is established on the edge of these systems, with the Dirac point located at the bulk gap, and their quantum spin Hall states are confirmed by a topological invariant, Z_2, of unity. Films on a BN substrate also maintain a non-trivial large-gap quantum spin Hall effect which harbors a Dirac cone lying within the band gap.[741]

Monolayer stanene, fabricated by molecular beam epitaxy, exhibits metallic features on a Bi_2Te_3(111) substrate. First-principles calculations show that varying of the substrate conditions tunes the electronic properties. Supported stanene has trivial or quantum spin Hall states, with significant Rashba splitting induced by inversion asymmetry. Large-gap (up to 0.3eV) quantum spin Hall states are obtained by growing stanene on substrates such as the anion-terminated (111)-surfaces of SrTe, PbTe, BaSe and BaTe.[742]

First-principles calculations, considering spin-orbit coupling effects and edge passivation, show that armchair nanoribbons are non-magnetic semiconductors with gap values which are a periodic function of ribbon-width. Zig-zag nanoribbons have anti-ferromagnetic ground states with opposite spin order between the two edges, and the gaps decrease as the ribbon width increases. The influence of dangling bonds is obvious in the ferromagnetic moments of zig-zag nanoribbons. Spin-orbit coupling effects can open the band-gap values of stanene sheets and zig-zag nanoribbons, but reduce the band-gap of armchair nanoribbons.[743]

First-principles calculations predict that cyanogen-saturated stanene should be a topological insulator, with a bulk gap as large as 203meV. This can be engineered by applying biaxial strains and an electric field. The band topology is identified by a Z_2

topological invariant, together with helical edge states. The mechanism is s-p_{xy} band inversion at the G-point, induced by spin-orbit coupling. Such systems have a robust topology with respect to chemical impurities.[744]

Suitable substrates are required in order to leave insulator non-trivial properties and their appreciable band-gaps (due to spin-orbital coupling) unaltered. *Ab initio* calculations predict that Ge(111) is a suitable candidate for supporting stanene sheets, because the band structure of (√3 x √3)stanene/Ge(111)-(2 x 2) surfaces has a typical Dirac cone at the Γ-point in the vicinity of the Fermi level. Such a system has been proved to maintain the non-trivial topological phase, as confirmed by edge-state calculations for stanene ribbons.[745]

Experimental results show that the strong interactions which arise from substrates will disturb the Dirac cones and cause a metallicity problem. Van der Waals density functional calculations indicate that stanene and stanane can form commensurate van der Waals heterosheets with InSe and GaTe layers, which exhibit quasi free-standing features with comparable spin-orbit coupling gaps. Supported stanane is still a large-gap quantum spin Hall insulator, in which the substrates induce a tunable Rashba splitting that can be modulated by changing the interlayer distance of heterosheets.[746] First-principles calculations demonstrate that stanane, stable hydrogenated stanene with a dumb-bell like structure, has large topological non-trivial band-gaps of 312meV (Γ-point) and 160meV for the bulk; characterized by a topological invariant, Z_2, of unity because of s-p_{xy} band inversion. Helical gap-less edge states appear in nanoribbon structures, with a Fermi velocity comparable to that of graphene. The non-trivial topological states are robust with respect to substrate effects.[747]

It is proposed that boron nitride sheet and reconstructed InSb(111)-(2 x 2) surfaces are ideal substrates for the experimental preparation of dumb-bell stanene while maintaining its non-trivial topology. When combined with standard semiconductor technologies, such as magnetic doping and electrical gating, the quantum anomalous Hall effect, Chern half-metallicity and topological superconductivity can be obtained on those substrates.[748]

The quantum spin Hall phase can be induced in stanene film on a h-BN substrate under a tensile strain of between 6.0 and 9.3%. For stanene on (5 x 5) h-BN, the quantum spin Hall phase can be preserved without strain but, for stanene on a AlN substrate, the quantum spin Hall phase cannot be found under compressive or tensile strains of less than 10%. For (2 x 2) stanene on (3 x 3) AlN, the compressive strain which is required to induce the quantum spin Hall phase is only 2%.[749]

A plane-wave based pseudo-potential approach to stanene, stabilized on a gold substrate, shows that the tin atoms bind to the gold support via strong chemical bonds and that significant electronic charge transfer occurs from the tin to the support. While a buckled structure is preferred for a layer of tin atoms in the free state, a planar graphene-like atomic arrangement is stabilized on the gold support. This structural change corroborates the metal-like band structure of planar stanene in contrast to the semi-metallic buckled configuration. The tunnelling current of the supported tin layer exhibits ohmic-like behavior and the calculated scanning tunnelling microscopic pattern of the supported tin layer contains distinct images of the holes which are characteristic of the hexagonal lattice.[750]

3.11 Thermal Conductivity

First-principles calculations, and self-consistent solutions of the linearized Boltzmann-Peierls equation within a three-phonon scattering framework, indicate that the single mode relaxation time approximation, which works quite well for bulk silicon, fails to describe properly the phonon thermal transport in stanene. Increasing strain hardens and stabilizes long-wavelength out-of-plane acoustic phonons, and leads to similar conductivity behaviors to those of silicene.[751] Combined first-principles calculations and phonon Boltzmann transport equation methods yield a thermal conductivity of 11.6W/mK for stanene. This value suggests a higher thermo-electric efficiency as compared with those of other two-dimensional materials. Detailed analysis of the phase-space for three-phonon processes shows that the phonon scattering channels, LA + LA/TA/ZA → TA/ZA, are restricted; leading to a predominant contribution of high group-velocity LA phonons to the conductivity.[752] Density functional theory shows that the phonon spectrum is positive for all modes in the first Brillouin zone, and that there is a phonon-energy band-gap between the acoustic and optical modes which is about 50/cm. Spin-orbit coupling opens up a direct energy band-gap at the Dirac point, softens the phonon spectrum and decreases the phonon group velocity.[753]

The introduction of a bond order potential for stanene, based upon an *ab initio* derived data set, accurately describes the energetics and properties of free-standing sheet and can be applied to various nanostructures such as tubes and ribbons. Using this potential, molecular dynamics simulation shows that the stanene structure is highly rippled; far more so than are other 2-dimensional materials such as graphene. This is due to its low in-plane stiffness, which is only about 25N/m in comparison with graphene's 480N/m.[754] The extent of the stanene rippling also exhibits a greater temperature dependence as compared to that of graphene. Above all, stanene-based nanostructures have significantly lower thermal conductivities than those of graphene-based structures, due to their low phonon group velocities, and a high anharmonic response.

3.12 Applications

Most surveys of the new two-dimensional materials focus on their applications in electronics, opto-electronic and magnetic devices[755,756,757], but the buckled structure of stanene can be a key structure with regard to chemical reactivity. First-principles calculations of the reactivity of stanene with atmospheric pollutants such as NO, NO_2, SO, SO_2, CO and CO_2 show that it is reactive towards them. The dissociation activation energies of those pollutants over stanene are lower than for previously reported catalysts. The origin of the low dissociation barriers is charge-transfer, from the stanene to the pollutants, resulting in bond-weakening.[758]

First-principles studies of the electronic structures of CO-, H_2O-, NH_3-, NO- and NO_2-adsorbed stanene monolayers show that the H_2O, NH_3 and CO molecules are physisorbed on the monolayer, while the NO and NO_2 molecules are chemisorbed, with quite a large charge transfer, considerable adsorption energy and strong covalent (Sn–O) bonding. Spin–orbit coupling calculations show that the band-gaps of the molecule-adsorbed monolayers can be effectively tuned: when biaxial strains and electric fields are applied, the adsorption energies and the charge transfer between the gas molecules and the monolayers change markedly.[759] Related first-principles calculations of the adsorption behavior of CO, NH_3, H_2O, H_2S, O_2, NO and NO_2 molecules on stanene sheet also indicate that the sensing performance of stanene is superior to that of other 2-dimensional materials such as silicene and germanene. The CO, O_2, NO and NO_2 molecules act as charge acceptors, whereas the NH_3, H_2O and H_2S molecules serve as charge donors. For non-polarized molecules, the molecule-stanene interaction is attributed mainly to the electrostatic attraction effect. In that case of polarized molecules, the covalent interaction plays a critical role in the adsorption process.[760] Work-function results reveal differing responses to the various molecules, thus indicating that the Schottky barrier height can be effectively tuned by selective adsorption of the small molecules. The non-trivial sensitivity and selectivity of stanene show that it has a potential application in the field of gas sensors and high performance catalysis.

First-principles simulations predict a topological to trivial insulating-phase transition induced by an out-of-plane electric field or by quantum confinement effects. The results highlight the potential use of stanene nanoribbons in gate-voltage controlled dissipation-less spin-based devices and can be used to set the minimum nanoribbon width for such devices. This is typically about 5nm.[761]

Density functional calculations and tight-binding simulations show that stanene on an α-alumina surface can have a sizeable topologically non-trivial band-gap of about 0.25eV at the Γ point. The stanene is atomically bonded to, but electronically decoupled from, the substrate; thus providing a high structural stability and largely isolated quantum spin Hall states.[762] The underlying physical mechanism is quite general, and lead to the opening of a new vista on the exploration of quantum spin Hall insulators for room temperature application.

Chapter 4

Antimonene

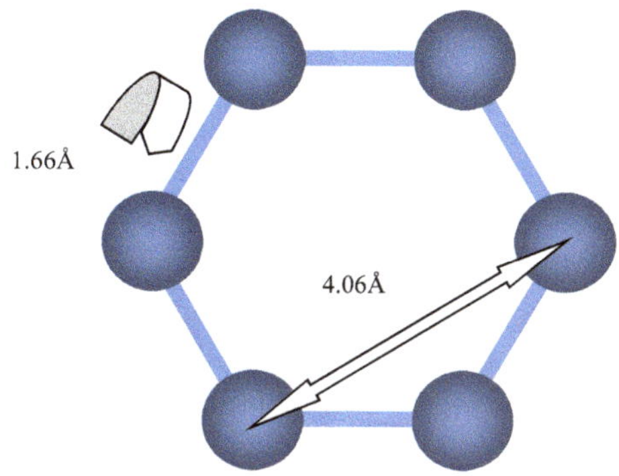

The pseudo-layer character of three-dimensional bulk crystals of antimony already suggests that its two-dimensional single-layer crystalline phase should have a buckled honeycomb structure like that of silicene. The antimony atoms also have an asymmetrical washboard-like structure, as does black phosphorene. *Ab initio* phonon and finite-temperature molecular dynamics calculations show that the single-layer phases are robust and can remain stable at high temperatures. Single-layer antimonene phases form bilayer and trilayer structures having wide interlayer spacings. This spacing is reduced in multilayers and the structure eventually changes into three-dimensional pseudo-layered bulk crystals. The phases are non-magnetic semiconductors, with band-gaps ranging from 0.3 to 1.5eV. Zig-zag and armchair nanoribbons of the antimonene phases have fundamental band-gaps arising from reconstructed edge states, and exhibit a range of magnetic and electronic properties, depending upon their width and edge geometry. Their band-gaps are tunable, using the width of the nanoribbon. The washboard form exhibits strongly directional mechanical properties, suggesting that strain may have a marked influence on electronic properties.[763] First-principles calculations indeed demonstrate that antimonene can be turned into a two-dimensional topological insulator by reducing the buckling height of the lattice, using tensile straining. A strain-driven band inversion in the vicinity of the Fermi level is responsible for the quantum phase transition. The buckled configuration permits the material to endure tensile strains of up to 18%, and the resultant bulk band-gap can be as large as 270meV. This tunable bulk band-gap makes antimonene a promising candidate material for achieving quantum spin Hall effects at high temperatures.[764]

First-principles calculations of the properties of antimonene show that the free-standing α and β allotropes are stable and semiconducting. The α-antimony form has a puckered structure, with two atomic sub-layers, while the β-antimony form has a buckled hexagonal lattice. Raman spectra and scanning tunnelling microscopic images have distinct features, which facilitates characterization of the allotropes. The β-antimony form has almost isotropic mechanical properties, while α-antimony form has strongly anisotropic characteristics. An indirect-direct band-gap transition is expected to occur when a moderate tensile strain is applied to monolayers.[765]

Investigation of the geometrical structure and properties of antimonene doped with 3d transition metals, using spin-polarized first-principles calculations, shows that strong orbital hybridization occurs between the transition metals and antimony atoms, where covalent bonds form in the antimonene. A spin-polarized semiconducting state appears in chromium-doped antimonene, while half-metallic states appear upon doping with titanium, vanadium or manganese.[766] These results indicate that, when combined with doping states, the bands of antimonene systems exhibit a variety of features: specific dopants can lead to half-metallic characteristics, with a high spin-polarization that has potential applications in spintronics.

4.1 Preparation

The synthesis of high-quality monolayer or few-layer antimonenes has however greatly hindered their development. High-quality few-layer monocrystalline antimonene polygons can be synthesized on various substrates, including flexible ones, by means of van der Waals epitaxial growth. Raman spectroscopy and transmission electron microscopy reveal that the polygons have a buckled rhombohedral atomic structure which is consistent with the theoretically predicted most stable β-phase allotrope. Very high stability of antimonenes has been observed following aging in air for 30 days. First-principles and molecular dynamics simulations confirm that, compared with phosphorene, antimonene is less likely to be oxidized and possesses a higher thermodynamic stability in an oxygen atmosphere at room temperature.[767] Antimonene polygons exhibit an electrical conductivity of up to 10^4S/m, and a good optical transparency in the visible light range.

A process which is assisted by sonification, but which does not require the addition of a surfactant, is a rapid and simple method for producing highly stable isopropanol/water (4:1) suspensions of few-layer antimonene via the liquid-phase exfoliation of antimony crystals. This method generates dispersions, of few-layer antimonene, which are suitable for on-surface isolation.[768] Atomic force microscopy, scanning transmission electron microscopy and electron energy loss spectroscopy have confirmed the formation of high-quality few-layer nanosheets having large lateral dimensions. These nanolayers are extremely stable under ambient conditions.

4.2 Composites

A novel type of seamless lateral heterostructure having an excellent stability can be formed within pristine arsenene and antimonene. Such heterostructures could exhibit direct and reduced energy-gaps, without any modulation, and type-II alignment plus a high carrier mobility would also be possible. Tensile strain can result in efficient band-gap engineering. Such predictions introduce a new vitality into the field of lateral heterostructures, and highlight the potential of lateral heterostructures as materials for future devices[769]. By combining antimonene with semimetal graphene, semiconducting arsenene and insulating hexagonal boron nitride, three new 2-dimensional van der Waals heterostructures can be created. The results indicate that BN/antimonene and arsenene/antimonene heterostructures are semiconductors with band-gaps of 0.14 and 0.31eV, respectively, while a graphene/antimonene heterostructure is metallic. The band-gaps of BN/antimonene and arsenene/antimonene can be markedly modulated by strong interlayer coupling. Biaxial straining can further tune the electronic properties of the arsenene/antimonene and BN/antimonene van der Waals heterostructures, resulting in an indirect-to-direct gap transition.[770] Biaxial straining produces band-gaps which are continuously variable from 1 to 0eV; corresponding to a spectra range of near-infrared and mid-infrared wavelengths. Density functional theory calculations of the electronic structures of antimonene/graphene van der Waals heterostructures demonstrate that weak van der Waals interactions predominate between antimonene and graphene, with their intrinsic electronic properties preserved. An n-type Schottky barrier to a p-type Schottky barrier transition occurs when the interlayer distance is decreased from 5.0 to 3.1Å; indicating that the Schottky barrier can be effectively tuned by varying that distance in van der Waals heterostructures.[771] Interfacial charge transfer and Fermi level shifts govern the Schottky barrier transition in the antimonene/graphene heterostructure.

Novel atomically thin nanocomposite systems, based upon germanene and antimonene, have been investigated by means of density functional theory. The germanene and antimonene monolayers are bound to each by orbital hybridization with an enhanced binding strength. The most important feature is that band-gap opening is feasible. One form of the composite has a direct band-gap of up to 391meV, while another 3 forms have indirect band-gaps which are tunable from 37 to 171meV. Changing the direction and strength of an external electric field can tune the energy-gap of the germanene/antimonene bilayer over a wide range; even encompassing a semiconductor-metal transition. The work function of heterobilayers, having the form which possesses a direct band-gap, can be varied from -3.21 to 12.33eV by changing the electric-field intensity.[772] The germanene/antimonene bilayer also exhibits a more pronounced optical conductivity.

Table 9
Binding energy between adatoms and a 5x5 antimonene super-cell

Adatom	**E(eV)**
Sb	1.19
As	1.22
H	1.25
P	1.30
Li	1.41
In	1.59
Al	1.66
N	1.87
Si	1.93
Cl	2.26
Ti	3.38
O	3.46
B	4.04
C	7.71

4.3 Substrate Interaction

The geometry, electronic structure and magnetic properties of the interfaces between cobalt (00•1) and monolayer or bilayer antimonene are such that the cobalt d and antimony sp orbitals exhibit marked hybridization near to E_F, while strong chemical bonds form between cobalt and antimonene at the interface, and the antimonene has a spin polarization of 12% at that interface.[773] Following contact, a high Schottky barrier forms whose height can be tuned by using various cobalt/antimonene stacking patterns in that the height decreases with increasing number of antimonene layers.

4.4 Surface Interaction

First-principles spin-polarized density functional calculations have been made of the chemisorption of adatoms, and of the physisorption of molecules, on two antimonene phases. Because adspecies/adspecies interaction is minimized by using large super-cells, these results mimic the effects of isolated single adatoms or molecules. Molecules such as H_2, O_2 and H_2O do not form strong chemical bonds, or dissociate. They are instead physisorbed, with a weak binding energy, without affecting the properties of antimonene. Adatoms such as hydrogen, lithium, boron, carbon, nitrogen, oxygen, aluminium, indium, silicon, phosphorus, chlorine, titanium, arsenic and antimony are chemisorbed with a significant binding energy, such that the atomic and electronic structures are locally modified. Boron and carbon adatoms are incorporated into the buckled antimonene crystal, leading to local reconstruction. Nitrogen gives rise to Stone-Wales type defects.[774] The lowest conduction and highest valence bands of antimonene in asymmetrical washboard structures have a very high curvature. When combined with adatom states, these bands offer a variety of features. Specific adatoms lead to spin polarization, acquire magnetic moments and can impart a half-metallic character to antimonene.

4.5 Defects

First-principles calculations, using five different models, have been made of monovacancy and divacancy antimonene. The various vacancies produce differing geometrical structures, with or without inversion symmetry, and influence the electronic structure. Monovacancy antimonene has a metallic character, while four divacancy models preserve a semiconducting nature. Inversion asymmetry and spin-orbit coupling lead to spin splitting in the monovacancy model and in two divacancy models (figures 12 and 13).[775] Zeeman-type spin-splitting appears, with out-of-plane spin-polarization occurring along M-K-Γ. Rashba and Dresselhaus effects induce spin-splitting at the Γ and M points of the monovacancy model.

First-principles calculations have also been used to study the structural and electronic properties of antimonene which contains Stone–Wales defects. The calculations show that the presence of a Stone-Wales defect changes the geometrical symmetry, and that the band-gap decreases with decreasing Stone-Wales defect concentration.[776] The formation energy and cohesive energy of the Stone-Wales defect explain its existence and good stability, respectively. The charge density is different near to the Stone-Wales defect.

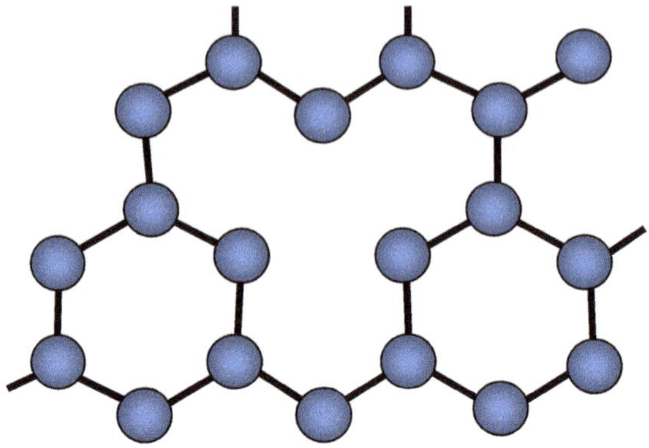

Figure 12. Model for the vacancy in antimonene

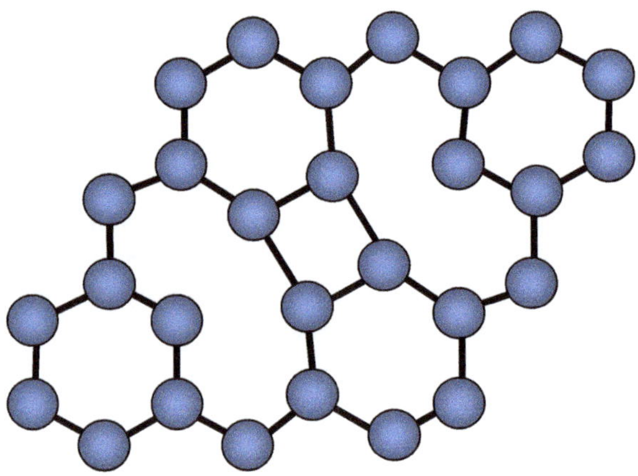

Figure 13. One of 5 models for the divacancy in antimonene

4.6 Thermal Conductivity

Use of the phonon Boltzmann transport method reveals that the semiconducting α-antimony and β-antimony phases have rectangular and hexagonal unit cells, respectively. The thermal conductivity increases with size, and decreases with edge roughness. A higher thermal conductivity is found for α-antimony, and distinct characteristics of the in-plane acoustic branches are associated with the anisotropic nature of the two-dimensional lattice.[777]

4.7 Optical Properties

The dielectric matrices, absorption coefficients, refractive indices, electron energy-loss spectra and optical reflectivities of the α and β allotropes of antimonene monolayers, between 0 and 21eV, have been calculated using the random phase approximation and density functional theory. The simulations predict that absorption begins in the infra-red, but peaks in the ultra-violet. The refractive index is 2.3 for α-Sb and 1.5 for β-Sb, in the zero-energy limit, and scales up to 3.6 in the ultra-violet region.[778] The reflectivity rises to 86% at ultra-violet energies, where antimonene behaves like a metal with regard to incident electromagnetic radiation. Antimonene is therefore suitable as a material for micro-electronic and opto-electronic nanodevices and solar-cell applications.

4.8 Semiconduction

The n- and p-type doping of antimonene will be essential for its use in opto-electronic devices. Tetrathiafulvalene and tetracyanoquinodimethane have been used as electron- and hole-dopants in order to produce n- and p-type semiconductors, respectively. Tetrathiafulvalene-modified antimonene is an n-type semiconductor with a deep donor state of 0.73eV. Tetracyanoquinodimethane functionalization produces a p-type semiconductor, with a shallow acceptor state of 0.27eV. Co-adsorption of tetrathiafulvalene and tetracyanoquinodimethane can significantly decrease the band-gaps to 0.15 and 0.12eV in the one- and two-sided configurations, respectively.[779] Angle-resolved photo-emission spectroscopic measurements show that single-bilayer Sb(111) ultra-thin film can be grown onto Bi_2Te_3 and Sb_2Te_3 substrates, and exhibits clear band dispersion. It is theoretically predicated to be a robust trivial semiconductor, but can be tuned to become a two-dimensional topological insulator by reducing the buckling height. The electronic structure has a similar nature on the two surfaces but, due to interfacial strain, the bands of antimonene on Bi_2Te_3 are flatter than those on Sb_2Te_3. This is attributed to the lattice constant of Bi_2Te_3 being larger than that of Sb_2Te_3. A charge-transfer effect is also seen in the core level shift, which influences the band dispersion.[780]

4.9 Applications

Although antimony is a semimetal when in bulk form, it is transformed into an indirect semiconductor with a band-gap of 2.28eV by thinning to one atomic layer. Under a small biaxial strain, it is transformed from being an indirect, to being a direct, band-gap semiconductor. Such a marked change in the electronic structure could permit the preparation of transistors having a high on/off ratio, of opto-electronic devices working with blue or ultra-violet light and of mechanical sensors.[781]

Detailed multiscale simulations which combine the accuracy of first-principles approaches to describing electronic properties with the efficiency of tight-binding Hamiltonians based upon maximally localized Wannier functions, of field-effect transistors based upon antimonene monolayers as channels provide initial estimates of upper limits on the electron and hole mobilities in the Takagi approximation.[782] Such simulations, which also include spin-orbit and multi-valley effects, demonstrate that devices of sub-10nm scale offer a level of performance that is compatible with industry requirements.

Chapter 5

Indiene

Very little attention has been paid to this material so far, and it is included here merely for the sake of completeness. For now, first-principles calculations of indium monolayers having a planar, puckered or buckled form have shown that the planar and buckled allotropes are stable and exhibit metallic and semiconducting behaviors, respectively. Their stability and electronic properties cannot be easily correlated with those of similar elemental monolayers. A van Hove singularity is observed in the electronic density of states and this could lead to an increase in the electronic conductivity. The planar allotrope remains metallic under both compressive and tensile strains, while the buckled allotrope changes from an indirect semiconductor to a metal.[783]

Chapter 6

Arsenene

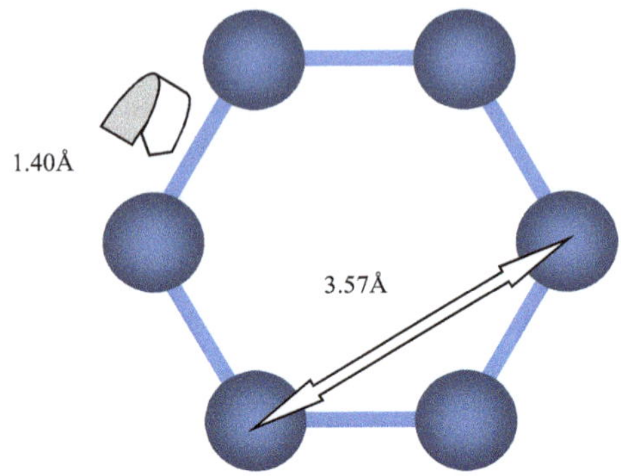

First-principles calculations indicate that two types of honeycomb structure, buckled and puckered, are stable; both possessing indirect gaps. The band-gaps of the puckered and buckled forms can be tuned by mechanical straining. Gap-closing occurs at 6% strain for the puckered form, where the bond angles between the nearest neighbors become almost equal. An indirect-to-direct gap transition occurs during straining, and 1% strain is sufficient to transform the puckered arsenene into a direct-gap semiconductor.[784] The emergence of a Dirac-like cone occurs with in-plane straining. Strain modulates not only the band-gap, but also the effective mass. The effective mass of electrons has a highly anisotropic dispersion in intrinsic puckered arsenene.[785] A bulk form of arsenic, termed gray arsenic, exists and can be used as a precursor for the buckled arsenene. Trions and bi-excitons in anisotropic two-dimensional materials can be investigated using effective mass theory. Explicit results have been obtained for arsenene, and for materials that share its features, such as a direct quasi-particle gap and anisotropic conduction and valence bands. Trions are predicted to have markedly high binding energies and an elongated electron-hole structure, with a preference for alignment along the armchair direction where the effective masses are lower. Bi-exciton binding energies are also high, being twice those of typical monolayer transition-metal dichalcogenides.[786]

First-principles density functional theory investigations of monolayer and bilayer arsenenes under in-plane biaxial straining show that, at sufficiently high tensile strains, the monolayer changes from the buckled honeycomb form to the planar honeycomb form. The interlayer distance of bilayers nevertheless remains at about 1.371Å. Both the monolayer and bilayer forms have indirect band-gaps and the electronic properties can be tuned, as already noted, by in-plane biaxial straining. Under compressive strains, the band-gap of both monolayer and bilayer arsenene initially increases but then rapidly decreases. The monolayer form exhibits an indirect-to-direct band gap transition when the compressive strain reaches -10%. Under tensile straining, the band-gap of monolayer arsenene decreases monotonically with strain, while the band-gap of the bilayer form decreases rapidly to zero under small tensile strains.[787]

6.1 Mechanical Properties

Theoretical calculations of phonon spectra and elastic constants have verified the dynamic and mechanical stability of arsenene. On the basis of the stable structure, the bulk modulus, shear modulus, Young's modulus and Poisson ratio have been calculated (table 9).[788] The shear anisotropy, Vickers hardness, Debye temperature, melting point and thermal conductivity have also been predicted, together with the electron mobility and relaxation time.

6.2 Semiconduction

First-principles density functional theory calculations demonstrate that monolayer arsenene is transformed from an indirect to a direct band-gap semiconductor by imposing an uniaxial tensile strain along the armchair and zig-zag directions. Compared with the biaxial tensile strain of 0.04, this transformation occurs at strains of 0.06 and 0.10 along the armchair and zig-zag directions, respectively. Spin-orbital coupling is available, to tune the band-gaps. The spin-orbit interaction opens up a 0.2eV band-gap at the Γ-point in unstrained monolayer arsenene. [789]A clear red-shift in the absorption properties is observed with increasing strain.

The intrinsic band-gap is indirect, and this is not favorable for realistic device applications. Arsenene can be transformed into a direct band-gap semiconductor by chemical doping with phosphorus, antimony or bismuth, and the effective mass of carriers can be modulated over a wide range.[790] Density functional theory calculations of alloyed and doped arsenene sheets show that AsN, AsP and SbAs alloys possess indirect band-gaps while BiAs has a direct band-gap. Although AsP, SbAs and BiAs sheets retain the semiconducting character of pure arsenene, they exhibit indirect-direct and semiconducting-metallic transitions upon applying biaxial strain. Boron- and nitrogen-doped arsenene is rendered p-type, while carbon- and oxygen-doped arsenene is metallic in character.[791] Carbon-doped arsenene is spin-polarization asymmetrical and can be turned into a bipolar spin-gapless semiconductor by applying an external electric field. The dopant concentration can affect the magnetism of the carbon-doped material.

Few-layer orthorhombic arsenene is an ideal semiconductor and, due the layer stacking, multilayer samples always behave as intrinsic direct band-gap semiconductors, with gaps of about 1eV. These band-gaps can be tuned in the case of nanoribbons. On the basis of the acoustic phonon-limited approach, the carrier mobilities are predicted to approach several thousand cm^2/Vs and also exhibit a high directional anisotropy.[792] First-principles calculations show that the band-gap of few-layer arsenic undergoes an indirect to direct transition as the number of arsenic layers, n, increases from 1 to 2. As n increases from 2 to infinity, the stacking of puckered honeycomb layers forms orthorhombic arsenic (ε-arsenic, arsenolamprite), which has a similar structure to that of black phosphorus and also has a direct band-gap. The indirect to direct transition arises from a distinct quantum-confinement influence upon the indirect and direct band-edge states with differing wave-function distributions. In-plane strains can induce a very different shift in the indirect and direct band edges, and thus also induce an indirect to direct band-gap transition. The band-gap dependence upon strain is non-monotonic, with both positive and negative deformation potentials. The gap opens between arsenic p-p bands, but the spin-orbit interaction decreases the gap by only 0.02eV. This is much smaller than the decrease in GaAs, with an s-p band gap. The calculated band-gaps of arsenene and ε-arsenic, using the hybrid functional, are estimated to be 1.4 and 0.4eV, respectively, and are comparable to those of phosphorene and black phosphorus.[793] First-principles studies of arsenene doped with non-magnetic elements show that dopants from groups III, V and VII, with odd numbers of valence electrons, maintain the semiconducting nature of the pristine material, while dopants from groups IV and VI, with even numbers of valence electrons, cause the metallic nature to change. The carbon- and oxygen-doped systems are spin-polarized, and can be changed into half-metals by applying an external electric field.[794]

First-principles calculations of arsenic nanoribbons, cut from arsenene nanosheets, show that the band features of arsenic nanoribbons depend upon the edge shape. All armchair arsenic nanoribbons retain the indirect band-gap feature, while the zig-zag ones change to direct semiconductors. Quantum confinement in nanoribbons enhances the gap size, and both the armchair and zig-zag ones have a gap which scales inversely as the ribbon width. The large deformation potential constants of arsenic nanoribbons lead to small carrier mobilities; of the order of 10^1 to $10^2 cm^2/Vs$.[795]

Nanostructuring arsenene into nanoribbons makes the band-gap a direct one. Direct band-gaps are found for narrow armchair, but wide zig-zag, nanoribbons; dominated by competition between in-plane and out-of-plane bonding. Straining the nanoribbons induces a direct band-gap and simultaneously modulates the transport properties. The gap energy is enhanced by tensile straining of armchair structures. In the case of zig-zag structures, tensile straining makes the effective mass of the holes much greater while compressive straining causes it to be much lower than that of electrons.[796] Density functional theory predicts that straining of honeycomb arsenene can result in an unique two-dimensional topological insulator with a bulk gap of up to 696meV. Band inversion occurs around the Γ-point, at tensile strains greater than 11.7%, indicating the occurrence of a quantum spin Hall effect at certain strains. The non-trivial topological phase nature is also confirmed by a topological invariant of unity, and edge states with a single Dirac-type crossing at the Γ-point. This raises the possibility of producing two-dimensional topological insulators with large bulk gaps from commonly used materials.[797]

Many of the reported two-dimensional semiconductors have band-gaps that are smaller than 2.0eV, and this greatly limits their application; especially as opto-electronic devices requiring a photoresponse in the blue and ultraviolet ranges. Monolayer arsenene is a semi-metal in the bulk but transforms into an indirect semiconductor with a band-gap of 2.49eV when thinned down to one atomic layer. Under a small biaxial strain, it transforms from an indirect into a direct band-gap semiconductor.[798]

First-principles studies of arsenene which is doped with boron, carbon, nitrogen, oxygen, gallium, germanium or selenium show that boron-, nitrogen- and gallium-substituted samples have spin-unpolarized semiconducting characteristics like those of pristine arsenene. Indirect-direct band-gap transitions are induced in boron- and nitrogen-substituted samples. On the other hand, oxygen-, selenium- and monovacancy-substituted samples are metallic, and carbon- or germanium-substituted samples have spin-polarized semiconducting characteristics with band-gaps of 1.1 and 1.3eV for the spin-up channels and 1.0 and 0.7eV for the spin-down channels. In arsenic-adsorbed samples, the Fermi level crosses the spin-up states, leading to metallic behavior, while the spin-down channel retains its semiconducting nature. In carbon-, germanium- and arsenic-adsorbed samples, strong hybridization between the dopant atoms and the arsenic atoms lead to energy-splitting near to the Fermi level and thus induce magnetic moments. By means of selective doping, therefore, the material can be changed from a spin non-polarized semiconductor into a spin-polarized semiconductor, into a half-metal or even into a metal.[799]

With regard to experimental methods, the cross-sectional layer structure at the nanoscale can be monitored by means of transmission electron microscopy, with an electron-beam evaporated SiO_2 cap on the surface to protect the thin film during sample preparation. The band-gap of multilayer arsenene can then be estimated by measuring the photoluminescence characteristics.[800]

6.3 Composites

Hetero-sheets composed of group-V (arsenene) and group-IV (graphene/silicene) layers exhibit typical van der Waals features, with small binding energies and soft interlayer elastic constants. The Dirac characteristics of the group-IV layer, and the semiconducting features of the group-V layer, are well preserved in the hetero-sheets, producing a Schottky contact at the metal/semiconductor interface. The Schottky barriers are always p-type in silicon-based hetero-sheets whereas, in carbon-based ones, the interfacial feature is sensitive to the interlayer distance. Tensile strain is expected to induce a p-type to n-type Schottky-barrier transition in As/C hetero-sheets.[801]

First-principles studies of arsenene/graphene van der Waals heterostructures show that a linear Dirac-like dispersion relationship around the Fermi level can be quite well preserved in those heterostructures. A p-type Schottky barrier (0.18eV), to n-type Schottky barrier (0.31eV), transition occurs when the interlayer distance is increased from 2.8 to 4.5Å. This suggests that the Schottky barrier can be tuned by modifying the interlayer distance in van der Waals heterostructures.[802] Density functional theory calculations of arsenene/graphene heterostructures, with an external electric field applied perpendicularly to the layers, demonstrate moreover that weak van der Waals interactions predominate between the arsenene and graphene, with their intrinsic electronic properties intact. An n-type Schottky contact is formed at the arsenene/graphene interface, with a barrier height of 0.54eV.[803] The vertical electric field can not only control the Schottky barrier height but also the Schottky contacts (n-type and p-type) and ohmic contacts (n-type) at the interface. Tunable p-type doping of graphene is produced by a negative electric field because electrons can transfer from the Dirac point of the graphene to the conduction band of the arsenene.

Ab initio simulations of monolayer van der Waals arsenene/$FeCl_2$ heterostructures reveal that spin splitting appears at the conduction-band minimum of arsenene, due to interfacial coupling between As-p and Cl-p, Fe-d states in the spin-down channel. The maximum splitting energy is 123meV among all of the stacking configurations. The splitting energy can be tuned by imposing a perpendicular electric field. In a field of 5V/nm, the spin-splitting direction can be inverted, and the splitting energy is -66meV. A positive electric field makes the system change from an ohmic contact to a Schottky one; achieving a continuous modulation of the barrier height. The electrical control of spin-splitting inversion can be converted into an electrical potential difference, according to the anomalous Hall effect.[804]

Table 10
Adsorption energies at 4x4 super-cell of arsenene

Adatom	Site	E(eV)
Sb	T	1.028
H	TH	1.185
As	BV	1.492
Ga	V	1.570
Al	V	1.598
Ge	T	1.622
N	BH	1.650
Li	V	1.720
Cl	T	1.799
P	BV	1.812
Si	VH	2.233
Se	T	2.428
Ti	B	3.405
O	B	4.130
C	V	4.398
B	V	4.511

B: bridge, BH: bridge hollow, BV: bridge valley, T: upper-layer As atom
TH: upper-layer hollow, V: lower-layer As atom, VH: valley hollow

6.4 Surface Interaction

First-principles calculations of the electronic structures of arsenene nanosheets with adsorbed light atoms (boron, carbon, nitrogen, oxygen, fluorine) show that most adatoms prefer to occupy the bridge site on the nanosheets, apart from carbon, which prefers the valley site. Defect states can be found in the middle gap of fluorine-adsorbed arsenene nanosheets, and nitrogen adatoms can induce n-type doping. The oxygen adatoms have a negligible effect upon the electronic structure. The boron, carbon, nitrogen and fluorine adatoms can all induce magnetism in arsenene nanosheets.[805] First-principles calculations of the electronic and topological properties of halogenated (fluorine, bromine, iodine) arsenene similarly show that the halogenated sheets exhibit Dirac type characteristics in the absence of spin-orbital coupling. An energy-gap is introduced by spin-orbital coupling, with values ranging from 0.194eV for fluorine-arsenene, to 0.255eV for iodine-arsenene.[806] These 2-dimensional systems are found to be quantum spin Hall insulators.

First-principles calculations of the adsorption of CO, CO_2, N_2, NH_3, NO and NO_2 molecules on pristine arsenene monolayers show that these gas molecules are held by an interaction that is intermediate between physisorption and chemisorption. The adsorption of NO and NO_2 can produce a noticeable modification in the density-of-states near to the Fermi level.[807] The adsorption of NO and NO_2 can alone lead to a magnetic moment of $1\mu B$.

The adsorption of various adatoms (hydrogen, lithium, boron, carbon, nitrogen, oxygen, aluminium, silicon, phosphorus, chlorine, titanium, gallium, germanium, arsenic, selenium, antimony) and the physisorption of H_2, O_2 or H_2O have been studied using large super-cells (table 10). The coupling between adspecies is thus minimized and the results mimic the effects of isolated adatoms or physisorbed molecules. The adatoms form strong chemisorption bonds and therefore locally modify the atomic structure and physical properties. Some of the adatoms cause significant local reconstruction of the atomic structure. The electronic states of some adatoms become spin-polarized and acquire net magnetic moments. They can even exhibit a half-metallic nature at high coverage. Most of the adsorbed atoms produce localized states in the fundamental band-gap.[808] The interactions between the H_2, O_2 and H_2O molecules and single-layer arsenene are rather weak and do not provoke any important changes in the physical properties of the molecules or the arsenene. Some of the molecules can however be dissociated at the edges of arsenene structures, with their constituents being adsorbed at edge atoms so as to cause local reconstruction.

First-principles calculations of the properties of arsenene monolayers under hydrogenation predict the existence of a new σ-type Dirac cone which is related to the $p_{x,y}$ orbitals of arsenic atoms and depends upon the in-plane tensile strain. Spin-orbit coupling opens up a quantum spin Hall gap of 193meV at the Dirac cone. A single pair of topologically protected helical edge states is established for the edges, and its quantum spin Hall nature is confirmed by a topological invariant, Z_2, of unity. A two-dimensional quantum-well, comprising hydrogenated arsenene with hexagonal BN sheets on each side, harbors a non-trivial quantum spin Hall state with the Dirac cone lying within the band-gap of the cladding BN substrate.[809] Calculations also show that the band-gap of arsenene decreases with increasing strain, and changes from indirect to direct. An s-p band inversion then takes place when the tensile strain is larger than 11.14%, leading to a non-trivially topological state. A single pair of topologically protected helical edge states is, as above, established for the edge of arsenene, and their quantum spin Hall states are confirmed by a non-trivial topological invariant, Z_2, of unity. High-dielectric BN is proposed to be an ideal substrate for the experimental synthesis of arsenene, while maintaining its non-trivial topology.[810]

On the basis of first-principles calculations, a new category of 2-dimensional topological insulator is proposed to exist in the form of chemically functionalized (-CH_3, -OH) arsenene films. Surface-decorated arsenene ($AsCH_3$, AsOH) films are intrinsic 2-dimensional topological insulators possessing a sizeable bulk gap. The bulk energy-gaps are 0.184eV and 0.304eV in $AsCH_3$ and AsOH films, respectively.[811] These large bulk gaps make them suitable for exploiting quantum spin Hall effects at experimentally accessible temperatures. Topologically helical edge states in these systems are desirable for obtaining dissipation-free transport. The present topological properties resist mechanical deformation caused by biaxial strain.

First-principles calculations of the adsorption of 15 metal adatoms onto arsenene show that all of these metal adatoms have binding energies which are larger than the cohesive energy of the bulk metal. This implies that stable adsorbates can form. Due to the localized states which originate from adatoms, the adsorption systems display a wide range of electronic properties: metallic, half-metallic, semiconducting and spin-semiconducting. Cobalt-doped arsenene possesses half-metallic properties.[812] The adsorption of copper, silver or gold adatoms turns semiconducting arsenene into a narrow-gap spin-semiconductor. Density functional theory investigation of arsenene with adsorbed transition-metal atoms (cobalt, copper, manganese, iron, nickel) shows that magnetism appears in the case of cobalt, manganese and iron. Among the magnetic cases, transition-metal atoms prefer the same adsorption site. Further study of the interaction in two-transition-metal adsorbed systems shows that different magnetic states occur.[813] Both non-magnetic and ferromagnetic states exist in a doubly cobalt-adsorbed system, and p-d hybridization results in a ferromagnetic state. In double-manganese and double-iron adsorbed systems, an interaction is found which can be explained in terms of a super-exchange mechanism.

6.5 Magnetism

First-principles calculations of the magnetic properties and electronic structures of arsenene, doped with 3 concentrations of a transition-metal, show that not all 3d transition-metal atoms can induce a magnetic state in arsenene. The dopant concentration hardly affects the magnetic moment, except in the case of iron or nickel doping. It is thought that the magnetism is predominantly controlled by the doped transition-metal atoms alone, and that no significant contributions arise even from the nearest arsenic atoms.[814] Density functional theory investigation of the properties of arsenene doped with transition-metal atoms (cobalt, copper, manganese, iron, nickel) shows that magnetism is observed in the case of copper, manganese, iron and nickel. Among the 4 magnetic systems, the nickel-doped one is the most easily formed. Study of the ferromagnetic interaction in doubly nickel-doped arsenene shows that p-d hybridization results in a ferromagnetic state.[815] The ferromagnetic interaction is however depressed with increasing Ni-Ni distance; a feature which can be well-explained in terms of the Zener-Ruderman-Kittel-Kasuya-Yosida theory. Exotic phenomena appear in doubly manganese-doped material. Both non-magnetic and ferromagnetic states are observed. First-principles investigation of substitutionally-doped 3d transition-metal atoms in an arsenene monolayer shows that, based upon the binding energy, transition-metal substituted arsenene is robust. Magnetic states are found for titanium, vanadium, chromium, manganese and iron doping.[816] A half-metallic state results from titanium and manganese doping, while a spin-polarized semiconducting state occurs for vanadium, chromium or Iron doping. First-principles density-functional theory study of 3d transition-metal doped nanosheets shows that scandium- and cobalt-doped arsenene exhibits non-magnetic semiconducting properties while arsensic-replacing titanium, chromium or copper atoms can produce a dilute magnetic semiconductor phase. Half-metallic properties are induced in vanadium-, manganese-, iron- or Ni-doped nanosheets.[817] Titanium-, vanadium-, manganese- or iron-doped nanosheets exhibit ferromagnetic coupling, while substitutional chromium-doping results in antiferromagnetic coupling in the most stable configuration.

6.6 Thermal Conductivity

By solving the Boltzmann transport equation for phonons, a highly anisotropic room-temperature thermal conductivity is predicted: with values of 30.4 and 7.8W/mK along the zig-zag and armchair directions, respectively. Phonons with mean free paths of between 20nm and 1μm make the main contribution to the large thermal conductivity in the zig-zag direction. The mean free paths of phonons contributing to heat transport in the armchair direction range from 20 to 100nm.[818]

Chapter 7
Phosphorene

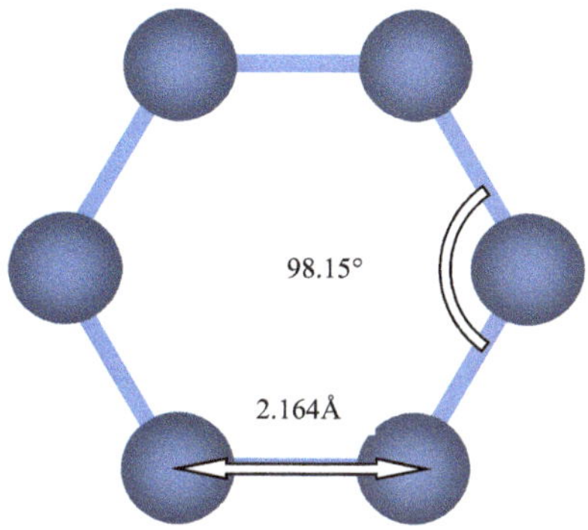

So-called black phosphorus has been known since 1960 to offer good carrier mobility as well as a fundamental band-gap. Phosphorene is the mono- or few-layer form of black phosphorus: a two-dimensional layered material offering great prospects in multidisciplinary fields such as electronics and opto-electronics.[819] There is a need for further investigations in order to establish controlled synthesis techniques.[820]

A single atomic layer of black phosphorus was experimentally prepared in 2014. It has anisotropic properties which arise from a puckered structure, quite unlike that of graphene or transition-metal dichalcogenides.[821,822,823] Group theory analysis of such 2-dimensional elemental systems with regard to dependence upon the number of layers and possible stacking arrangements shows that, departing from the most symmetrical D_{6h1} space-group, the structures have a group-subgroup relationship. Analysis of the irreducible representations of their lattice vibrations makes it possible to distinguish between the various allotropes. The analysis can be used to study the effect of strain, to understand structural phase transitions and to characterize the number of layers, crystallographic orientation and non-linear behaviors.[824]

7.1 Preparation

Single crystals of orthorhombic black phosphorus can be grown, by transport reaction, from red phosphorus; with Sn/SnI_4 as a mineralization additive. High-quality pure crystals which are several millimeters in size can be obtained; thus making the large-area preparation of single or multilayer phosphorene possible. *In situ* neutron diffraction studies show that black phosphorus is formed directly via the gas phase, without the appearance of any other intermediate phase. Crystal growth is initiated upon cooling the starting materials from 500C.[825] Monolayer phosphorene can be prepared via mechanical cleavage and subsequent Ar^+ plasma-thinning. The thickness of the phosphorene can be determined by means of optical contrast spectra and atomic force microscopy, while Raman spectroscopy is used to characterize the pristine and plasma-treated samples. The Raman frequency of the A_{2g} mode stiffens, and the intensity ratio of the A_{2g} to A_{1g} modes exhibits a discrete monotonic increase, with decreasing phosphorene thicknesses down to a monolayer. All of these measurements can be used to identify the thickness of the two-dimensional semiconductor.[826]

The stability of phosphorene is degraded by water, but the presence of a small amount of water permits the synthesis of high-quality material by the liquid exfoliation of black phosphorus, using dimethyl sulfoxide as a solvent. A phosphorus/water molar ratio of between 1.5 and 14 maximizes the quality of the phosphorene flakes and their stability.[827] Liquid exfoliation of black phosphorus has also been performed using N-methyl-2-pyrrolidone.[828] Liquid-phase exfoliation in dimethyl sulfoxide, dimethylformamide, isopropyl alcohol, N-methyl-2-pyrrolidone and N-cyclohexyl-2-pyrrolidone has been modelled using an atomistic force field, based upon *ab initio* calculations and lattice dynamics, which accurately reproduces the experimentally observed mechanical properties. Planar molecules such as N-methyl-2-pyrrolidone preferentially orient themselves parallel to the interface. The exfoliation efficacy of a solvent is enhanced when molecular planarity sharpens a molecular wedging effect, or when strong phosphorene-solvent adhesion stabilizes newly exposed phosphorene surfaces. Examination of the colloidal stability of exfoliated flakes shows that dispersion is favored when the cohesive energy between the molecules in the solvent monolayer confined between phosphorene sheets is high. This is true of dimethyl sulfoxide. It is hindered when adhesion between these molecules and the phosphorene is strong. Molecular planarity in solvents such as dimethylformamide enhances the cohesive energy.[829]

The renaissance of black phosphorus indeed depends largely upon the mechanical exfoliation method, which is able to produce few-layer forms from the bulk. The exfoliation mechanism can be simulated in terms of a sliding process of bilayer phosphorene. Interlayer Coulomb interactions dictate the optimal sliding pathway, revealing minimum energy barriers as low as about 60meV and a surface energy of about $59mJ/m^2$. Black phosphorus can thus be exfoliated by using the sliding approach. Considerable band-gap modulation also occurs along these sliding pathways.[830] Suitable liquid exfoliation conditions permits exfoliation to be carried out at the 10g scale. A rapid method for quantifying the thickness of two-dimensional phosphorus shows that the monolayer and few-layer flakes which are produced are crystalline and unoxidized. Air-exposure leads to rapid oxidation and to the production of acid. When layered black phosphorus crystals are prepared by chemical vapor transport, the use of scanning electron microscopy, energy-dispersive X-ray spectroscopy, X-ray photo-electron spectroscopy and X-ray diffraction reveals that they exhibit an inherent electrochemistry and are oxidizable (about 560mV vs Ag/AgCl).[831]

Density functional theory calculations of the Raman spectra of few-layer phosphorene show that, due to the interlayer van der Waals interactions, the low-frequency rigid layer A_g breathing mode and B_{1g} shear mode can shift by as much as 45.1 and 38.5/cm, respectively, as the layer numbers increase from 2L to 5L. A typical characteristic for the experimentally observable mode (~460/cm in the bulk) is identified and shows that this mode changes from coupled in-plane and out-of-plane vibrations, in a single layer, to pure in-plane vibrations in a few-layer sample. The corresponding frequencies vary by up to more than 10/cm. This Raman-frequency variation could be used to characterize experimentally the thickness of the 2-dimensional layered material.[832]

A highly-controllable method for fabricating high-quality air-stable phosphorene films having a number of layers, ranging from a few down to a monolayer, is to use oxygen-plasma dry-etching to produce thick-exfoliated phosphorene flakes, layer by layer, and with atomic precision. In such a stabilized phosphorene monolayer, it is possible to engineer defects precisely for the first time, leading to efficient photon emission at new frequencies in the near-infrared at room temperature. An electrostatic gate can also be used to tune the photon emission from defects in monolayer phosphorene.[833]

By using first-principles approaches, a novel kinetic pathway can be established for fabricating blue phosphorene via epitaxial growth. Systematic energetic studies reveal that both black and blue phosphorene monolayers can be easily stabilized on Cu(111), Au(111) and GaN(001) substrates, while GaN(001) is best for fabricating blue phosphorene via a half-layer by half-layer growth mechanism. The GaN(001) surface is here first preferentially covered with a half layer of phosphorus adatoms, followed by the other half.[834] Once formed, such a blue phosphorene monolayer is thermodynamically stable, as checked using *ab initio* molecular dynamics simulations. This half-layer by half-layer growth mechanism may permit the mass production of high-quality blue phosphorene, and might aid the epitaxial growth of black phosphorene and other 2-dimensional materials.

7.2 Structure

A certain amount of artistic licence has obviously been used in the diagram at the head of this chapter, particularly with regard to the bond-angle. The non-planar structures of the various layered phosphorene allotropes can be mapped onto a two-color two-dimensional triangular tiling pattern. In the buckled structure of a phosphorene monolayer, atoms in the so-called top positions are mapped onto dark tiles while atoms in so-called bottom positions are mapped onto light tiles. Optimum sp^3 bonding is seen throughout the structure when each triangular tile is surrounded by the same number, N, of like-colored tiles; where $0 \leq N \leq 2$. *Ab initio* density functional calculations indicate that both the relative stability and the electronic properties depend mainly upon N. This mapping approach can also be applied to phosphorene structures having non-hexagonal rings, and to two-dimensional quasicrystals having no translational symmetry. These are predicted to be almost as stable as the hexagonal network.[835] In all of the reported phosphorus monolayers, the phosphorus is 3-fold coordinated, but its flexible chemistry permits it to have a coordination number of up to 6. Three new phosphorus monolayers (α-P_6, β-P_6, 558-P_6), having 2-, 3- and 4-fold coordination, can be observed in epitaxial growth when flakes merge to form ridges at the boundaries. Simulations show that these 3 new allotropes are thermally and dynamically stable, and have an energetic stability which is comparable to that of monolayers such as δ-P, γ-P and ε-P. Because of their special atomic configurations, they exhibit an extremely high electron mobility, anisotropic Young's moduli, and optical absorbance in the visible and ultra-violet regions.[836] Also on the basis of *ab initio* density functional calculations, γ-P and δ-P have been proposed to be two additional stable structural phases of layered phosphorus besides the layered α-P (black) and β-P (blue) phosphorus allotropes. Monolayers of some of these allotropes have a wide band-gap while others, including γ-P, undergo a metal-insulator transition which is caused by in-layer strain or a change in the number of layers. One benefit is the possibility of connecting different structural phases at no energy cost. This will be particularly useful for assembling heterostructures having well-defined metallic and semiconducting regions in the same contiguous layer.[837] First-principles calculations also predict the existence of 7 novel phases by including 6 types of enantiomer corresponding to three kinds of structures which exhibit chirality. The Poisson ratios exhibit conventional behavior, rather than the negative values of monolayer or bulk black phosphorus; due to the structures being non-puckered. The phase transformations of these enantiomers reveal that the possibility of transformation between them exists because the energy barriers are low.[838] A new class of monolayer phosphorus allotropes, ε-P, ζ-P, η-P, and η-P, has been independently predicted. They are completely different to monolayer α-P (black) and to the previously predicted forms, β-P, γ-P and δ-P, and have a buckled honeycomb lattice. The new allotropes are composed of P_4 (square) or P_5 (pentagon) units that favor tri-coordination of phosphorus atoms. Their stabilities are confirmed by first-principles calculations and, in particular, η-P is shown to be as stable as α-P (black) and to be more stable than all previously reported phosphorene polymorphs.[839]

A two-dimensional allotrope of phosphorus called red phosphorene has also been reported. This results from re-structuring segments of blue and black phosphorenes. First-principles calculations indicate that red phosphorene is dynamically stable and exhibits a marked thermodynamic stability which is comparable to that of black phosphorene. Because of sp^3-hybridization and the formation of a localized lone pair, red phosphorene is a semiconductor with an indirect band-gap of about 1.96eV. As usual, this can be modulated by in-plane straining, due to its wave-like configuration. Red, black and blue phosphorenes have distinct layer thicknesses, surface work functions and colors.[840]

Based upon crystal structure predictions, a new allotrope, ψ-phosphorene, has been proposed which has a porous structure that is thermally and dynamically stable as compared with previously reported allotropes. Due to its unique atom configuration, ψ-phosphorene possesses highly orientation-dependent mechanical properties, with excellent flexibility. Other calculations predict that it is semiconducting, with an indirect band-gap of 1.57eV, and has anisotropic transport properties. The electron mobility along the x-direction is up to $1.3 \times 10^4 cm^2/Vs$; comparable to that of black phosphorene. It is demonstrated that ψ-phosphorene could be used for highly-selective hydrogen separation from a mixture of CH_4, CO_2, N_2, CO and H_2. By combining a suitable band-gap with high carrier mobility, a $MoSe_2$/ψ-phosphorene van der Waals heterojunction is predicted to be a good solar-cell material, with a power-conversion efficiency of up to 20.26%.[841] The Au(110) surface is expected to be a suitable substrate for the synthesis of ψ-phosphorene.

7.3 Mechanical Properties

When compared to graphene and other two-dimensional materials, phosphorene has a relatively low Young's modulus and fracture strength. This will limit its possible applications, due to possible structural failure. Some studies of its mechanical properties have been based upon first-principles calculations at 0K. Molecular dynamics simulations for finite temperatures show that temperature has a significant effect upon the mechanical properties. The fracture strength and strain are reduced by more than 65% when the temperature is increased from 0 to 450K. The fracture strength and strain in the zig-zag direction are more sensitive to temperature than those in the armchair direction. Failure cracks prefer to propagate along the groove in the puckered structure when uniaxial tension is applied in the armchair direction. In contrast, when uniaxial tension is applied in the zig-zag direction, multiple cracks with rough fracture surfaces propagate.[842]

Density functional theory investigations of phosphorene nanoribbons show that their ideal strength is lower than that of two-dimensional phosphorene. The Young's modulus of armchair nanoribbons exhibits a marked size effect, due to edge relaxations. The stress-strain curves indicate that chemisorbed oxygen atoms greatly reduce the ideal strength and Young's modulus of two-dimensional phosphorene oxide along the zig-zag direction; especially at high oxidation ratios. The strain and oxidation have a marked effect upon the phonon dispersion.[843] Molecular dynamics simulations indicate that fracture under uniaxial tension along the armchair direction is due to a break in the interlayer bond angles, while failure in the zig-zag direction is provoked by a break in both intra-layer angles and bonds. A modified Griffith criterion for the energy release-rate, and its dependence upon the strain-rates and crack orientations, indicates that the energy release-rate remains almost unchanged in the armchair direction but fluctuates wildly in the zig-zag direction. The strain-rate plays a negligible role in the energy release rate. The geometrical factor in the Griffith criterion is almost constant when the crack orientation is smaller than 45°, regardless of the crack orientation and loading direction.[844]

Density functional tight-binding method studies of armchair and zig-zag nanotubes, subjected to uniaxial tensile strain, show that the deformation and failure of the nanotubes are very anisotropic. In the case of zig-zag nanotubes, 3 deformation stages are found. A primary linear elastic stage is associated with interactions between neighboring puckers. This is followed by a bond-rotation stage in where the puckered configuration is smoothed-out by bond-rotation. In the third stage, bond-elongation occurs in which the P-P bonds are directly stretched up to the maximum limit and failure is initiated by rupture of the most highly-stretched bonds, which are located in specific rows. In the case of armchair nanotubes, the applied stress stretches the bonds up to the limit and causes failure. For both armchair and zig-zag nanotubes, the failure strain and stress are sensitive to the tube diameter while the Young's modulus, flexural rigidity, radial Poisson's ratio and Poisson ratio are relatively insensitive.[845] In armchair nanotubes, the failure strain decreases from 0.40 to 0.25 and the failure stress increases from 13 to 21GPa as the tube diameter increases from 13.3 to 32.8Å. In zig-zag nanotubes, the failure strain decreases from 0.66 to 0.55 and the failure stress increases from 4 to 9GPa when the tube diameter increases from 13.2 to 31.1Å. The Young's modulus, flexural rigidity, radial and thickness Poisson ratios are 114.2GPa, 0.019eVnm^2, 0.47 and 0.11 for armchair nanotubes and 49.2GPa, 0.071eVnm^2, 0.07 and 0.21 for zig-zag nanotubes, respectively.[846]

In situ transmission electron microscopy and *in situ* micromanipulation have been combined in order to visualize directly the nanomechanical behaviour of individual nanoflakes. The nanoflakes can be easily bent, scrolled and stretched; thus demonstrating their remarkable mechanical flexibility without fracturing. An out-of-plane plate-like bending mechanism and an in-plane tensile strain of up to 34% are observed.[847] An easy liquid-phase shear exfoliation method can produce monolayer and few-layer nanoflakes in organic solvents, using only a kitchen blender.

Two-dimensional materials tend to crumple according to the Mermin-Wagner theorem. The resultant ripple deformation can markedly affect the electronic properties. First-principles calculations of a new highly-anisotropic ripple pattern in phosphorene, where compression-induced ripple deformation occurs only along the zig-zag direction at strains of up to 10%, show that this direction-selective ripple deformation arises from its puckered structure, with its coupled hinge-like bonding configurations, and from the resultant anisotropic Poisson ratio.[848] Phosphorene exhibits considerable structural flexibility along the armchair direction, permitting it to support high curvatures. Semiconducting and a direct band-gap are retained during buckling along the armchair direction. The band-gap decreases, and transforms into an indirect band-gap, with buckling along the zig-zag direction. This structural flexibility and electronic robustness along the armchair direction can facilitate the fabrication of devices having complex shapes (folded, nano-scrolled).[849]

First-principles studies of monolayer and bilayer phosphorene show that, under sufficiently high isotropic tensile strains, the monolayer will lose its puckered structure and transform into a flat hexagonal plane. Two inner sub-layers of bilayer phosphorene can be bonded due to its interlayer distance contraction. Under uniaxial tensile straining along a zig-zag direction, the pucker distance of each layer in bilayer phosphorene can exhibit a negative Poisson ratio. The properties of monolayer and bilayer phosphorene depend upon the magnitude and direction of the applied strain. Their band-gaps decrease more rapidly under isotropic compressive strain than under uniaxial strain. Their direct-indirect band-gap transitions occur at higher isotropic tensile strains than that under uniaxial strain. Under isotropic compressive strain, bilayer phosphorene undergoes a transition from a direct-gap semiconductor to a metal. On the other hand, monolayer phosphorene initially undergoes a direct-indirect transition, and then changes into a metal. Under isotropic tensile straining, both the bilayer and monolayer phosphorene undergo a direct-indirect transition and finally change into a metal.[850] When an uniaxial stress is applied along an arbitrary direction, there is an enhancement of the inherent anisotropy.[851]

First-principles calculations show that phosphorene exhibits a markedly anisotropic elasticity with an anisotropy factor of 15.5. Phosphorene sheet has a high tensile ductility along the armchair direction, characterized by an elastic strain limit of 0.31. In the zig-zag direction, the critical strain is dictated by the phonon instability and an in-plane soft mode occurs beyond a strain of 0.22. Under uniaxial strain, the band-gaps of phosphorene can be continuously modulated. A Dirac-like band structure appears under sufficient straining along the zig-zag direction. These Dirac cones are clearly anisotropic, with Fermi velocities of up to 7 x 10^5m/s along the armchair direction. They drop to zero along the zig-zag direction.[852] Strain is already marked in as-prepared samples, and serves as a useful tool for engineering the material's properties, even though quantifying the strain and characterizing its spatially inhomogeneous distribution is difficult. First-principles simulations of the lattice vibrational modes and corresponding Raman spectra of strained black phosphorus (phosphorene) monolayers show that the frequencies of the vibrational modes and their Raman scattering peaks undergo substantial and distinct shifts, depending upon the type and degree of strain. When combined with high spatial-resolution Raman scattering measurements, these calculated results can quantify arbitrary strain distributions in phosphorene.[853]

By calculating the Grüneisen parameters, the frequency shift of the Raman and infra-red active modes under symmetrical biaxial strain can be deduced, as can the occurrence of a strain-induced semiconductor-metal transition, the gap size, and the effective mass of carriers in various strain configurations. A peculiar Dirac-shaped dispersion is noted for specific conditions, including zigzag-oriented tensile straining. The linear energy spectrum has distinct velocities which correspond to each of its linear branches, and is limited to the Γ-X direction in the first Brillouin zone.[854]

When density functional tight-binding simulations are used to optimize the two main edge configurations, armchair and zig-zag, it is found that the energy relaxation of nanoribbons leads to noticeable changes in the edge atomic configurations. The effective width of the edge region, which includes all of the atoms involved in edge relaxation, contains some 3 atomic rows near to the edge for both armchair and zig-zag nanoribbons. The edge stresses are positive for both, indicating the presence of tensile stress at the edges. In addition, the moduli of both are positive. The edge elastic modulus and edge stress of zig-zag nanoribbons are about three times larger than those of armchair nanoribbons. The tensile edge stresses along zig-zag and armchair edges are able to cause distortion of free-standing phosphorene nanoribbons.[855]

Using a density functional based potential and realistic boundary conditions, the flexibility of individual α, β and γ phosphorene allotrope layers along an arbitrary bending direction has been investigated. Correlation of these results with the in-plane elasticity shows that, although a monolayer thickness cannot be defined in the normal continuum sense, an orthotropic plate with a direction-dependent thickness can describe the out-of-plane deformation of α and γ allotropes.[856] First-principles calculations indicate that δ-phosphorene nanosheets are good auxetic materials, with a high negative Poisson ratio. The Young's modulus and Poisson ratio are anisotropic. The negative Poisson ratio value along the grooved direction has values of up to -0.267. The auxetic nature of δ-phosphorene arises from its puckered structure and is insensitive to the number of layers, due to weak interlayer interactions.[857] The δ-phosphorene possesses a good flexibility because of its relatively low Young's modulus and high critical crack strain.

7.4 Diffusion

Density functional theory calculations of lithium diffusion reveal that lithium atoms form strong bonds with phosphorus atoms, and exist in a cationic state. The shallow energy barrier, 0.08eV, for lithium diffusion on monolayer phosphorene along the zig-zag direction leads to ultra-high diffusivity. This is of the order of 100 times higher than that on MoS_2 and 10,000 times higher than that on graphene at room temperature. A larger energy barrier, 0.68eV, along the armchair direction results in almost no diffusion. No such marked diffusion anisotropy is observed in graphene or MoS_2. A notably high average voltage of 2.9V is predicted for a phosphorene-based lithium-ion battery, and a semiconducting to metallic transition - induced by lithium intercalation of phosphorene - gives rise to good electrical conductivity.[858] Density functional theory shows that intrinsic point defects, including vacancy and Stone-Wales types, block the directional ultra-fast diffusion of lithium. On defect-free phosphorene, the diffusion of lithium along the zig-zag direction is 1.6 billion times faster than that along the armchair direction, and 260 times faster than that in graphite. Upon introducing intrinsic vacancies and Stone-Wales defects, the energy barrier to lithium diffusion along the zig-zag direction increases sharply, to between 0.17 and 0.49eV; blocking the ultra-fast migration of lithium along the zig-zag direction. The formation energies of defects in phosphorene are considerably lower than those in graphene or silicene sheet. The open-circuit voltage increases with the appearance of defects; an unsuitable quality for anode material.[859] Density functional theory calculations show that both armchair and zig-zag nanoribbons have a significantly enhanced lithium binding strength, without sacrificing Li mobility, due to the presence of unique edge states. Zig-zag nanoribbons with a width of 21.5Å have a moderate working voltage (0.504 to 0.021V), a high capacity (541mAh/g) and a rapid charge/discharge rate. An inferior voltage is found for armchair nanoribbons, due mainly to its low stiffness that cannot support the structural deformation during lithiation. It is very necessary therefore to avoid undesirable structural expansion in the armchair nanoribbons.[860]

Density-functional theory calculations, and experimental measurements, of lithium diffusion show that the lithium mobility along the zig-zag direction in the phosphorene valley is some 7 to 11 orders of magnitude faster than that along the other directions; thus reflecting its ultra-fast and anisotropic diffusivity. Lithium intercalation has been studied by considering various Li_nP_{16}, n = 1 to 16, configurations; including single-side and double-side adsorption. Phosphorene can accommodate up to one lithium atom per phosphorus atom ($Li_{16}P_{16}$). Even at high Li concentrations (x = 1 in Li_xP), there is no lithium-clustering and the phosphorene structure, when fractured, is reversible during lithium intercalation. The theoretical value of the lithium capacity of monolayer phosphorene is predicted to be above 433mAh/g; depending upon whether the lithium atoms are adsorbed on one side or both sides. Experimental measurement of the lithium capacity of few-layer phosphorene indicates a reversible stable value of about 453mAh/g, even after 50 cycles.[861]

First-principles calculations of diffusion in sodium-doped phosphorene indicate that sodium can be stabilized on the surface, without clustering. Comparison of the adsorption of sodium atoms on one, and both, sides of phosphorene shows that Na-Na exhibits strong repulsion at separations of less than 4.35Å. The sodium intercalation capacity is estimated to be 324mAh/g, and the calculated discharge curve indicates a relatively low Na^+/Na voltage. The energy barrier to sodium diffusion on the phosphorene surface, at both low and high sodium concentrations, can be as low as 40 to 63meV. This implies a high mobility of sodium during the charge/discharge process.[862]

7.5 Magnetism

Although magnetism in phosphorene nanoribbons is essential for practical applications, edge reconstruction and Peierls distortion prevent the nanoribbons from becoming highly magnetized. First-principles calculations show that relaxed oxygen-saturated diagonal nanoribbons exhibit spin-polarized antiferromagnetic coupling, and the magnetism is entirely localized at the saturated edges. The antiferromagnetic state is quite stable under tensile and limited compressive straining. The irreversible Wilson transition occurs upon applying strain, but the non-magnetic metal phase (a new ground state) becomes more stable than the antiferromagnetic state when the compressive strain exceeds -4%. Dual tuning of the geometrical and electronic structures is seen as a deviation from a honeycomb to orthorhombic-like structure and by the formation of P-p_y bonding (P-p_z non-bonding) from P-p_z non-bonding (P-p_y anti-bonding) because of an increase in the proportion of P-p_y (P-p_z) orbitals.[863]

Ab initio investigation of zig-zag edge phosphorene nanoribbons with various widths has identified a stable magnetic state in pristine nanoribbons by permitting the system to be spin-polarized. The ground state of such nanoribbons prefers ferromagnetic order in the same edge, but antiferromagnetic order between two opposite edges. The magnetism arises from dangling bond states as well as from edge-localized π-orbital states. The presence of a dangling bond is however crucial to the magnetism of zig-zag edge phosphorene nanoribbons. Hydrogenated samples become non-magnetic semiconductors with a direct band-gap, while oxygen-saturated samples have magnetic ground states due to the weak P-O bond in the ribbon plane between the p_z-orbitals of the edge oxygen and phosphorus atoms.[864] Density functional theory calculations have also identified an antiferromagnetic insulating state in zig-zag phosphorene nanoribbons. When compared with other one-dimensional systems, this magnetism displays several particular characteristics in that the magnetic moments are arranged antiparallel at each zig-zag edge, the magnetism is quite stable in energy (some 29meV/magnetic-ion), the band-gap is large (about 0.7eV), the electronic and magnetic properties are almost independent of the nanoribbon width and a moderate compressive strain will induce a magnetic to non-magnetic as well as a semiconductor-to-metal transition. All of these phenomena arise due to the electronic instability which is induced by half-filled one-dimensional bands which cross the Fermi level at about $\pi/2a$.[865]

First-principles calculations of the magnetic properties of phosphorene doped with 3d transition-metal atoms predict values of 1.02, 2.00, 3.00, 4.04, 1.01 and $0.93\mu_B$ for titanium-, vanadium-, chromium-, manganese-, iron- and nickel-doped material, respectively. There is no spin-polarized state in scandium- and cobalt-doped material. Such doped materials can become magnetic half-metals or magnetic semiconductors, depending upon the type of impurity atom. Straining is a promising means for tuning the magnetic moment of transition-metal doped phosphorene.[866] A small biaxial strain can induce a transition from a state possessing a magnetic moment, to another state, in material which is doped with iron or manganese. In manganese-doped material, the magnetic moment changes from 2 to $4\mu_B$. In iron-doped material, the magnetic moment changes from 1 to $3\mu_B$.

Density functional theory study of 3d transition-metal atoms (scandium to zinc) adsorbed on phosphorene sheet shows that, due to the existence of lone pair electrons on phosphorus atoms in the phosphorene, all of the transition-metal atoms, except for the closed-shell zinc atom, can bond strongly to the phosphorene and have correspondingly high binding energies. For transition-metal atoms ranging from scandium to cobalt, the systems exhibit interesting magnetic properties which arise from exchange splitting of the transition-metal 3d orbitals. Strain is an effective means of controlling the magnetism of these systems, by tuning the interaction of the transition-metal with phosphorene and thus the relative positions of in-gap 3d orbitals. A small biaxial strain can induce a magnetic transition from a low-spin to a high-spin state in phosphorene decorated with scandium, vanadium or manganese.[867]

In a two-dimensional dilute magnetic semiconductor, consisting of a phosphorene monolayer doped with cobalt atoms in substitutional and interstitial defects, those defects can be stabilized and are electrically active. By controlling holes or electrons by a potential gate, the exchange interaction and magnetic order can be engineered and may induce a ferromagnetic-to-antiferromagnetic phase transition in p-doped samples. At a cobalt concentration of 2.7%, the Curie temperature is estimated to be 466K in the mean-field approximation.[868]

Full-potential linearized augmented plane wave analysis of non-magnetic (aluminium, silicon, sulfur, chlorine) impurity-doped phosphorene layers reveals that the binding-energy (between 5.1 and 8.3eV) tends to decrease with increasing atomic number. The electronic band structure and magnetic properties depend strongly upon the specific impurity atom. The aluminium-doped system thus exhibits semiconducting properties while, in all of the other systems, metallic band structures are found. It is predicted that aluminium and silicon act as donor elements whereas sulfur and chlorine act as acceptor elements. A magnetic state appears in silicon-, sulfur- and chlorine-doped systems, while aluminium induces no spin-polarized state. In the silicon-doped system, the major contribution to magnetic moment originates from the silicon impurity itself, while the presence of dangling bonds at a neighboring phosphorus atom near to the impurity site plays a crucial role for magnetism in sulfur- and chlorine-doped systems. A half-metallic feature is found in sulfur-doped layers. It is concluded that the magnetic state can be manipulated even by non-magnetic impurity-doping of phosphorene layers.[869] Another systematic, first-principles, investigation of phosphorene monolayers doped with non-metallic atoms (hydrogen, fluorine, chlorine, bromine, iodine, boron, carbon, silicon, nitrogen, arsenic, oxygen, sulfur, selenium) showed that although substitutional doping with hydrogen, fluorine, chlorine, bromine, iodine, boron, nitrogen, oxygen, sulfur or selenium results in substantial structural deformation at the doping sites, all of the neutral non-metal doped systems are stable. Calculated formation energies indicate that substitutional doping with many non-metal atoms is possible under suitable experimental conditions. The charged dopants, C^-, Si^-, S^+ and Se^+, are stable. Substitutional doping with hydrogen, fluorine, chlorine, bromine, iodine, boron, nitrogen, arsenic, C^-, Si^-, S^+ or Se^+ cannot induce magnetism in the monolayers, due to a saturation or pairing of valence electrons of the dopant and its neighboring phosphorus atoms. The ground states of neutral carbon-, silicon-, oxygen-, sulfur- or selenium-doped systems are magnetic, all[870] with a moment of $1.0\mu_B$, due to the appearance of an unpaired valence electron of carbon and silicon or the formation of a non-bonding 3p electron of a neighboring phosphorus atom around oxygen, sulfur and selenium. The magnetic coupling between the moments induced by two silicon, oxygen, sulfur or selenium atoms are long-range antiferromagnetic and the coupling is attributed to a hybridization interaction involving polarized electrons. The coupling between the moments induced by two carbon atoms is weak.[871] A further study showed that a magnetic moment can be induced in monolayer phosphorene when it is doped with the non-magnetic impurities oxygen, sulfur and selenium. Oxygen-doping causes it to become a spin-gapless semiconductor, while sulfur and selenium produces half-metallic characteristics. The p-orbital of the dopant undergoes strong hybridization with the surrounding phosphorus atoms, thus inducing spin splitting of the phosphorus and impurities. Carbon induces metallic characteristics, while boron acts as an n-type dopant and fluorine like an p-type dopant. The magnetism of phosphorene which is doped with two oxygen atoms depends strongly upon the doping sites and the distance between the two oxygen atoms.[872]

First-principles modelling of nitrogen- and boron-doped material reveals a tendency to form highly ordered structures: nitrogen doping leading to the formation of -N-P-P-P-N- lines. Further transformation to -P-N-P-N- lines, across the chains of phosphorene, occurs with increasing band-gap and increasing nitrogen concentration; coinciding with a decreasing chemical activity of nitrogen-doped phosphorene. In contrast to nitrogen, boron prefers to form -B-B- pairs; with the further formation of -P-P-B-B-P-P- patterns along the phosphorene chains. A low concentration of boron dopants changes the material from a semiconductor to a semimetal, with a simultaneous enhancement of its chemical activity. Co-doping with both boron and nitrogen starts with the formation of -B-N- pairs, which provide flat bands and further transformation of the pairs into hexagonal BN lines and ribbons across the phosphorene chains.[873]

Wannier-function based interpolation techniques provide firm numerical evidence for the existence of a saddle-point Van Hove singularity at the Γ-point near to the phosphorene Fermi energy. In the proximity of the Van Hove singularity, the spin susceptibility exhibits the logarithmic temperature dependence which is typical of Liftshitz phase transitions. The critical temperature for the ferromagnetic transition can be increased if strain is applied along the zig-zag ridges. The spin susceptibility due to the Van Hove singularity persists at up to about 97K.[874]

An *ab initio* calculation of the possibility of long-range magnetic ordering in two-dimensional self-passivated porous phosphorene layers predicts a non-magnetic state. A pore geometry with a dangling bond at two zig-zag edges at a distance of 7.7Å exhibits antiferromagnetic ordering. Pore-to-pore magnetic interaction, with a distance of 13.5Å between two pores, is remarkably long-ranged and arises from interactions between the magnetic tails of the edge states in the armchair direction. The antiferromagnetic state is maintained by oxidation of the edge. The long-range antiferromagnetic ordering changes to long-range ferromagnetic ordering in an external electric field.[875]

First-principles studies of armchair black phosphorene nanoribbons show that, in the unrelaxed nanoribbons, antiferromagnetic coupling is found between two phosphorus atoms in the same edge. The edge magnetic moment vanishes after structure relaxation, and all of the nanoribbons exhibit semiconducting features. Charge-doping greatly alters the band structures of the nanoribbons because metallic states reappear in the charge-doped nanoribbons. A magnetic moment is found in charge-doped systems, and the Stoner condition neatly explains the magnetic moment at edge atoms.[876]

First-principles study of the tunnel magnetoresistance and spin-injection efficiency in phosphorene with nickel electrodes, under mechanical tension and bending, shows that both the tunnel magnetoresistance and spin-injection efficiency are largely improved by the mechanical deformation. Uniaxial tensions of 0 to 15%, applied in the armchair transport direction, increase the tunnel magnetoresistance ratio to a maximum of 107% at 10%. The spin-injection efficiency increases monotonically from 8 to 43% with increasing strain. Under out-of-plane bending, the overall tunnel magnetoresistance increases from 7 to 50% within the bending ratio of 0 to 3.9%. The spin-injection efficiency is improved to about 70%, as compared to that (30%) of flat phosphorene[877]. The tunnel magnetoresistance and spin-injection efficiency are mediated by the transmission of spin-up electrons in the parallel configuration which, in turn, is highly dependent upon the applied mechanical tension and bending.

First-principles density functional theory and plane pseudo-potential method calculations of phosphorene, with some phosphorus atoms replaced by carbon, oxygen and sulfur, show that sulfur-doped phosphorene exhibits less geometrical distortion than does phosphorene doped with oxygen or carbon. Although pure phosphorene is non-magnetic, all of the doped systems are magnetic; each with a magnetization of μB per cell.[878]

7.6 Semiconduction and Band Structure

A comprehensive first-principles study of phosphorene nanoribbons, phosphorus nanotubes, multilayer phosphorene sheets and hetero-bilayers of phosphorene and two-dimensional transition-metal dichalcogenide monolayers shows that bare zig-zag phosphorene nanoribbons are metals, regardless of the ribbon width. Bare armchair phosphorene nanoribbons however are semiconductors with indirect band-gaps, and the latter decrease with increasing ribbon width.

Compressive and tensile strains can reduce or enlarge the band-gap of bare armchair nanoribbons, respectively. An in-plane electric field can significantly reduce the band-gap of bare armchair nanoribbons, leading to a semiconductor-to-metal transition at above a certain electric-field strength. Upon edge-passivating nanoribbons using hydrogen, zig-zag nanoribbons become semiconducting, with almost direct band-gaps.

Armchair nanoribbons remain semiconductors, but with direct band-gaps. The effect of tensile strain and electric fields on edge-passivated nanoribbons is similar to that for bare armchair nanoribbons. In the case of single-walled phosphorus nanotubes, both armchair and zig-zag nanotubes are semiconductors with direct band-gaps. Under tensile straining or transverse electric fields, a band-gap modulation similar to that in armchair nanoribbons can occur. Multilayer phosphorene sheets are semiconductors whose band-gaps decrease with increasing number of multilayers. In the presence of a vertical electric field, the band-gaps of multilayer phosphorene sheets decrease with increasing electric field, and the band-gap modulation is more significant with more layers.

Heterobilayers of phosphorene (p-type) with an n-type MoS_2 or WS_2 monolayer are still semiconductors and their band-gaps can again be reduced by applying a vertical electric field. Phosphorene/MoS_2 heterolayers can be an efficient solar cell material, with an estimated maximum power conversion efficiency of 17.5%.[879]

The band structure is highly anisotropic, with the k_x direction being Dirac-like and the k_y direction being Schrödinger-like. A notable feature is the presence of a quasi-flat edge band that is entirely detached from the bulk band in nanoribbons. The mechanism of the emergence of the quasi-flat band can be explained by using the topological argument which was invoked in order to explain successfully the flat band in graphene. The quasi-flat band can be controlled by applying an in-plane electric field perpendicular to the ribbon direction, and the conductance is switched off above a critical electric field.[880] Hybrid density functional theory calculations, and the deformation potential approximation, show that bilayer phosphorene - under a slight compression perpendicular to its surface - exhibits an extraordinary room-temperature electron mobility, of the order of $7 \times 10^4 cm^2/Vs$. This is some 2 orders of magnitude higher than that widely reported for ground-state phosphorenes, and is the result of the emergence of a new conduction-band minimum which is decoupled from the in-plane acoustic phonons that dominate carrier scattering.[881]

First-principles studies demonstrate that the monolayer material exhibits a giant phononic anisotropy which is precisely opposite to its electronic counterpart and can be effectively tuned by strain engineering. Upon sampling the whole Brillouin zone for the monolayer several, so-called hidden, directions are found along which small-momentum phonons are frozen by strain and exhibit the smallest amount of anharmonicity. The hidden directions are intrinsically different from the usual armchair and zig-zag directions. The anisotropy of the interlayer coupling of few-layer samples can be explored by examining the rigid-layer vibrations.[882]

A symmetry analysis of the electronic band structure including spin-orbit interaction close to the insulating gap edge and expressions for the energy dispersion relationship and spin-dependent eigenstates for electrons and holes, found via simplification of a perturbative expansion in wave-vector away from the zone center using elementary group theory, exposes the underlying symmetries that give rise to a substantial anisotropy in optical absorption, charge and spin transport properties. This reveals the mechanism responsible for valence-band distortion and the possible lack of a true direct gap.[883]

The availability of these products makes it possible to reveal that the band-gap of black phosphorus increases from 0.33eV in the bulk to 1.88eV in bilayers. This range is larger than that of any other two-dimensional material. A higher-energy optical transition also occurs, which changes from 2.0eV in the bulk to 3.23eV in bilayers.[884]

Chemical species such as hydrogen, OH, fluorine and chlorine can act as scissors to cut phosphorene and give phosphorus nanochains and nanoribbons. This scissors effect results from strong bonding between the chemical species and the phosphorus atoms. Species such as oxygen, sulfur and selenium fail to cut phosphorene nanostructures, due to their weak bonding with phosphorus. The electronic structures of the resultant P-chains reveal that the hydrogenated chain is an insulator while the pristine chain is a one-dimensional Dirac material in which the charge carriers are massless fermions travelling at an effective speed of about 8×10^5m/s. The resultant zig-zag phosphorene nanoribbons exhibit either metallic or semiconducting behavior, depending upon the treatment of the phosphorus edge atoms.[885]

Edge-induced gap states in finite layers, as examined using analytical models and density functional theory, are such that their nature depends upon the direction of the cut. Armchair nanoribbons are insulating, whereas nanoribbons cut in the perpendicular direction (with zig-zag or cliff-type edges) are metallic. That is, unless they undergo reconstruction or distortion with cell-doubling; which opens up a gap. All stable nanoribbons with unsaturated edges have gap states that can be removed by hydrogen passivation. Armchair nanoribbon edge-states decay exponentially with distance to the edge, and can be described by a nearly-free electron model. This then proves the existence of gap states in the limit of a semi-infinite plane.[886] First-principles calculations of the electronic structures and edge properties of phosphorene nanoribbons with and without edge hydrogenation show that, unlike the direct band-gap semiconducting characteristics observed in phosphorene sheet, bare armchair phosphorene nanoribbons are semiconductors with an indirect band-gap, while bare zig-zag nanoribbons are conductors. Edge hydrogenation of either armchair or zig-zag nanoribbons produces semiconductors with a direct band-gap. The electronic properties of the edge-hydrogenated nanoribbons are independent of their edge orientation. This behaviour contrasts with that of bare nanoribbons, which exhibit a significant edge orientation-dependence. Edge-energy and stress analyses show that hydrogenated nanoribbons have an extremely low edge energy and also a low edge stress, thus indicating that they are energetically and mechanically stable.[887] So phosphorene is stable and, unlike graphene, has a natural direct band-gap. *Ab initio* calculations indicate that the band-gap depends upon the number of layers and the in-layer strain, and is significantly wider than the bulk value of 0.31 to 0.36eV. A photoluminescence peak observed in the visible optical range for single-layer phosphorene confirms that the band-gap is larger than that of the bulk. Transport studies indicate a hole mobility that reflects the structural anisotropy of phosphorene. At room temperature, few-layer phosphorene field-effect transistors with a 1.0μm channel length exhibit an on-current of 194mA/mm, a hole field-effect mobility of $286cm^2/Vs$ and an on/off ratio of up to 10^4.[888]

A density functional theory study of the thermodynamic and electronic properties of nanoribbons having various terminations and reconstructions of the ribbon edges, with and without hydrogen passivation, leads to the construction of a phase diagram. This identifies the energetically preferred edges as a function of the temperature and of the hydrogen partial pressure. Detailed electronic structure calculations show further that ribbons having hydrogenated edges are typically direct-gap semiconductors with fundamental gaps that vary inversely with the ribbon width. Ribbons having bare or partially passivated edges instead either have metallic edges, or are semiconducting with band-gaps that are smaller than those of the hydrogenated counterparts due to the appearance of mid-gap edge states. These observations offer a means of tailoring the electronic properties of phosphorene nanoribbons by controlling the edge termination state via control of processing conditions such as the temperature and hydrogen partial pressure; not to mention the confinement of carriers by adjusting the ribbon width.

Density functional theory calculations of phosphorene bilayers having three different stacking orders show that their direct band-gap can vary from 0.78 to 1.04eV. A vertical 0.5V/Å electric field can further reduce the band-gap to 0.56eV. When a MoS_2 monolayer covers p-type AA- or AB-stacked bilayers, the combined trilayer can be an effective solar-cell material with type-II heterojunction alignment. The power conversion efficiency is predicted to be about 18 and 16% for AA- and AB-stacked bilayers, respectively. These values are better than those reported for rival state-of-the-art trilayer graphene/transition metal dichalcogenide solar cells.[889]

Density functional theory shows that zig-zag phosphorene nanoribbons exhibit the greatest quantum size effect, changing the band-gap from 1.4 to 2.6eV when the width is reduced from 26 to 6Å. Straining modifies the charge-carrier transport, leading to a sudden increase in electron effective mass at +8% strain for armchair nanoribbons or hole effective mass at +3% strain for zig-zag nanoribbons. Straining of diagonal nanoribbons results in a direct-to-indirect band-gap transition at -7% or +5% strain and creates degenerate energy valleys.[890]

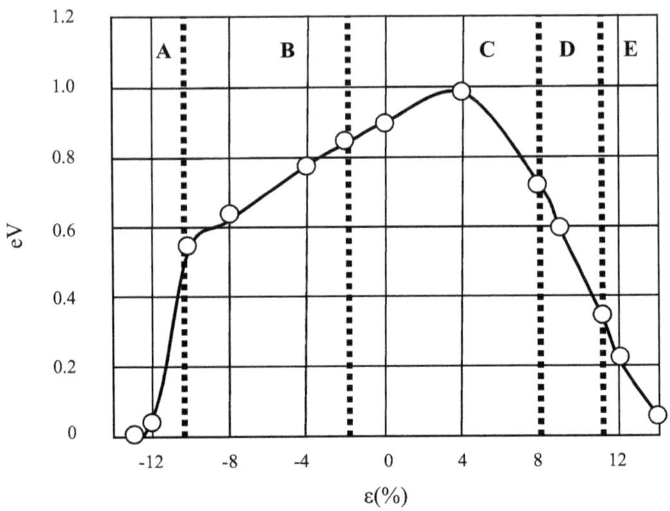

Figure 14. Band-gap of phosphorene as a function of strain applied in the zig-zag direction, A: direct, B: indirect, C: direct, D: indirect, E: direct. The boundaries are at -10.2, -2, 8 and 11.3%

A study of the effect of strain upon the electronic band structure shows that phosphorene can withstand a tensile stress and strain of up to 10N/m and 30%, respectively. The band-gap undergoes a direct-indirect-direct transition when an axial strain is applied. A moderate -2% compression in the zig-zag direction can trigger this gap transition. Sufficient tension (11.3%) or compression (10.2%) changes the gap from indirect to direct again. Strain-zones having distinct electronic band structures can be identified (figures 14 and 15). Although density functional theory is known to underestimate the band-gap of semiconductors, it here correctly predicts the strain effect upon the electronic properties, as validated by using a hybrid functional method. A general mechanism explains the energy-shifts with straining in terms of the bond nature of near band-edge electronic orbitals. The effective masses of carriers in the armchair direction are an order of magnitude smaller than those for the zig-zag axis; indicating that the former direction is favored for carrier transport. The effective masses can be markedly altered by straining, and a sharp jump/drop occurs at the zone boundaries of the direct-indirect gap transition.[891] First-principles calculations, based upon density functional theory, of the electronic properties of monolayer and bilayer material, subjected to uniaxial and biaxial strain, show that strain engineering has a clear effect upon the electronic properties. Biaxial strain is more effective in tuning the band-gap than is uniaxial strain. Dirac-like cones emerge upon applying zig-zag tensile strain to the monolayer and bilayer systems. In the latter material, anisotropic Dirac-like dispersion is induced upon applying suitable armchair or biaxial compressive strains.[892]

A simple model successfully reproduces the isolated edge state in the bulk gap of monolayer phosphorene but, in the case of bilayer phosphorene, it shows that edge states are almost buried in the valence band and that there is no isolated mid-gap edge state at the zig-zag edge. Because the buried edge state functions as an acceptor, holes are doped from the edge state into the bulk. This can explain the p-type conduction of undoped black phosphorus. In a vertical electric field, the intrinsic hole doping is reduced because some of the edge states move into the gap.[893]

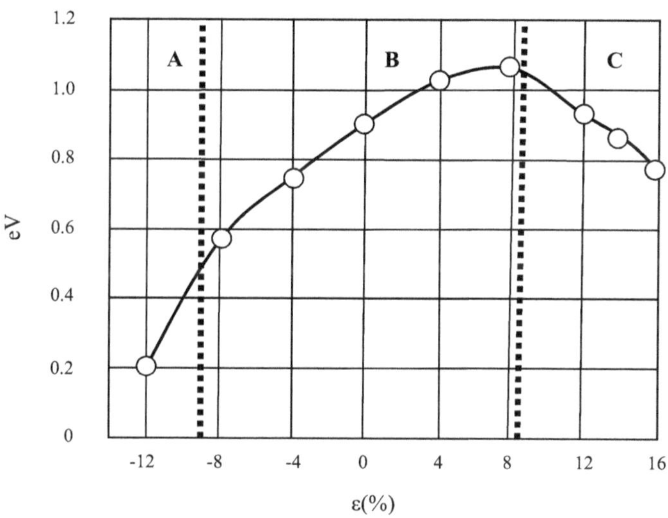

Figure 15. Band-gap of phosphorene as a function of strain applied in the armchair direction, A: indirect, B: direct, C: indirect. The boundaries are at -9 and 6.8%

On the basis of first-principles calculations, a mechanism can be proposed for changes in the band-gap of a blue phosphorene nanoribbon, with armchair or zig-zag edges and hydrogen saturation, as a function of its width. Both types of nanoribbon are indirect semiconductors. An energy gap of about 1eV can be achieved when the ribbon-width is decreased to about 10Å, and this is attributed to a quantum confinement effect. The parameters which are required to describe the quantum confinement can be obtained by fitting the calculated band-gaps to the ribbon widths. The results show that the quantum confinement in armchair nanoribbons is stronger than that in zig-zag ones.[894] Density functional theory investigation of the edge passivation effect in phosphorene nanoribbons shows that hydrogen and fluorine passivate the metallic edge states of nanoribbons and can open the band-gap up to 2.25eV. The two configurations of passivated atoms can exist at two edges and affect the band-gap of narrow nanoribbons. The band-gap of each type of zig-zag nanoribbon decreases as the ribbon width increases, this again being attributed to the quantum confinement effect.[895]

Zig-zag blue phosphorene nanoribbons with widths of 1.5 to 5nm can exhibit an intra-edge antiferromagnetic ground state with a semiconducting band-gap of about 0.35eV, and this is insensitive to the edge structure relaxation effect. The edge magnetism disappears upon H-passivation, and the band-gap is increased to about 1.77eV; almost the same as that of a two-dimensional blue phosphorene layer. Oxygen-passivated zig-zag nanoribbons also exhibit an intra-edge antiferromagnetic state. Both unpassivated and oxygen-passivated samples have almost the same band-gap.[896]

First-principles density functional theory calculations of edge and quantum confinement effects upon the electronic properties of phosphorene nanoribbons show that armchair nanoribbons are semiconductors for all edge groups. The latter were: hydrogen, fluorine, chlorine, hydroxyl, oxygen, sulfur and selenium - in addition to the pristine case – and the ribbon widths ranged up to 3.5nm. Zig-zag nanoribbons however exhibit either semiconductor or metallic behavior, depending upon the edge chemical species. One group (hydrogen, fluorine, chlorine, hydroxyl) forms saturated bonds with phosphorus atoms in armchair and zig-zag nanoribbons, and the edge states keep far away from the band-gap. The other group (pristine, oxygen, sulfur, selenium) forms weak unsaturated bonds with the pz orbital of the phosphorus atoms, and brings edge states within the band-gap of the ribbon. In zig-zag nanoribbons, the edge states of the former group are present around the Fermi level within the band gap, which close up the band-gap of those nanoribbons. In armchair nanoribbons, these edge states are located at the bottom of the conduction band and result in a reduced band-gap.[897]

When blue phosphorene is doped with light non-metallic atoms (boron, carbon, nitrogen, oxygen, fluorine), all of the substituted systems are very stable. The boron-substituted system is a direct band-gap semiconductor with a band-gap size of about 1.5eV. Magnetism is found in the case of carbon and oxygen substitution, but no magnetic moment is detected in the case of the other impurities.[898]

The oxidation of phosphorene can lead to the formation of a new family of two-dimensional (planar) and one-dimensional (tubular) oxides and sub-oxides; the majority being insulators. The band-gap of a phosphorene oxide depends upon the oxygen concentration, so that controlled oxidation may be used as a means of modifying the band-gap at will. In the case of the oxygen-saturated composition, P_2O_5, the planar and tubular phases both have a band-gap energy of about 8.5eV and are transparent in the near-ultraviolet. These two forms of phosphorene oxide are predicted to have the same formation enthalpy as O'-P_2O_5.[899]

Study of the shear-induced direct-to-indirect gap transition and anisotropy decrease, using first-principles methods, reveals that phosphorene can sustain an applied shear strain of up to 10%. The band-gap undergoes a direct-to- indirect transition when a shear strain of 5% is applied, leading to a change from 1.54eV under ambient conditions to 1.22eV at a shear strain of 10%. The anisotropy decrease is analysed by calculating the maximum sound velocities, effective mass and decomposed charge density which signals the onset of the undesirable shear-induced direct-to-indirect gap transition. On the other hand, the shear-induced electronic anisotropy suggests that phosphorene can be used as a switch in nano-electronic applications.[900]

First-principles calculations of the evolution of band-edges in few-layer phosphorene, as a function of the number of phosphorus layers, predict that monolayer phosphorene is an indirect band-gap semiconductor, and that its valence-band edge is extremely sensitive to strain. The band-gap can undergo an indirect-to-direct transition under a lattice expansion which is as small as 1% along the zig-zag direction. A semi-empirical interlayer-coupling model closely reproduces the evolution of the valence-band edges. The interlayer coupling plays a predominant role in the evolution of the band edges by decreasing both the band-gap and the effective carrier mass with increasing phosphorene thickness.[901]

Density functional theory analysis of the energy levels of phosphorene nanoribbons shows that the ribbons are small band-gap semiconductors with gaps, at absolute zero, of less than 0.81eV. The electronic structure depends upon the direction, with respect to the unit cell, in which the ribbons are formed. The unit cell is predicted to be a ground state triplet, suggesting that the ribbons can be ferromagnetic semiconductors.[902]

First-principles calculations, based upon density functional theory, of phosphorene which is co-doped with vanadium and various non-metallic atoms (boron, carbon, nitrogen, oxygen) show that the properties are clearly affected by these dopants. Marked local lattice distortions are observed in all of the co-doped systems. All of the substituted systems retain their semiconducting nature, but the band-gaps become smaller than that of pristine phosphorene, indicating that impurity-doping is an effective means of modulating the band-gap. Obvious covalent interactions exist between the impurities and their neighbouring phosphorus atoms.[903] Transition-metal doped layers are expected to have dilute magnetic semiconductor properties. A spin-polarized semiconducting state is obtained by substitutional doping (titanium, chromium, manganese) while a half-metallic state is obtained by vanadium and iron doping. The most promising dopants are suggested to be chromium and manganese, because a spin polarized state is achieved for a band-gap larger than 0.5eV.[904]

First-principles density functional theory studies of phosphorene which is substitutionally doped with groups III, IV, V and VI elements, show that the electronic properties are sharply modified by the number of valence electrons in the dopant atoms. Those with an even number of valence electrons enable the doped phosphorene to have metallic features, while dopants with an odd number of valence electrons retain semiconducting features. This behavior is attributed to the peculiar bonding characteristics of phosphorene and to a strong hybridization of sp orbitals between the dopants and phosphorene. Calculated formation energies indicate that such doped systems can be thermodynamically stable.[905] When phosphorene is doped with semiconducting impurities (gallium, germanium, arsenic, selenium), the gallium and arsenic doped layers exhibit semiconducting behavior while germanium-doped layers are metallic. A magnetic state appears in the case of selenium-doped material. Deep donor or acceptor states are found upon doping with semiconducting impurities. With external electron carrier doping, the donor level induced by selenium becomes an effective shallow donor level. Due to Fermi-level shifting, a metallic to semiconductor conversion or *vice versa* is found for germanium and arsenic. With external hole carrier doping, all of the systems have metallic band structures, apart from the selenium one. The latter impurity creates a shallow acceptor state with an extra hole. The most promising dopant is selenium because it creates a magnetic state, and extra electron- or hole-doping can induce n-type or p-type behaviour, respectively.[906] First-principles studies of phosphorene in the presence of typical monovalent (hydrogen, fluorine) and divalent (oxygen) impurities suggest a decomposition of phosphorene into weakly-bonded one-dimensional chains as a result of single- and double-sided hydrogenation and fluorination. In spite of a considerable quasi-particle band-gap of 2.29eV, fully hydrogenated phosphorene is dynamically unstable. Complete fluorination of phosphorene instead gives rise to a stable structure which is an indirect band-gap semiconductor with a gap of 2.27eV. Fluorination from the gas phase is also much more likely than hydrogenation due to the relatively low energy barrier to the dissociative adsorption of F_2 (0.19eV) as compared to H_2 (2.54eV). At low concentrations, monovalent impurities tend to form regular atomic rows even though such patterns do not seem to be easily achievable due to high migration barriers, being equal to 1.09eV for H_2 and 2.81eV for F_2. Instability occurs upon oxidation, leading to the formation of disordered amorphous-like structures at high impurity concentrations.[907]

Under normal compressive straining, a structural transformation occurs from pristine Z-phosphorene to A-phosphorene at a strain of 48%. Anisotropic biaxial in-plane straining produces the same effect at -16% and 54%. When the puckered structure is flattened into a plane, the structure is quite unstable at finite temperatures and transforms into a new phase, H-phosphorene. The anisotropic structure of A-phosphorene gives rise to direction-dependent mechanical properties, but H-phosphorene exhibits isotropic mechanical properties. Both the A- and H-phosphorenes are semiconductors, having indirect band-gaps of about 0.42 and 1.94eV, respectively. The electronic properties of these two phases are sensitive to the magnitude and direction of the applied strain.[908]

Molecular orbital theory predicts that interactions between lone-pair electrons will give rise to van der Waals forces between layers, due to non-equivalent hybridization in bulk black phosphorus. First-principles calculations show that phosphorene nanoribbons exhibit high activity and can be easily bonded by oxygen atoms and hydroxyl groups, indicating that the nanoribbons can be easily oxidized. The so-called cliff nanoribbon configuration can be maintained when it is passivated with hydroxyl groups and can be stable in a strongly alkaline environment. By oxidation of zig-zag, armchair and cliff edges, the nanoribbons can be changed from semimetallic to semiconducting, and the band-gap can be changed from direct to indirect. Nanoribbons passivated with OHO-[(OH+O)-] and OH- [(O+H)-] have intrinsic spin magnetic moments of about $2\mu_B$, originating from edge unsaturation electrons and symmetry reduction.[909] Density functional theory calculations of the effect of edge functionalization using O, OH and alternate terminations (OHO) of blue phosphorene nanoribbons indicate that O-termination is the most stable, with OHO-termination being more stable than OH-termination. *Ab initio* dynamic simulation shows that they are all thermodynamically stable at room temperature. The ground structures of O- and OH-nanoribbons are spin-polarized semiconductors, while OH-functionalized nanoribbons are non-magnetic semiconductors.[910] As the ribbon width increases, the band-gap of O-nanoribbons tends to 1.04eV while that of OH-nanoribbons tends to 1.97eV. This is comparable to the band-gap of single-layer blue phosphorene, since it is dominated by p_z electrons of the inner phosphorus atoms.

A trion (charged exciton) binding energy of 162meV is found for few-layer phosphorene at room temperature. This is some one to two orders of magnitude higher than those (20 to 30meV) in two-dimensional transition-metal dichalcogenide semiconductors and quasi two-dimensional quantum wells (1 to 5meV). Such a large binding energy is usually observed only in truly one-dimensional materials such as carbon nanotubes, where opto-electronic applications are greatly limited by intrinsically small optical cross-sections. Phosphorene instead permits quasi one-dimensional excitonic and trionic behaviors over a large two-dimensional area, thus allowing opto-electronic integration. Experiments validate the quasi one-dimensional nature of excitonic and trionic dynamics in phosphorene by demonstrating completely linearly polarized light emission from excitons and trions in few-layer samples.[911] A theoretical study of the exciton binding energy of anisotropic two-dimensional crystals yields analytical expressions upon considering variational wave functions at various limits of the screening-length to exciton-size ratio and comparing them with variational and exact numerical solutions. In the case of phosphorene, and aided by density-functional theory evaluation of the 2-dimensional polarizability, the analytical solution for the exciton binding energy gives results which compare well with experimental data.[912]

Strain is a powerful tool for band-structure engineering. In-plane straining in the armchair or zig-zag direction changes the effective mass components along both directions, while vertical straining has a significant effect upon the effective mass only in the armchair direction. The band-gap is narrowed by compressive in-plane straining and tensile vertical straining. For certain strain configurations, the gap is closed and the energy band changed into one of semi-Dirac type. That is, the dispersion is linear in the armchair direction and is gap-less quadratic in the zig-zag direction. The band-edge optical absorption is completely polarized along the armchair direction and the polarization rate is reduced when the photon energy increases. Strain changes not only the absorption edge, but also the absorption polarization.[913] The electronic band structure of strained phosphorene, within the range of experimentally accessible carrier densities and uniaxial strains, is described well by using a strain-dependent decoupled electron-hole Hamiltonian. This shows that a uniaxial tensile strain in the normal direction enhances the anisotropy of the carrier mobility, while a uniaxial strain in the zig-zag direction acts inversely. On the other hand, a uniaxial compressive strain in the normal direction weakens the anisotropy of the carrier mobility, while a uniaxial strain in the zig-zag direction again acts inversely. Application of a uniaxial strain in the armchair direction is ineffective in affecting the anisotropy of the carrier mobility.[914] The room-temperature carrier mobility in atomically thin two-dimensional materials is usually far below the intrinsic limit imposed by phonon scattering, as a result of scattering by remote charged impurities in the environment. Simulation of the charged impurity-limited carrier mobility, μ, in bare and encapsulated monolayer samples reveals a significant temperature-dependence of the carrier mobilities: $\mu \propto T^{-\gamma}$. This results from the temperature variability of charge screening, and varies with crystal orientation. The anisotropy in the effective mass leads to an anisotropic carrier mobility, with the mobility in the armchair direction being about an order of magnitude higher than that in the zig-zag direction. This mobility anisotropy is particularly increased at low temperatures and high carrier-densities. When encapsulated using a high-κ overlayer, the mobility increases by up to an order of magnitude. The temperature-dependence and anisotropy are reduced however.[915]

Numerical results show that the size-dependent carrier mobility of passivated phosphorene nanoribbons is not very sensitive to various passivation groups, such as hydrogen, fluorine and chlorine, but is strongly associated with the orientation of the nanoribbon. Passivated armchair nanoribbons have a much higher carrier mobility than do passivated zig-zag nanoribbons of similar width. With increasing width, the electron mobility of passivated nanoribbons can be further increased. The anisotropy of carrier mobility in the latter nanoribbons can be reversed by applying a tensile strain.[916]

A five-hopping-parameter tight-binding model shows, in the case of phosphorene, how breakage of the bipartite lattice structure gives rise to an electron-hole asymmetry of the energy spectrum. The finite-size phosphorene plaquette exhibits a Hofstadter-type spectrum made up of two unequal butterflies separated by a gap, where a quasi-flat band composed of zig-zag edge states is located. The transport properties can be investigated by simulating a four-lead Hall device (with all of the leads attached on the same zig-zag side) and by using the Landauer-Büttiker formalism. The chiral edge states due to the magnetic field yield quantum Hall plateaux, but the topological edge states in the gap do not support the quantum Hall effect, and exhibit dissipative behavior. Calculation of the complex eigen-energies of the non-Hermitian effective Hamiltonian that describes the open system (plaquette+leads) proves the super-radiance effect in the energy range of the quasi-flat band.[917] Using the Keldysh non-equilibrium Green's function method, and again applying the Landauer-Büttiker formula, the spin-dependent electron transport properties of a zig-zag nanoribbon at zero temperature have similarly been studied by depositing 2 ferromagnetic strips onto the edges of the nanoribbon. This induces exchange fields having parallel and anti-parallel configurations. An external perpendicular electric field is also applied in order to manipulate the spin-dependent electron conductance properties. Due to the order of the zig-zag nanoribbon edge bands near zero Fermi energy, parallel and anti-parallel configurations result in differing properties. In the anti-parallel configuration, spin filtering occurs only in the presence of a perpendicular electric field. In the parallel case, spin filtering takes place in both the absence and presence of an electric field. Although the spin filtering properties are invariant with regard to inversion of the direction of the electric field in a parallel configuration, the inversion changes the spin filtering type in an anti-parallel configuration.[918]

There are two types of zig-zag edge and two types of armchair edge for the phosphorene lattice. Study of the winding number of various types of edges of phosphorene ribbons suggests that, as well as the typical zig-zag edge, a flat zero-energy edge band can be found in ribbons of another non-typical armchair edge. Every edge state of the atypical armchair edge decays to the bulk condition at two different decay rates.[919]

First-principles calculations of the electronic properties of wrinkled phosphorene, and their response to charge injection and an external vertical electric field, show that small-sized wrinkled systems have a lower energy than do wrinkle-free monolayers; thus suggesting that free-standing phosphorene spontaneously forms small protrusions on its nanosheets. The ratio of wrinkle-height to curvature radius increases with increasing height. Injected charges are distributed mainly at peaks and valleys. A direct-to-indirect band-gap transition is found for zig-zag wrinkles with a height of 14.81Å, and the band-gaps of wrinkled nanosheets decrease almost linearly with increasing field.[920]

A high-resolution scanning tunnelling spectroscopic study of the light and heavy effective-mass directions reveals atomic-scale electronic variations which are related to strain-induced anisotropic deformation of the puckered honeycomb structure of freshly-cleaved black phosphorus. A proposed model for edge reconstruction self-passivates most dangling bonds by switching the coordination number of phosphorus from 3 to 5 or from 3 to 4.[921]

The discontinuous Galerkin density functional theory method permits large-scale Kohn-Sham calculations for nanoribbons with armchair edges containing a few thousands to ten thousand atoms. The use of the Galerkin method permits here the achievement of conventional plane-wave basis set accuracy with only some 15 adaptive local basis functions per atom. The relatively small number of degrees of freedom required to represent the Kohn-Sham Hamiltonian, together with the pole expansion inversion technique which avoids the need to diagonalize the Hamiltonian, results in a highly efficient and scalable scheme for computing the electronic structures of nanoribbons having armchair edges; together with their dynamics. The time required to calculate the electronic structures of large-scale nanoribbons containing 1080 to 10800 atoms is then only 10 to 25s per self-consistent field iteration. The accuracy is still comparable to that found using conventional plane wave density functional calculations. In these nanoribbons with armchair edges, the Galerkin method can scale up to 50000 processors. The use of Galerkin *ab initio* molecular dynamics calculations to study the thermodynamic stability of nanoribbons with armchair edges reveals that a (2 x 1) edge reconstruction occurs at room temperature.[922]

A strain-induced topological phase transition from the normal to the quantum-spin Hall state is detected in bilayer phosphorene, accompanied by a band-inversion that changes the $\mathbb{Z}2$ number from 0 to 1. This is highly dependent upon the interlayer stacking. When the bottom layer is shifted by half of a unit cell along zig-zag/armchair direction with respect to the top layer, the maximum topological band-gap of 92.5meV is sufficiently large to produce the quantum-spin Hall effect, even at room-temperature.[923]

Five-parameter tight-binding approximation analysis of monolayer phosphorene, zig-zag nanoribbons and armchair nanoribbons shows that, in zig-zag nanoribbons, the ratio of the two dominant hopping parameters indicates the possibility of a relativistic dispersion relationship and of the existence of a pair of separate quasi-flat bands at the Fermi level. The corresponding states are edge-localized if their bands are well-separated from the valence and conduction bands. In armchair nanoribbons, a transverse electric field along the ribbon width increases band-gap closure by shifting the energies of the valence and conduction band edge states. In zig-zag nanoribbons, a gap occurs at the middle of the relatively degenerate quasi-flat bands.[924]

A theoretical tight-binding investigation of rectangular, hexagonal and triangular monolayer phosphorene quantum dots shows that edge states appear in the band-gap of all types of quantum dot, regardless of their shape and edge configuration, due to anisotropic electron hopping in monolayer phosphorene. The charge density of any edge state is localized only in specific edges of a monolayer quantum dot, unlike the case of graphene quantum dots. The magnetic levels of monolayer phosphorene quantum dots exhibit a Hofstadter-butterfly spectrum and approach the Landau levels of monolayer phosphorene as the magnetic field increases. A so-called flat band appears in the magneto-energy spectrum, which is entirely different to that of monolayer phosphorene. The electronic and optical properties can be tailored by choosing the dot-size, the type of boundary edge and the external magnetic field.[925]

The theoretical Landau levels and magneto-transport properties of phosphorene under a perpendicular magnetic field have been studied within the framework of the effective kp Hamiltonian and tight-binding model. In the low-field regime, the Landau levels depend linearly upon both the Landau level index and the magnetic field in a manner which is similar to that of a conventional semiconductor 2-dimensional electron gas. The Landau splittings of the conduction and valence bands are different, and the wave-functions which correspond to the Landau levels are strongly anisotropic; due to the differing anisotropic effective masses. An analytical expression for the Landau levels in the low-energy regime can be obtained by solving the decoupled Hamiltonian, and agrees well with numerical calculations. In the high magnetic regime, a self-similar Hofstadter butterfly spectrum is obtained upon using the tight-binding model. This butterfly spectrum is consistent, with the Landau level results deduced from the effective kp theory, over a wide range of magnetic fields. The Landau levels of the phosphorene nanoribbon depend strongly upon the ribbon orientation, due to the anisotropic hopping parameters. The Hall, and longitudinal, resistances clearly reveal the structure of the Landau levels.[926]

The general electronic properties of monolayer phosphorene are exotic due to its puckered structure and large intrinsic direct band-gap, and the band structure is affected by a perpendicular magnetic field. The magneto-optical Hall and longitudinal optical conductivities, as functions of temperature, magnetic field and Fermi energy, are strongly influenced by the magnetic field. The imaginary part of the former, and the real part of the latter, exhibit regular interband oscillations as a function of frequencies of between 1.5 and 2eV. Strong intraband responses in the latter, and weak ones in the former, occur at much lower frequencies. The magneto-optical response can be tuned in the microwave-to-terahertz and visible frequency ranges; unlike graphene, where the response is limited to the terahertz regime.[927]

An analysis of the effects of the anisotropic energy bands of phosphorene on the magnetoroton branches for electrons and holes in the two Landau levels close to the band edges shows that the fractional quantum Hall effect gap in the lowest Landau level in the conduction band is slightly larger than that for conventional semiconductor systems. Conversely, the fractional quantum Hall-effect gap in the highest Landau level in the valence band is slightly larger than that for conventional semiconductors. Both effects should be experimentally observable in phosphorene. The magnetoroton mode for both electrons and holes consists of two branches, with two minima due to the anisotropy. In the long-wavelength limit, a second mode with upward dispersion which is well separated from the magnetoroton mode appears that is entirely a result of that same anisotropy. These novel features of the collective mode, which are unique to phosphorene, can be detected in resonant inelastic light-scattering experiments.[928]

A low-energy Hamiltonian for single-layer and bilayer black phosphorus can be devised which describes the electronic states in the vicinity of the Γ-point. It is based upon a tight-binding description of electron and hole bands close to the Fermi level. Expressions can be calculated for the Landau-level spectrum as a function of magnetic field and, in the case of bilayer black phosphorus, an investigation of the effect of an external bias upon the electronic band-gap demonstrates the highly anisotropic character of black phosphorus. In particular, in bilayer material the presence of bias permits a field-induced semiconductor-metal transition.[929] Another Hamiltonian, for silicene and phosphorene, can be devised by using the method of invariants. In the case of phosphorene, the bands near to the Brillouin-zone center have terms only in even powers of the wave-vector. It is predicted that the energies change quadratically in the presence of a perpendicular external electric field, but change linearly in a perpendicular magnetic field. The energies for silicene vary linearly in both cases.[930]

A comprehensive first-principles computational study of the electronic properties of 1-dimensional nanotubes, taking account of strain effects upon the mechanical and electrical properties, including elastic modulus, energy band structure and carrier effective mass, demonstrates that armchair nanotubes exhibit semiconducting properties along the axial direction and that the carrier mobility can be significantly improved by compressive strain. The hole mobility increases from 40.7 to 197.0cm^2/Vs as the compressive strain is increased to 5% at room temperature. Investigations of size effects in armchair nanotubes indicate that the conductance increases appreciably with increasing diameter.[931]

7.7 Surface Interaction

Ammonia adsorption on pristine and doped single-layer black phosphorus has been investigated using density functional theory. From a geometrical point of view, doping with hetero-atoms causes internal distortion of the single-layer phosphorus. This distortion is large for boron-, carbon-, nitrogen- or oxygen-doped black phosphorus sheet. Doping with hetero-atoms also decreases the E_g value. The highest adsorption energies for ammonia on single-layer black phosphorus occur for boron-, silicon and nitrogen-doped material, with energy values of −43.3, −35.3 and −17.05kcal/mol, respectively. The adsorption energies for pristine and carbon-, nitrogen- or oxygen-doped single-layer material are insignificant: E_{ad} < 8.6kcal/mol. It is suggested that metal dopants (boron, nickel, silicon) improve the reactivity of the phosphorus sheet to the ammonia molecule; unlike pristine and non-metal (carbon, nitrogen, oxygen) doped phosphorus. Whereas the substitution of most dopants has a significant effect upon the gap width of the doped system, nitrogen-doping has no great influence on the gap-width or overall shape of the density-of-states curve. The adsorption of ammonia on nitrogen-, carbon- or oxygen-doped systems has no significant effect upon the electronic structure, but slightly changes the density-of-states curves of silicon- and nitrogen-doped systems.[932] In the case of boron-doped single-layer black phosphorus, ammonia absorption leads to a marked change in the density-of-states, accompanied by fading of a small peak in the gap region.

Density functional theory combined with a non-equilibrium Green's function formalism, applied to the optimum enhancement of the p-type conductance of monolayer phosphorene under surface charge-transfer doping of electrophilic molecules such as 2,3,5,6-tetrafluoro-7,7,8,8-tetracyanoquinodimethane, NO_2 and MoO_3, indicates that such enhancement originates from charge transfer between the adsorbed molecule and the phosphorene layer. The dynamic transport behavior demonstrates that additional channels for hole transport in the host monolayer are generated upon adsorption of the molecule.[933]

When laser-pruning is used to reduce multilayer- to few-layer black phosphorus under ambient conditions, phosphorene oxides and sub-oxides are formed and the degree of laser-induced oxidation is governed by the laser power. Because the band-gaps of the phosphorene sub-oxide depend upon the oxygen concentration, this technique is able to achieve localized band-gap engineering of thin material. Micro-patterns of few-layer phosphorene sub-oxide flakes with unique optical and fluorescence properties are created. Some of these sub-oxide flakes exhibit long-term (up to 340h) stability under ambient conditions. It is possible to develop a so-called calibration-map to choose the laser power required to tune the band-gap of the sub-oxide flakes. The latter exhibit a higher drain-current modulation and mobility than does pristine material. The surface of the laser-patterned region can also be altered so as to be sensitive to toxic gas by means of fluorescence contrast.[934] *Ab initio* density functional theory calculations, combined with the equivalent cores approximation, can be used to determine the core-level binding-energy shifts at phosphorus sites caused by the oxidation of phosphorene. The presence of oxygen increases the core-level binding energies of phosphorus atoms and leads to binding-energy shifts of up to 6eV in highly defective geometries. Identification of the likely binding geometries of oxygen helps to interpret the observed core-level photo-emission spectra at various stages of oxidation and thus permits the determination of the fractions of specific local geometries.[935]

First-principles calculations of the interaction of physisorbed small molecules such as CO, H_2, H_2O, NH_3, NO, NO_2 and O_2 with phosphorene show that CO, H_2, H_2O and NH_3 molecules act as weak donors whereas O_2 and NO_2 act as strong acceptors. The NO molecule, in particular, donates electrons to graphene, but receives electrons from phosphorene. Among the above molecules, NO_2 exhibits the strongest interaction, by hybridizing its outer orbitals with the 3p orbital of phosphorene. The non-trivial and distinct charge transfer between phosphorene and these physisorbed molecules makes phosphorene a promising material for application as a gas sensor. It also provides an effective route for modulating the polarity and density of carriers. The intermediate binding energy of hydrogen molecules on phosphorene also implies that stable hydrogen storage under ambient conditions and subsequent release will be possible.[936]

Table 11
Adsorption energies for black phosphorene

Adatom	Site	E(eV)
Zn	H	0.15
Mg	H	0.66
Ag	H	1.14
Na	H	1.35
Au	H	1.61
P	H	1.77
Be	H	1.79
Li	H	1.93
Al	H	2.00
Cu	H	2.15
Si	H	2.52
Fe	H	2.98
N	B	3.49
B	H	3.58
Co	H	3.75
Ni	H	4.41
Pt	H	4.82

B: bridge, H: hollow

Surface reaction with oxygen is a fundamental cause of the degradation of phosphorene. First-principles calculations show that, for each oxygen atom adsorbed, there is an energy release of about 2eV. Although the most stable forms of adsorbed oxygen are electrically inactive and lead to only minor distortions of the lattice, there are low-energy metastable forms which introduce deep donor and/or acceptor levels into the gap. It is suggested that reactive dangling oxygen atoms increase the hydrophilicity.[937] Thermodynamic analysis of oxy-(O) or imine-(NH) functionalized monolayers with single-sided or double-sided functionalization shows that oxy-functionalized material can be formed under conditions ranging from ultra-high vacuum to high concentrations of molecular O_2. Imide-functionalized phosphorene can be formed at relatively high concentrations of molecular N_2H_2. Born-Oppenheimer molecular dynamics simulations show that, under ambient conditions, both O_2 and N_2H_2 can etch phosphorene away.[938] Systematic density functional theory computations of the behaviour of tetra-cyanoquinodimethane, tetra-cyanoethylene and tetra-thiafulvalene, on the surface of phosphorene monolayers, and thicker layers, reveals considerable charge transfer and a strong non-covalent interaction between these molecules and phosphorene. The band-gap decreases dramatically due to molecular modification and can be further modified by applying an external electric field. Surface molecular modification is also an effective means for enhancing the light-harvesting ability of phosphorene in various directions.[939] Analysis of phonon dispersion curves of phosphorene oxide after its formation by oxygen adsorption on bare phosphorene reveals that stoichiometric and non-stoichiometric oxide configurations are stable under ambient conditions, thus suggesting that oxygen adsorption does not degrade phosphorene. The nature of the band-gap of the oxides depends upon the degree of functionalization of the phosphorene. An indirect gap is predicted for non-stoichiometric configurations, while a direct gap is predicted for stoichiometric oxide. An applied mechanical strain or external electric field permits modification of the band-gap of the oxide. Unlike the case of bare phosphorene, a dependence of the diode-like asymmetrical current-voltage response upon the degree of stoichiometry is predicted for the oxide.[940]

Table 12
Adsorption energies for blue phosphorene

Adatom	**Site**	**E(eV)**
Zn	T	0.04
Mg	H	0.30
Ag	H	0.99
Be	V	1.16
Na	V	1.30
P	T	1.32
Au	H	1.43
Al	V	1.60
Li	V	1.78
Cu	H	1.80
Si	H	2.00
N	T	2.18
C	B	3.23
S	T	3.27
Co	V	3.43
Pt	H	4.13
Ni	V	4.24
O	T	5.46

B: bridge, H: hollow, T: top, V: valley

First-principles calculations of the adsorption of alkali, alkaline-earth, non-metallic, transition-metal and noble-metal adatoms on black and blue phosphorus nanosheets show that both types of sheet exhibit a good adsorption capability with regard to foreign atoms (tables 11 and 12). The binding energies of adatoms here are stronger than those on BN, SiC, MoS_2 or graphene. On black-type sheet, most adatoms prefer to adsorb at the hollow sites. In the case of blue-type sheet, the favored adsorption site is element-dependent. The majority of alkali, alkaline-earth and transition-metal adatoms prefer the valley site, noble-metal adatoms prefer the hollow site and non-metallic adatoms prefer the bridge and top sites. The semiconducting behavior of phosphorene is modified by the adatoms, which can cause p-type and n-type doping or can induce mid-gap states in the sheets. Surface adsorption effectively functionalizes the phosphorene system to exhibit versatile spintronic features: N-/P-decorated blue sheets are half-metals, B-/Fe-decorated sheets become bipolar semiconductors and Co-/Au-decorated blue and black sheets turn into spin-gapless-semiconductors.[941] Density functional theory study of the absorption of transition-metals on pristine and defected single-layer phosphorene predicts that an increased transition-metal diffusivity on pristine material would impair any possibility of controlled magnetism. On the other hand, point defects can anchor the metals and exponentially reduce their diffusivity. Metals bind strongly to pristine and defected material, and a microscopic description of the bonding can qualitatively explain trends in the diffusivity with increasing number of valence electrons. Divacancy complexes are essential to permitting practical application due to their greater thermodynamic stability in comparison with monovacancies.[942] Defect-plus-metal complexes retain intrinsic semiconduction properties, and induce local magnetic moments which exhibit marked exchange-splitting and high spin-flip energies. Metal absorption can completely alter the intrinsic nature of the semiconduction, and give rise to half-metallic and metallic composite states.

A systematic first-principles calculations of 20 different adatoms including s and p valence metals, 3d transition metals, noble metals, semiconductors, hydrogen and oxygen adsorbed on phosphorene shows that the adsorbed adatoms produce a wide diversity of structural, electronic and magnetic properties. Phosphorene forms strong bonds with all of the adatoms while preserving its structural integrity. The adsorption energies of adatoms on phosphorene are more than twice as high as on graphene. The greatest distortions of phosphorene are about 0.1 to 0.2Å. The charge carrier type in phosphorene can be tuned by adatom adsorption.[943] A similar first-principles density-functional theory investigation of the behaviour of lithium, sodium, magnesium, aluminium, chromium, iron, cobalt, nickel, molybdenum, palladium, platinum and gold, adsorbed on phosphorene, shows that all of the adatoms prefer the hollow sites. The magnesium, chromium, molybdenum and gold adatoms can exhibit a three-dimensional growth mode on the phosphorene substrate, whereas all of the other adatoms exhibit a two-dimensional growth mode. A metallic state is observed in the case of lithium-, sodium-, aluminium- and chromium-doped systems, and chromium-doped phosphorene exhibits a magnetic state. The other eight metal-doped systems maintain a semiconducting band-gap. Spin-polarized band structures exist in iron-, cobalt- and gold-doped systems with a band-gap. The iron-doped material is a potential candidate for a dilute magnetic semiconductor material, because no clustering problem will occur. There is a variation in the dipole moment and work functions for each impurity-doped layer and dipole moment and a shift in the Fermi level can account semi-quantitatively for the change in work function.[944]

Density functional theory studies of single-layer phosphorene which is functionalized with the electrophilic molecule tetracyano-p-quinodimethane as an electron acceptor, or the nucleophilic molecule tetrathiafulvalene as an electron donor, show that the former molecule introduces shallow acceptor states into the gap of phosphorene close to the valence band edge. This makes the doped system into a p-type semiconductor. When the tetrathiafulvalene molecule is adsorbed on phosphorene, the occupied molecular states which are introduced into the gap are of deep donor type and so effective n-doping for transport cannot be achieved. This can be corrected by applying an external out-of-plane electric field which is directed from the phosphorene towards the tetrathiafulvalene. An in-plane tensile strain can also be used, or a combination of tension and electric field. Under these conditions the conduction band edge of the phosphorene moves closer to the tetrathiafulvalene-derived donor states. The tetrathiafulvalene-adsorbed phosphorene system then becomes an n-type semiconductor. The out-of-plane electric field and the in-plane strain can considerably modify the band-gap of the tetrathiafulvalene-adsorbed phosphorene.[945]

7.8 Substrate Interaction

Study of the interaction between phosphorene and SiO_2, Al_2O_3 or h-BN surfaces demonstrates that phosphorene forms strong bonds with oxygen-terminated SiO_2 (00•1) and aluminium- and oxygen-terminated Al_2O_3 (00•1) surfaces. The structure of the phosphorene changes markedly. Phosphorene adsorbs onto h-BN surfaces via van der Waals interactions. The two-dimensional planar structure of h-BN is free from dangling bonds, which provide an ideal substrate for phosphorene. The band-gap of the phosphorene/h-BN system decreases monotonically with increasing vertical electric field. A semiconductor-to-metal transition occurs at about 6V/nm.[946]

Because of its similar in-layer hexagonal lattice to that of MoS_2, the latter is expected to be a suitable substrate for growing blue phosphorene. Van der Waals heterostructures are constructed theoretically by stacking blue phosphorene on top of MoS_2, and the thermal stability and electronic structures are evaluated on the basis of first-principles calculations and a Van der Waals corrected exchange-correlation functional. Formation of the heterostructures is found to be exothermic and the most stable stacking configuration is confirmed. Blue-phosphorene/MoS_2 heterostructures preserve the properties of both blue phosphorene and MoS_2, but exhibit relatively narrower band-gaps due to the interlayer-coupling effect. The band structures can be further engineered by applying an external electric field. An indirect-direct band-gap transition in the bilayer is controlled by the symmetry property of the built-in electric dipole fields.[947]

The influence of thermal effects upon the Raman phonons of few-layer phosphorene formed on hafnium-dioxide high-k dielectric is such that, when annealed at up to 200C, the phosphorene film exhibits a blue-shift in both the out-of-plane (A_g^1) and in-plane (B_g^2 and A_g^2) phonon modes as a result of a compressive strain effect. This in turn is attributed to out-diffusion of hafnium atoms from the underlying HfO_2 dielectric, which compresses the phosphorene in both the zig-zag and armchair directions. Upon further increasing the thermal energy beyond 250C, strain relaxation within the phosphorene eventually takes place. The phosphorene is then unable to retain its intrinsic crystallinity prior to annealing, as revealed by a broadening of the full-width at half maximum of the Raman phonons.[948]

Direct measurement indicates a strong layer-dependent surface potential, for mono- and few-layer phosphorene on gold, which is consistent with theoretical predictions. When photoluminescence spectroscopy is used to probe charge transfer in the phosphorene/gold hybrid system, highly anisotropic and layer-dependent photoluminescence-quenching is observed in the phosphorene/gold hybrid system and is attributed to highly anisotropic layer-dependent interfacial charge transfer.[949]

7.9 Optical Properties

The optical response of phosphorene nanostructures, studied using time-dependent density functional theory, shows that – when compared with the absorption spectrum of graphene - that of the nanostructure exhibits a high absorbance in the ultra-violet region; indicating high light absorptivity. In the low-energy resonance zone, a spectral band extends over the entire near-infrared region. When impulse excitation polarizes in the armchair-edge direction, the low-energy plasmons in few-layer phosphorene nanostructures exhibit an apparent long-range charge-transfer excitation. This is significantly less pronounced along the zig-zag edge direction. The edge configuration significantly affects the absorption spectrum of monolayer nanostructures. The armchair edge and zig-zag edge serve different purposes in the absorption spectrum. The absorption spectrum of few-layer nanostructures changes according to the number of layers when the impulse excitation polarizes in the armchair edge direction. The change in the low-energy resonance zone is also appreciably different to that in the high-energy resonance zone.[950]

The band-gap of phosphorene nanoribbons is strongly enhanced by quantum confinement, but differently orientated nanoribbons exhibit distinct scaling laws for describing the expression for band-gap versus the ribbon width, w. The band-gaps of armchair nanoribbons scale as $1/w^2$ whereas zig-zag nanoribbons exhibit a 1/w behavior. The distinct scaling laws reflect a significant aspect of the band dispersion of phosphorene. That is, electrons and holes behave as non-relativistic particles along the zig-zag direction but instead resemble relativistic particles along the armchair direction. This peculiar merging of non-relativistic and relativistic properties in a single material is expected to impart novel electrical and magnetotransport properties to few-layer black phosphorus and its ribbon structures. The respective nanoribbons host electrons and holes having markedly different effective masses and optical absorption spectra.[951]

Strong and highly layer-dependent photoluminescence is observed in few-layer (2 to 5) phosphorene. This confirms the theoretical prediction that few-layer phosphorene has a direct, and layer-sensitive, band-gap. The few-layer material is more sensitive to temperature-modulation, than are graphene or MoS_2, with regard to Raman scattering. The anisotropic Raman response in few-layer material permits the use of an optical method for determining quickly the crystalline orientation; that is, without the use of tunnelling electron microscopy or scanning tunnelling microscopy. Such results provide essential experimental information on the band structures and exciton behavior in few-layer phosphorene.[952] The optical response of phosphorene can be gradually changed, by applying moderate uniaxial compression, because the material undergoes a transition to an indirect-gap semiconductor and eventually to a semimetal. The straining tunes not only the gap between the valence band and conduction band local extrema, but also the effective masses and thus the exciton anisotropy and binding strength. Theoretical considerations suggest how the exciton stability, and resultant luminescence energy, evolve under uniaxial strain. The exciton binding energy can be as high as 0.87eV in vacuum for a 5% transverse strain; thus placing it among the highest values for two-dimensional materials. The large shift in the luminescence peak, and its linear dependence upon strain, suggest that it can be used to probe the strain-state of single layers.[953] Lattice vibrational Raman and infra-red modes in strained few-layer phosphorene have been studied using density functional theory techniques. Significant frequency shifts, of up to some 100/cm in the applied strain range of ±6%, of the Raman activities in both the high- and low-frequency regions and of the infra-red activities in the low-frequency region occur. The type of applied strain, armchair or zig-zag, selectively affects specific vibrational modes.[954] When combined with high spatial-resolution Raman and infra-red scattering experiments, such results can aid the understanding of complicated inhomogeneous strain distributions in few-layer phosphorene and the preparation of materials, having desired electronic properties, by straining or layering.

Certain configurations have a semiconducting nature, and the states around the Fermi energy are contributed mainly by the p-orbitals of atoms. With regard to optical properties, the reflective spectrum is dispersed approximately within the visible range, thus suggesting that this 2-dimensional nanostructure can be considered to be shielding for the visible region. Due to anisotropy of the electronic structure of black phosphorene, device performance is substantially better in the armchair direction. Zero-bias transmission exhibits no conductance channel near to the Fermi level but, further away, prominent spectra are observed. Some configurations possess non-linear current-voltage characteristics. There is a sensitivity to NH_3 and NO_2 gas molecules which can be explored using electronic and current-voltage characteristics. This shows that black phosphorene has a strong affinity for electron seeking NO_2 molecule; thus offering possibilities for sensor application.[955]

7.10 Defects

Due to low formation energies, it is very easy to create atomic defects in phosphorene during its fabrication process. How these atomic defects affect its mechanical behavior, however, remain unknown. The defect formation energy in phosphorene is generally much lower than that in graphene, and the defect configuration strongly affects the electronic structure. The band-gap vanishes in single-vacancy layers, but reappears in divacancy layers. A single vacancy behaves like a p-type impurity with regard to transport properties. A vacancy defect can contribute greatly to increasing the current. Along the zig-zag direction, the current in the most stable single vacancy structure is significantly increased when compared to that found in pristine layers. The current along the armchair direction is always greater than that along the zig-zag direction.[956] Density functional theory, combined with semi-empirical van der Waals dispersion corrections, shows that a vacancy or a self-interstitial defect is more stable in the outer layer than in the inner layer of defective few-layer phosphorene The formation energy and transition energy of a vacancy or a self-interstitial phosphorus defect decrease with increasing film thickness. This is due mainly to an upward shift of the host valence band maximum with reference to the vacuum level. Both vacancies and self-interstitials can thus act as shallow acceptors, and this explains the experimentally observed p-type conductivity of few-layer phosphorene. As these native point defects have moderate formation energies, and are stable in the negatively charged state, they can also serve as electron-compensating centers in n-type few-layer phosphorene.[957]

Two possible monovacancies and 14 possible divacancies have been identified in phosphorene. Monovacancy-(5|9), with a pair of pentagon-nonagons, is the ground state of monovacancies and divacancy-(5|8|5), with a pentagon-octagon-pentagon structure, is the most stable divacancy. The 14 divacancies can be further sub-divided into four basic types, depending upon their topological structures, and they transform between these configurations via bond-rotation. The diffusion of monovacancy-(5|9) is anisotropic, in that it prefers to migrate along the zig-zag direction in the same half-layer. The introduction of monovacancies and divacancies affects the vibrational properties and introduces localized defect modes. These can be used to distinguish between the various vacancy structures. Phosphorene with monovacancy-(5|9) is a ferromagnetic semiconductor with a magnetic moment of 1.0μ_B and a band-gap of about 0.211eV. The divacancy induces a direct-indirect band-gap transition.[958] First-principles calculations of 10 types of point defect, including the Stone-Wales (SW-1, SW-2), monovacancy (SV-[5|9], SV-[55|66]) and divacancy (DV-[5|8|5]-1, DV-[5|8|5]-2, DV-[555|777]-1, DV-[555|777]-2, DV-[555|777]-3, DV-[4|10|4]), show that all of them are quite easily created in phosphorene. Most of them are easy to distinguish from one another, and their defective atomic structures can be correlated with simulated scanning tunnelling microscopic images. The SW, DV-(5|8|5)-1, DV-(555|777) and DV-(4|10|4) defects have little effect upon the electronic properties, and defective monolayers are still semiconducting; with similar band-gap values to those of perfect phosphorene. The SV-(5|9) and DV-(5|8|5)-2 defects can introduce unoccupied localized states into the fundamental band gap. In particular, the SV-(5|9) and SV-(55|66) defects can introduce hole-doping, and SV-(5|9) can produce local magnetic moments.[959]

First-principles calculations of the hopping rate of vacancies predict that direct observation of these monovacancies, with their high mobility and anisotropic motion, is possible only at temperatures of about 70K, where the thermal activity is largely suppressed. Because of the low diffusion barrier (0.3eV), the motion of a monovacancy at room temperature is 16 orders of magnitude faster than it is in graphene. The intrinsic strain associated with the vacancies extends far along the zig-zag direction, but falls off rapidly along the armchair direction. The motion of divacancies, via multiple dissociation-recombination processes involving monovacancies, exhibits novel features due to the small energy-cost; about 1.05eV for the splitting of a divacancy into two monovacancies.[960] An uniaxial tensile strain, applied along the zig-zag direction, can promote the motion of monovacancies while a tensile strain applied along the armchair direction has the opposite effect. These features of vacancies, and the puckering structure, facilitate bond reorganization.

Vacancies introduce local stress concentrations and provoke early bond-breaking, leading to a marked degradation of the mechanical properties. A 2% concentration of randomly distributed monovacancies can reduce the fracture strength by about 40%. An increase in temperature from 10 to 400K further degrades the fracture strength by about 60%. The fracture strength of defective phosphorene is affected by the defect distribution.[961] When the defects are arranged in a line, the reduction in fracture strength depends markedly upon the tilt-angle and the loading direction. Divacancies cause an even larger reduction in fracture strength than do monovacancies when loading is applied in the armchair direction.

First-principles calculations of the stress-strain relationship of phosphorene which contains vacancies indicate that, due to the presence of the vacancies, mechanical properties such as the ideal strength and Young's modulus are greatly degraded in the zig-zag direction.[962] The tensile strain produces a marked reduction in the band-gap, and Li diffusion through the defected material is more energetically favorable here than in pristine phosphorene. This effect is further increased by applying biaxial tension.

A monovacancy creates an in-gap state in the band-gap of pristine phosphorene and induces a magnetic moment, even though pristine phosphorene is non-magnetic. Planar and staggered divacancies do not change the magnetic properties, although a staggered divacancy creates states in the gap. Density functional theory calculations show that the adsorption of non-metallic elements (carbon, nitrogen, oxygen) and transition-metal elements (Fe, cobalt, Ni) can change the magnetic properties, with or without vacancies. Non-magnetic pristine phosphorene thus becomes magnetic following the adsorption of nitrogen, iron or cobalt adatoms and magnetic phosphorene with a monovacancy becomes non-magnetic following the adsorption of carbon, nitrogen or cobalt atoms. In the case of an oxygen- or iron-adsorbed monovacancy structure, the electronic and magnetic properties are tunable via control of the charge on the system.[963] Density functional theory examination of the role played by extrinsic point defects, including surface adatoms (boron, carbon, nitrogen, oxygen, fluorine) with a [He] core electronic configuration, shows that boron and carbon, with an electronegativity close to that of P, prefer to break the sp^3 bonds of phosphorene and reside at the interstitial sites of the two-dimensional lattice by forming sp^2-like bonds with the native atoms. The nitrogen, oxygen and fluorine atoms, which are more electronegative than P, prefer surface sites and attract the lone pairs of phosphorene. The boron, nitrogen and fluorine adsorption also introduces local magnetic moments into the lattice. The boron, carbon, nitrogen and fluorine adatoms modify the band-gap to yield metallic transverse tunnelling characteristics. Oxygen does not modify the band-gap, and a diode-like tunnelling behavior is observed.[964] First-principles calculations based upon density functional theory show again that a vacancy or external strain alone do not induce magnetism in phosphorene. On the other hand, the interplay of a vacancy and external strain can indeed lead to magnetism. When biaxial or uniaxial strain along the zig-zag direction of samples containing phosphorus vacancies reaches 4%, the system favors a spin-polarized state with a magnetic moment of about $1\mu_B$ per vacancy site. This is due to the spin-polarized p-states of under-coordinated phosphorus atoms next to the vacancy. They are bonded in the absence of external strain or when phosphorene is subjected to small strains.[965]

Ab initio calculations show that non iso-electronic foreign atoms form quasi-bound states at various energy levels and create differing polarization states, depending upon the number of valence electrons between the phosphorus and dopant atoms. Distortions of the electronic bands, induced by the external species, lead to strong back-scattering effects on the propagating charge carriers. Depending upon the energy of the charge carrier and the type of doping, the conduction may range from diffusive to localized. Interstitial defects at vacant sites lead to homogeneous transport for different types of dopant atom. The relatively low reported values of charge mobility may originate from the presence of defects.[966]

Molecular dynamics simulations incorporating a Nosé–Hoover Langevin thermostat show that the Young's modulus, shear modulus, specific heat capacity and coefficient of thermal expansion of armchair and zig-zag sheets decrease with increasing vacancy defect ratio. The specific heat capacity and linear coefficient of thermal expansion increase with increasing temperature, at low temperatures, and converge to a constant value when the temperature is greater than 600K.[967] The calculated specific heat capacity of the two types of phosphorene sheet closely obeys the Debye T^3-law at low temperatures.

During the synthesis of 2-dimensional materials, line defects are often generated at the grain boundaries between domains of differing orientation. Twelve types of line-defect structure in monocrystalline phosphorene have been studied using first-principles calculations. These line defects are usually formed via the migration and aggregation of intrinsic point defects such as Stone-Wales, monovacancy or divacancy. The results show that the migration of point defects in phosphorene is anisotropic. The lowest migration energy barriers are 1.39 and 2.58eV for Stone-Wales defects in zig-zag and armchair directions, respectively, while the lowest migration energy barriers are 0.40 and 0.49eV for monovacancies in zig-zag and armchair directions, respectively.[968] The aggregation of point defects into lines is energetically favored over separate point defects. The axis of line defects is direction-selective, depending upon the type of point defect of which it is composed. The presence of line defects effectively modulates the electronic properties, making defect-containing material metallic or semiconducting with a tunable band-gap. A monovacancy-based line defect can behave like a metallic wire, thus offering the possibility of creating circuits of sub-nanometer width for nanoscale electronic applications. Density functional theory investigation of fault-line defect models shows that such low-energy line defects could modulate the electronic band-gap energies of single-layer two-dimensional phosphorene[969].

It is advantageous to use blue phosphorene in order to prepare small nanotubes, and it is possible to create low-energy nanotubes via the inclusion of defect lines. Such defect-induced faceted nanotubes, made using 5 different types of defect line, have negligible bending stresses. This leads to a reduction in the formation energy with respect to round nanotubes. Armchair faceted nanotubes have similar formation energies to those of multiphase faceted nanotubes, but have a wider range of possible structures. Zig-zag faceted nanotubes have lower formation energies than do round tubes or multiphase faceted nanotubes. The electronic properties of defect-induced faceted nanotubes are governed by the defect lines. They control the band-gap, and the shape of the electronic states at the band edges. The band-gaps increase with the radius of the nanotubes, and converge to equal those of isolated defect lines.[970]

First-principles calculations predict the existence of two typical grain boundaries in phosphorene. The grain-boundary region is reactive, and carbon or oxygen impurity atoms prefer to be incorporated into the grain boundary region rather than into the bulk. The formation of OP along the grain boundaries is an exothermic process. Both carbon and oxygen doping of grain-boundary defects give rise to magnetism. The band structures can also be tuned by carbon or oxygen doping.[971]

High-angle tilt boundaries with a higher density of (5|7) defect pairs oriented along the armchair direction are stronger than are low-angle tilt boundaries with a lower defect density. High-angle boundaries with a higher density of (4|8) defect pairs oriented along the zig-zag direction are similarly stronger than are low-angle boundaries having a lower defect density. Failure is due to the rupture of the most pre-strained bonds in the heptagons of the (5|7) defect pair, or octagons of the (4|8) pair. High-angle grain boundaries are more able to accommodate pre-strained bonds in heptagon and octagon defects, leading to higher failure stresses and strains.[972] These results cannot be explained in terms of a Griffith-type fracture-mechanics criterion because that does not take account of bond pre-stretching. Such anomalous mechanical and failure characteristics of phosphorene tilt grain boundaries are shared by graphene and hexagonal boron nitride. One conclusion is that they may be universal features of 2-dimensional materials.

In order to emphasize the close relationship which exists between material properties and local geometry, phosphorene having a conical structure was examined. A decrease in the semiconducting gap is observed which is directly linked to its non-planar shape. This geometrical effect occurs regardless of which phosphorene allotrope is involved, and thus constitutes an unique optical means for identifying local structural defects in this 2-dimensional material.[973]

7.11 Composites

Combining the electronic structures of two-dimensional monolayers in ultra-thin hybrid nanocomposites is expected to produce new properties which go beyond those of the individual components. First-principles calculations of hybrid graphene and phosphorene nanocomposites show that weak van der Waals interactions predominate between the graphene and phosphorene, with their intrinsic electronic properties being preserved. As the interfacial distance decreases, the Dirac point of the graphene moves from the conduction band to the valence band of phosphorene in the hybrid nanocomposites, thus inducing a transition from an n-type Schottky contact to a p-type Schottky contact at the graphene/phosphorene interface.[974] Other studies of single-layer and bilayer phosphorene with graphene again show that the properties of both graphene and phosphorene are preserved in the heterostructure. By applying a perpendicular electric field, it is also possible to tune the position of the band structure of phosphorene with respect to that of graphene, thus permitting control of the Schottky barrier height and doping of phosphorene.[975] Investigations of the electric field-dependent band structures and optical properties of graphene/phosphorene bilayers show that the effective electron Fermi velocity is comparable to that of a pure graphene monolayer. This situation is unaltered even in the presence of a finite external electric field. The electronic band structure changes from semiconducting to metallic under an external electric field strength of 0.5V/Å. The electric field enhances the structural stability of a bilayer system because the binding energy is increased by 20 to 30% as compared with that for a zero electric field. The optical anisotropy of electric polarization along the armchair and zig-zag directions disappears, but the anisotropic optical properties survive with regard to the electric polarization for in-plane and perpendicular directions.[976] Phosphorene and graphene have a tiny lattice mismatch, along the armchair direction, which can result in an atomically sharp in-plane interface. First-principles investigations of the electronic properties of lateral phosphorene/graphene heterostructures demonstrate that the electronic properties of this type of heterostructure can be highly tunable via quantum size effects and an externally applied electric field. Under a strong externally applied electric field, Dirac fermions can be created; with Fermi velocities which are about an order or magnitude smaller than that of graphene. Undoped and hydrogen-doped configurations exhibit three markedly different electronic phases, revealing the strongly tunable nature of this type of heterostructure. Graphene is a naturally good electrode for phosphorene. The transport properties of two-probe graphene/phosphorene/graphene devices exhibit tunnelling transport characteristics.[977] The interlayer thermal conductance at the phosphorene/graphene interface has been systematically investigated using classical molecular dynamics simulations. The transient pump-probe heating method predicts the interfacial thermal resistance of the phosphorene/graphene interface to be 8.41×10^{-8} Km^2/W at room temperature. Temperature, contact pressure, vacancy defects and chemical functionalization can all effectively reduce the thermal resistance at the interface. Numerical predictions of the thermal resistance reduction as a function of temperature, interfacial coupling strength, defect ratio and hydrogen-coverage yield figures of 56.5, 70.4, 34.8 and 84.5%, respectively.[978]

Vertical integration of two-dimensional materials has thus emerged as being an interesting means of designing novel electronic and opto-electronic devices. Density functional theory investigation of the graphene/phosphorene and hexagonal boron nitride/phosphorene heterostructures shows that the interlayer distance, binding energy and charge transfer in the two systems are similar. Interlayer non-covalent bonding is attributed to a weak coupling between the p_z orbital of phosphorene and the π-orbital of graphene or BN. A small amount of electron transfer from graphene and BN, depending upon the vertical strain, makes phosphorene slightly n-doped in both heterostructures. Some attractive characteristics of phosphorene, including the direct band-gap and linear dichroism, are preserved but a large redistribution of electrostatic potential across the interface occurs which might significantly renormalize the carrier dynamics and affect the excitonic behavior of the phosphorene. It is suggested that graphene and BN can be used as an effective capping layer, to protect phosphorene from structural and chemical degradation, while maintaining its useful electronic characteristics and acting as an active layer which tunes the carrier dynamics and optical properties of phosphorene.[979] Phosphorus suffers from the problem of degradation due to oxidation, but the properties can be preserved by encapsulation in hexagonal boron nitride; which also enhances the band-gap of the phosphorene layer. In comparison with the band-gap of a pristine phosphorene layer, the band-gap of a phosphorene/BN-monolayer system is increased by 0.15eV. It is further increased by 0.31eV in a BN(monolayer) /phosphorene/ BN(monolayer) trilayer structure. The band-gap does not increase further when the thickness of the h-BN layer is increased up to 4 monolayers. The anisotropic effective mass and optical properties are maintained by BN/phosphorene/BN heterostructures.[980] Measurements of phosphorene/hexagonal-BN heterostructures with one-dimensional edge contacts show that these transistors are stable under ambient conditions for up to

300h. They exhibit ambipolar behavior, a gate-dependent metal-insulator transition and mobilities of up to 4000cm^2/Vs. At low temperatures, gate-tunable Shubnikov de Haas magneto-oscillations and Zeeman splitting are observed in a magnetic field, with an estimated g-factor of about 2. The cyclotron mass of few-layer phosphorene holes increases from 0.25 to 0.31m_e as the Fermi level moves towards the valence-band edge.[981]

When the electronic structures of black-phosphorene/XT_2 (X = Mo, W; T = S, Se, Te) two-dimensional) van der Waals heterostructures are calculated using first-principles methods, the electronic band structures of both black phosphorene and XT_2 are preserved in the combined van der Waals heterostructures. The WSe_2/black-phosphorene van der Waals heterostructure exhibits type-I band alignment, but heterostructures involving MoS_2, $MoSe_2$, $MoTe_2$, WS_2 or WTe_2 exhibit type-II band alignment. Strong spin-splitting appears in all of the heterostructures when considering spin orbital coupling.[982]

7.12 Superconductivity

First-principles calculations predict that electron-doped phosphorene is a potential BCS-like superconductor. The stretching modes at the Brillouin-zone center are markedly softened by electron-doping, which results in strong electron-phonon coupling. Superconductivity can be induced by a doped electron density of above 1.3 x 10^{14}/cm^2, and may exist above liquid-helium temperatures when the doped electron density is greater than 2.6 x 10^{14}/cm^2. The superconductivity can be appreciably tuned and enhanced by tensile straining. The maximum critical temperature of electron-doped phosphorene is predicted to be higher than 10K.[983]

First-principles calculations also predict that bilayer phosphorene will become a quasi two-dimensional superconductor under a certain degree of interlayer compression. A decreasing interlayer distance may provoke a transition from a semiconducting phase to a metallic phase, while high vertical compression could make the lattice become dynamically unstable. In the stable metallic phase of bilayer phosphorene, interlayer phonon modes dominate electron-phonon coupling. The coupling can be greater than unity and the superconducting temperature can be higher than 10K.[984] The bilayer material can be transformed from a direct-gap semiconductor to a BCS superconductor by intercalating lithium atoms. In lithium-intercalated samples, the electron occupation of the lithium-derived band is small and the superconductivity is intrinsic. With increasing intercalation of lithium atoms, the increased metallicity and strong electron-phonon coupling favour increased superconductivity. As above, the electron-phonon coupling can be greater than unity and the superconducting temperature can be increased up to 16.5K.[985] The electron-phonon coupling comes mainly from low-frequency optical phonon modes, and biaxial strain gives rise to a more obvious increase in density of states around the Fermi level, and phonon softening in the low-frequency regime, as compared to the other two types of uniaxial strain. The electron-phonon coupling is therefore more significantly enhanced by biaxial strain than by uniaxial strains and the superconducting transition temperature increases sharply from 3 to 16K for a typical dopant concentration of $3.0 \times 10^{14}/cm^2$ when the biaxial strain reaches 4.0%.[986]

7.13 Thermal Conductivity

First-principles calculations of the thermal conductivity of black phosphorene and blue phosphorene show that the former exhibits a thermal conductivity anisotropy ratio of three, with predicted values of 110 and 36W/mK along the armchair and zig-zag directions, respectively, at 300K. Blue phosphorene is isotropic, with a zig-zag structure, and the predicted value is 78W/mK. These allotropes exhibit strikingly different thermal conductivity behaviours, with phonons having mean-free-paths of between 10nm and 1μm predominating in black phosphorene. A much narrower band (50 to 200nm) of mean-free-paths predominates in blue phosphorene.[987] A combination of density functional calculations and the Peierls-Boltzmann transport equation suggests the coexistence of size-dependent and size-independent thermal conductivities in single-layer black phosphorus along the zig-zag and armchair directions, respectively. The thermal conductivity of single-layer phosphorene is thought to be highly anisotropic because of orientation-dependent group velocities and phonon relaxation times. Thus the thermal conductivities of 10μm single-layer black phosphorus at 300K are 83.5 and 24.3W/mK along the zig-zag and armchair directions, respectively.[988] A Stillinger-Weber interatomic potential, parameterized for phosphorene, reproduces well the crystal structure, cohesive energy and phonon dispersion predicted by first-principles calculations. At room temperature, the intrinsic thermal conductivities along the zig-zag and armchair directions are about 152.7 and 33.0W/mK, respectively, with an anisotropy ratio of five. The marked directional-dependence of the thermal conductivity is attributed mainly to the strong anisotropy of phonon group velocities and to the weak anisotropy of phonon lifetimes. The effective phonon mean-free-paths in the zig-zag and armchair directions are about 141.4 and 43.4nm, respectively.[989]

First-principles calculations and non-equilibrium Green's function methods show that a significant crystallographic orientation-dependence of the thermal conductance exists in two-dimensional monolayer phosphorene sheet. The room-temperature thermal conductance along the zig-zag direction is 40% higher than that along the armchair direction. As usual, the conductance anisotropy can be tuned by straining. The zigzag-oriented conductance is enhanced when zigzag-oriented strain is applied, but decreases when armchair-oriented straining is applied. The armchair-oriented conductance decreases when either zigzag- or armchair-oriented strain is applied.[990]

Investigation of the in-plane and cross-plane thermal conductivities of single- and multi-layer phosphorene reveals a marked anisotropy in the in-plane thermal conductivity, with the value for the zig-zag direction being much higher than that for the armchair direction. The in-plane thermal conductivity of multi-layer phosphorene is insensitive to the layer number. This contrasts sharply with the case of graphene, where interlayer interactions strongly influence thermal transport. Tensile straining leads to an anomalous increase in the in-plane thermal conductivity; particularly in the armchair direction. Both the in-plane and cross-plane thermal conductivities can be modulated by such external strain, but strain modulation along the cross-plane direction is more effective and more tunable than that along the in-plane direction.[991]

A comparative analysis of thermal transport in graphene and phosphorene shows that the thermal conductance of phosphorene nanoribbons is greater than that of graphene nanoribbons. The graphene/phosphorene ratio of thermal conductivities decreases with decreasing temperature and decreasing nanoribbon width and *vice versa.* The greater thermal conductance and thermal conductivity in phosphorene nanoribbons arises from the lower cut-off frequencies of the acoustic modes.[992] When the thermal conductivity of monolayer phosphorene is calculated using large-scale classical non-equilibrium molecular dynamics simulations, the predicted thermal conductivities for infinite-length armchair and zig-zag phosphorene sheets are 63.6 and 110.7W/mK, respectively. The markedly anisotropic thermal transport is attributed to the distinct atomic structures and direction-dependent group velocities. When the thermal conductivities of two-dimensional graphene sheets having the same dimensions are calculated for comparison, the extrapolated thermal conductivities of two-dimensional graphene sheet are 1008.5 and 1086.9W/mK in the armchair and zig-zag directions, respectively. These are clearly an order of magnitude higher than those in phosphorene. In comparison with graphene, the vibrational frequencies that can be excited in phosphorene are severely limited. The thermal conductivities of phosphorene and graphene sheets exhibit a monotonically decreasing trend in both structures.[993]

When the orientation-dependent thermal transport in phosphorene sheets is studied for crystal chiralities ranging from 0° to 90°, using the Boltzmann transport equation and first-principles calculations, it is found that the intrinsic thermal conductivity is a smooth monotonically decreasing function of the chirality, which exhibits a sinusoidal behavior that is bounded by terminal values of 48.9W/mK (0°) and 27.8W/mK (90°). The optical modes make unusually large contributions to heat transfer, and account for almost 30% of the total thermal conductivity. This is because the optical phonons have group velocities and relaxation times which are comparable to those of the acoustic phonons.[994]

7.14 Applications

7.14.1 Solar Power

Edge-modified phosphorene nanoflakes have been proposed for use as donor and acceptor materials for heterojunction solar cells. Density functional theory calculations show that heterojunctions which consist of hydrogen- and fluorine-passivated nanoflakes have several desirable opto-electronic properties that are relevant to solar-cell use. The calculations also predict that the maximum energy conversion efficiency of this easily-fabricated type of heterojunction can be as high as 20%. This makes them extremely competitive with other types of two-dimensional heterojunction.[995] Using first-principles calculation studies of the electronic properties of few-layer phosphorene, such as the layer-dependent behavior of the band-gap, work-function band-alignment and carrier effective mass, shows that the material exhibits a robust direct band-gap behavior. The band-gap decreases, with the number of layers, according to a power-law relationship. The work function decreases rapidly from the monolayer (5.16eV) to a trilayer (4.56eV) and then falls more slowly with further increases in layer number. When compared with monolayer material there is seen to be a sharp decrease in the hole effective mass along the ridge (zig-zag) direction in bilayer material. This indicates the existence of strong interlayer coupling, and a screening effect. It is suggested that few-layer material having a layer-dependent band-gap and a robust direct band-gap nature is a promising material for efficient solar-energy harvesting. The few-layer material performs better, than monolayer samples, in terms of a lower carrier effective mass, a higher carrier density and a weaker scattering due to enhanced screening. The layer-dependent band edges and work functions of few-layer material allow for the modification of Schottky barriers to give increased carrier-injection efficiency.[996]

7.14.2 Sensors

First-principles calculations of the adsorption of CO, CO_2, NH_3, NO and NO_2 gases on monolayer phosphorene predict a superior sensing performance that may surpass that of other two-dimensional rivals, such as graphene and MoS_2. Molecular doping (charge transfer between gas molecules and phosphorene) is identified as being the driving mechanism for the high adsorption strength. Calculations of the current-voltage relationship, using the non-equilibrium Green's function formalism, reveal a one- to two-orders of magnitude anisotropy along the armchair and zig-zag directions. This is consistent with the anisotropic electronic band structure. The current-voltage relationship exhibits distinct responses, with a marked change in the relationship along the armchair or zig-zag directions, depending upon the type of gas molecule. Such a selectivity and sensitivity to adsorption thus makes phosphorene a superior gas sensor and promises wide-ranging applications.[997] A single-electron transistor device, made of a material such as phosphorene, has been proposed as a future low-power device for sensing purposes. The operation and performance of such a device has been investigated by using density functional theory-based *ab initio* calculations to predict their relative sensitivities to various gas molecules. The adsorption of CO, CO_2, NH_3 and NO_2 on monolayers of phosphorene has been investigated with regard to their most stable configurations and relative orientations on the host layer, and it seems that the structural and electronic properties of the host layers are unaffected by the adsorption processes. Phosphorene is also sensitive to nitrogen-based molecules, and magnetism may be induced in the presence of a paramagnetic molecule. The current results suggest that the charge stability diagram of a single-electron transistor is unique to a specific gas molecule on the two-dimensional layer, and this is sensitive up to the addition or removal of a single molecule from the island. The wide temperature-range of operation, extreme detection sensitivity and the versatility of two-dimensional materials for gas-sensing will make these devices leading candidates for practical application.[998] A field-effect transistor sensor device can also be based upon two-dimensional phosphorene nanosheets. Such a sensor exhibits an ultra-high sensitivity to NO_2 in dry air, and the sensitivity depends upon the thickness. The maximum response is observed for 4.8nm-thick nanosheet, with a room-temperature sensitivity of up to 190% for 20ppb. First-principles calculations, combined with statistical thermodynamics modelling, predict that the adsorption density is about $10^{15}/cm^2$ for 4.8nm-thick nanosheets when exposed to 20ppb of NO_2 at 300K. Modelling also suggests that the dependence of sensitivity upon nanosheet thickness is dictated by the band gap for sheets of less than 10nm, and by the effective thickness for thicker sheets.[999] First-principles calculations, based upon density functional theory, of phosphorene which is substitutionally doped with 17 different atoms reveal that anionic X (X = O, C, S) dopants have a large binding energy and highly dispersive electronic states, indicating the formation of covalent X-P bonds and strong structural stability. Alkali atom (lithium, sodium) dopants donate most of the electrons in the outer s-orbital to form ionic bonds with P, and the band-gap is decreased by pushing down the conduction band. This suggests that the optical and electronic properties of doped phosphorene can be tailored. In the case of doping with VIIIB-group (Fe, cobalt, Ni) elements, a strong affinity is predicted and the binding energy and charge transfer are correlated strongly with their electronegativity. Examination of NO molecule-adsorption shows that metal-doped phosphorenes generally exhibit a significantly enhanced chemical activity, as compared with pristine phosphorene. Substitutionally-doped phosphorene thus exhibits many interesting electronic and optical properties which differ from those of pristine phosphorene, and metal-doped phosphorenes offer great promise in chemical applications which involve molecular adsorption and desorption processes: such as, materials growth, catalysis, gas sensing and storage.[1000]

First-principles calculations show that the CrO_3 molecule, with its acceptor electron nature, is chemisorbed at the surface with minimal geometrical distortion. Following adsorption, a partial charge gradient is established between the phosphorus atoms located in the upper and lower phosphorene planes. Variations in the CrO_3 concentration lead to differing interaction strengths: at high concentrations of adsorbed CrO_3, interaction with the surface becomes stronger due to an increased steric effect between neighboring molecules.[1001] This effect, together with geometrical distortions produced in the structure due to the large number of adsorbed molecules, leads to a decrease in the adsorption energy. It is concluded that phosphorene is a promising material for use as a gas sensor. First-principles calculations of the Van der Waals binding mechanism between pristine or defected phosphorene and PH_3 and AsH_3 show that both gas molecules interact weakly with the phosphorene sheet; with the binding being slightly stronger for AsH_3 than for PH_3. The creation of monovacancies and divacancies within the phosphorene sheet significantly enhances the adsorption, and the adsorption energies of both PH_3 and AsH_3 are increased by factors of 4 and 3, respectively, over the values for pristine phosphorene.[1002] These findings again show that phosphorene which contains vacancies can be considered to be an efficient sensor for toxic gases.

7.14.3 Catalysis

Density functional theory identifies strain-engineered phosphorene as being a possible photocatalyst for use in water-splitting hydrogen production[1003]. Lattice dynamic calculations demonstrate the stability of such artificial materials under various strains. The phosphorene lattice is unstable under compressive strains, but is very stable under tensile strains. A tunable band-gap ranging from 1.54eV under ambient conditions, to 1.82eV under tensile strains, is determined using parameter-free hybrid functional calculations. The appropriate band-gaps and band-edge alignments at certain pH levels demonstrate the potential application of phosphorene as an efficient photocatalyst for visible-light water-splitting. Strained phosphorene exhibits significantly improved photocatalytic properties under visible-light irradiation. The negative splitting energy of absorbed H_2O indicates that water-splitting on phosphorene is energetically favorable, with and without strain.[1004]

Like graphene, single-layer phosphorene is flexible and can be mechanically exfoliated. Phosphorene catalysts have been prepared by exfoliation and exposed to 4 levels (0.5, 1,

2, 4ppm) of arsenic concentration.[1005] The most appropriate adsorption rate for As^{III} was 0.0329/min and the sorption capacity was 4.67mg/g at a pH of 7.

7.14.4 Energy Storage

Pristine or defect-containing phosphorene is considered to be a promising anode material for lithium-ion batteries. Calculations suggest that the binding energies of lithium adsorption at various sites vary within a narrow range, and that the binding between lithium atoms and pristine phosphorene is relatively weak. Defects can greatly improve the performance of phosphorene in terms of binding energy and lithium diffusion. The binding energy of lithium atoms around a vacancy is higher, by about 1eV, than that of perfect phosphorene. The lithium atoms can also diffuse between two adjacent grooves, over an energy barrier of 0.13eV; opening up a novel channel for lithium diffusion and markedly improving rapid charge/discharge capabilities. These properties indicate that defect-containing phosphorene has great potential as an electrode material in lithium-ion batteries.[1006] Density functional theory studies of the possibility of using phosphorene nanoribbons as anode materials for lithium-ion batteries show that both phosphorene and its nanoribbons are direct band gap semiconductors, and undergo a semiconductor-metal transition upon lithiation; together with significant charge transfer. The estimated energy barrier to lithium diffusion on phosphorene monolayers, along the zig-zag direction, is 0.09eV, thus promising ultra-fast charging/discharging. The fast lithium ion mobility is retained in phosphorene nanoribbons.[1007] The structure of phosphorene is not greatly influenced by lithiation, and the volume change is only 0.2%. The barriers to lithium diffusion are calculated to be 0.76 and 0.72eV on monolayer and double-layer phosphorene, respectively, and the theoretical specific capacity of a phosphorene monolayer is 432.79mAh/g. This is larger than those of other commercial anode materials. Lithium-ion adsorption and diffusion in one-dimensional armchair-type nanotubes with lithium intercalation on the inside or outside surfaces has been compared with the behavior in 2-dimensional phosphorene sheet. The maximum lithium storage capacity (for the $Li_{22}P_{44}$intercalation state) is, as above, evaluated to be 432mAh/g.[1008] The inside-surface system has a higher adsorption energy and lower lithium-diffusion energy-barrier than that of phosphorene sheet or the outside-surface system. The results suggest that nanotubes can serve as potential anode materials for lithium-ion batteries.

Implantation with silicon significantly improves the binding of lithium atoms, while the diffusion barrier is unaffected. The high capacity, low open-circuit voltage, small volume change and electrical conductivity of lithiated phosphorene indeed make it a good candidate for application as an electrode material in batteries.[1009] When phosphorene/graphene heterostructures are examined using first-principles calculations, the binding energy of lithium on pristine phosphorene is found to be relatively weak (about 1.9eV) while phosphorene/graphene heterostructures can have a greatly improved binding energy (2.6eV) without this affecting the high lithium mobility within the layers. The electronic structures show that the large lithium adsorption energy and the rapid diffusivity in the heterostructures arise from an interfacial synergy effect. The heterostructures also exhibit ultra-high stiffnesses of 350N/m in the armchair direction and 464N/m in the zig-zag direction. This can avoid distortion of the pristine phosphorene following the insertion of lithium. The use of heterostructures can therefore greatly improve the cycle life of a battery and, due to the high capacity, good conductivity, excellent lithium mobility and ultra-high stiffness, phosphorene/graphene heterostructures are very promising anode materials for lithium-ion batteries.[1010]

First-principles calculations of the sodium adsorption energy, specific capacity and sodium diffusion barriers on monolayer phosphorene indicate a favorable sodium-phosphorene interaction and a high theoretical sodium storage capacity. The sodium-phosphorene system undergoes a semiconductor-metal transition at high sodium concentrations. The sodium diffusion on phosphorene is fast and anisotropic, with an energy barrier of 0.04eV. Due to its high capacity, good stability, excellent electrical conductivity and high sodium mobility, monolayer phosphorene is deemed to be a very promising anode material for sodium-ion batteries.[1011] The main challenge to competitive sodium-ion battery technology is the development of suitable anode materials. A hybrid material consisting of a few phosphorene layers, sandwiched between graphene layers, has a specific capacity of 2440mAh/g (calculated using the mass of phosphorus alone) at a current density of 0.05A/g, and an 83% capacity retention after 100 cycles while operating between 0 and 1.5V. The large capacity of the anode is attributed to the intercalation of sodium ions along the x-axis of the phosphorene layers, followed by the formation of a Na_3P alloy. The graphene layers in the hybrid material act as both a mechanical backbone and an electrical highway, ensuring that a suitable elastic buffer-space accommodates the anisotropic expansion of phosphorene layers along the y and z axial directions.[1012]

7.14.5 Energy Harvesting

A nanogenerator has been proposed which is powered by cyclically doping phosphorene with ambipolar organic molecules. Benzyl viologen is used as an n-type donor and tetracyanoquinodimethane is used as a p-type acceptor. Density functional theory calculations show that the device is able to achieve an open-circuit voltage of about 2.7V. The output power can be modulated by adjusting the concentration and cyclic frequency of the molecular doping. The proposed design is simple, is experimentally feasible and is promising for applications in energy-harvesting and solid-state lighting.[1013]

7.14.6 Thermoelectricity

Thermoelectric devices which exploit the Seebeck effect in order to convert heat-flow into electrical energy are highly desirable for the development of portable solid-state passively-powered electronic systems. The conversion efficiency of such a device depends upon the dimensionless thermoelectric figure-of-merit, which is proportional to the ratio of the electrical conductance to the thermal conductance. First-principles and model calculations show that phosphorene possesses a spatially anisotropic electrical conductance, and that its lattice thermal conductance also exhibits a pronounced spatial anisotropy. The main electrical and thermal conduction directions are mutually orthogonal, thus enhancing the ratio of these conductances. The thermoelectric figure-of-merit along the armchair direction of phosphorene at 500K may consequently meet the criterion for commercial use. It is close to unity even at room temperature for moderate doping (about $2 \times 10^{12}/cm^2$). It is hoped that phosphorene will prove to be an environmentally sound thermoelectric material possessing unprecedented qualities. It is also a mechanically flexible material that very efficiently converts heat energy at about 300K, and its exploitation does not require sophisticated engineering techniques.[1014] Density functional theory studies of phosphorene nanoribbons having various widths and edge configurations show that armchair nanoribbons are semiconducting while zig-zag nanoribbons are metallic. The band-gaps of armchair nanoribbons decrease monotonically with increasing ribbon width. Upon passivating the edge phosphorus atoms with hydrogen, the zig-zag form also become semiconducting, while the armchair form exhibits a larger band-gap than the pristine counterpart. Boltzmann theory and relaxation-time approximations show that all of the semiconducting nanoribbons exhibit very large Seebeck coefficients and these can be further increased by hydrogen passivation of the edges. On the basis of the significantly enhanced Seebeck coefficient and decreased thermal conductivity, it is expected that at least one type of phosphorene nanoribbon can be optimized so as to exhibit a very high figure-of-merit at room temperature and thus offer great promise for thermoelectric applications.[1015]

First-principles calculations based upon density functional theory show that the thermo-electric performance of phosphorene is markedly enhanced upon applying a tensile strain along a particular direction. Material which is subjected to 10% strain exhibits a very large power factor, exceeding $10mW/mK^2$. This is almost two orders of magnitude higher than the power-factors of conventional flexible thermo-electric materials. The extreme improvement in the thermo-electric performance is explained in terms of the concept of a strain-induced energy valley.[1016] First-principles calculations, combined with semi-classical Boltzmann theory, of further enhancement of the thermo-electric performance by using strain-induced band convergence via the effect of uniaxial strain shows that, when zigzag-direction strain is applied, the Seebeck coefficient and electrical conductivity in the zig-zag direction are both simultaneously enhanced at a critical strain of 5%; where band convergence is achieved. The highest figure-of-merit, of 1.65 at 300K, could be conservatively estimated by using the bulk lattice thermal conductivity. When an armchair-direction strain of 8% is applied, the room-temperature figure-of-merit can attain 2.12 in the armchair direction.[1017] First-principles calculations of the electron-phonon interaction in phosphorene, based upon density functional perturbation theory and Wannier interpolation, reveal that the high anisotropy provides extra phase space for electron-phonon scattering and that optical phonons make appreciable contributions. Neither effect can be captured by deformation potential calculations. Simulation predicts carrier mobilities of about 170cm^2/Vs for both electrons and holes at 300K, and a thermo-electric figure-of-merit of up to 0.14 in p-type impurity-free material at 500K.[1018]

The generalized Boltzmann theory has been used to describe the thermo-electric transport properties of monolayer phosphorene in the presence of short-range and long-range charged impurity interactions. The electrical conductivity is highly anisotropic, while the Seebeck coefficient and figure-of-merit are almost isotropic, without being affected by the presence or absence of a coupling term.[1019] The conductivity for n-type doping is more affected by the coupling term than is that for p-type.

7.14.7 Electronics

First-principles studies of the magnetic properties and spin caloritronics of zigzag-type blue phosphorene nanoribbons show that bare, or monohydrogenated, samples exhibit spin-semiconducting properties which arise from the edge electronic states. The p_y orbitals of the edge phosphorus atoms make the main contributions to these states. The spin-semiconducting property imparts a natural advantage on the fabrication of perfect thermospin devices having a stronger spin Seebeck effect, than charge Seebeck effect, at the Fermi level. When a temperature difference is applied, the electric current with the different spin index exhibits bipolar behavior, and the spin-filtering efficiency can attain 1200%. By changing the widths of bare and monohydrogenated samples, the ratio of the spin Seebeck coefficient to the charge Seebeck coefficient at the Fermi level can be equal to about 10 at room temperature.[1020]

Self-consistent atomistic quantum transport simulations of bilayer phosphorene field-effect transistors show that 10nm-channel transistors can exhibit excellent device characteristics such as a large current ratio (greater than 7) and a small sub-threshold swing of 66mV/dec; with a double-gate device structure. Scaling of the gate dielectric monotonically improves the overall performance, but the channel-length can be scaled down only to about 8nm due to significant short-channel effects. Phosphorene field-effect transistors have more favorable switching characteristics than those of other similar two-dimensional bilayer semiconductor devices, thus making both monolayer and bilayer phosphorene attractive for future switching applications.[1021] Self-consistent atomistic quantum transport simulations of ultra-thin two-dimensional phosphorene with a direct band-gap suitable for tunnelling field effect transistors shows that the latter has a sub-threshold slope of below 60mV/dec. It is predicted that devices which are oriented in the small effective mass direction can yield a much larger on-current, for a given on-current/off-current ratio, than can monolayer $MoTe_2$ tunnelling field effect transistors. A gate underlap structure is required for the scaling-down of phosphorene tunnelling field effect transistors, in the small effective mass direction, in order to suppress the source-to-drain direct tunnelling leakage current.[1022] Combined density functional and non-equilibrium Green's function studies of monolayer black phosphorus nanoribbons reveal a giant Stark effect which can switch on transport channels in semiconducting nanoribbons under low bias and induce an insulator-metal transition. Unlike the cases of graphene and MoS_2 nanoribbons, the carrier transport channels under low bias are located mainly in the interior of both armchair and zig-zag nanoribbons and are immune to a small number of edge defects. A device made from a phosphorene nanoribbon-based dual-gate field-effect transistor, with an on/off ratio of 1000, has been proposed on the basis of the giant electric-field tuning effect.[1023] Phosphorene offers the promise of the sought-after high-mobility large direct band-gap two-dimensional layered crystal that will be ideal for opto-electronics and flexible devices. Its instability in air is a drawback to practical application. Air-stable devices, with dielectric and hydrophobic encapsulation, have however been demonstrated. An aging mechanism, which can initiate from the surface of bare samples or from the edges of samples with thin dielectric coatings, highlights the ineffectiveness of conventional scaled dielectrics. Studies indicate that a double-layer capping of Al_2O_3 and hydrophobic fluoropolymer produces transistors which exhibit indefinite air-stability.[1024] Study of the transistor behavior on mechanically exfoliated few-layer phosphorene indicates an on-current of 144mA/mm and a hole mobility of $95.6cm^2/Vs$. Atomic layer deposition of Al_2O_3 on the top shows that the polarity of the transistor then alternates from p-type to ambipolar. This transition is attributed to changes, in the effective Schottky barrier heights for both electrons and holes at the metal contact edges, which originate from fixed charges in the atomic layer deposited dielectric.[1025] A few-layer phosphorene MOSFET, with Al_2O_3 surface passivation and a Ti/Au top gate, is stable in the ambient atmosphere for at least 100h. Following annealing (200C, 1h, dry nitrogen), the drain current increases by an order of magnitude, to about 100mA/mm. This is attributed to a reduction in trapped charge in the Al_2O_3 and/or Schottky barrier at the source and drain contacts. The drain current is stable, between -50 and 150C, for up to at least 2000h. Such results suggest that the environmental protection of phosphorene should not be a major concern, and that its passivation should focus on its effect upon electronic control and transport. With cut-off frequencies now approaching the GHz range, such phosphorene MOSFETs will be able to satisfy the speed and stability requirements of most flexible electronics. Phosphorene here has a natural advantage due to its corrugated lattice structure.[1026]

Phosphorene is a promising material for studying spin transport, due to its low spin-orbit coupling and its 1/2 nuclear spin; which can permit the study of hyperfine effects. By making permalloy and cobalt contacts to few-layer material, the Schottky barrier height can be determined as a function of gate bias. Flat-band barrier heights, relative to the valence band edge, of 110 and 200meV are found for permalloy and cobalt, respectively.[1027]

It can be demonstrated experimentally that the transport gap can be tuned monotonically, from about 0.3 to about 1.0eV, when the flake thickness is decreased from bulk dimensions to a single layer. As a result, the ON current, the OFF current and the current ON/OFF ratios of phosphorene field effect transistors are found to be significantly affected by the layer thickness. The transport gap can be deduced from the transfer characteristics of phosphorene field effect transistors by using a robust technique. By scaling the thickness of the gate oxide, it is possible to demonstrate an enhanced ambipolar conduction in both monolayer and few-layer phosphorene field effect transistors. The asymmetry of the electron and hole currents depends upon the layer thickness, and can be explained in terms of dynamic changes in the metal Fermi level with the energy band of phosphorene, depending upon the layer number. The Schottky barrier heights for both electron and hole injection are functions of the layer thickness.[1028]

Non-equilibrium Green's function simulations of Schottky barrier field-effect transistors, based upon few- and mono-layer phosphorene, show that scaling-down of the gate-oxide thickness results in marked ambipolar I-V characteristics and to a significant increase in the minimal leakage current. The problem of leakage is particularly great when the gate insulator is thin and the number of layers is large. It can be suppressed by reducing the phosphorene to a mono- or bi-layer. Unlike graphene and layered dichalcogenide materials, both the on-current of the phosphorene Schottky barrier field-effect transistors and the metal-semiconductor contact resistance between metal and phosphorene depend strongly upon the crystalline transport direction.[1029] Quantum transport simulations of the ballistic transport characteristics of metal-oxide semiconductor field effect transistors based upon phosphorene reveal that, at the 10nm channel length scale, phosphorene

MOSFETs out-perform monolayer HfS_2 MOSFETs in terms of on-state current. On the other hand, monolayer HfS_2 MOSFETs can offer comparable, but slightly degraded, device performances when compared with phosphorene MOSFETs at the 5nm channel length.[1030] The stacking-order of bilayer black phosphorene has an effect upon the properties of p- and n-MOSFETs when two layers of black phosphorus are stacked in three different orders and used as channel materials. The results show that a roughly 1.37 times and 1.49 times increase in on-current is possible along armchair and zig-zag directions, respectively. A 55.11% variation in transconductance is possible along the armchair direction, by changing the stacking order (AA, AB, AC) and a roughly 8 times increase in on-current can be achieved by changing the channel orientation (armchair, zig-zag) in the p-MOSFET. A roughly 14.8mV/V drain-induced barrier decrease is observed in both p- and n-MOSFETs, indicating a certain immunity to short-channel effects.[1031] Self-consistent multiscale quantum transport simulations show that the MOSFET structure can effectively suppress ambipolar conduction and decrease the leakage current. It is thus more scalable than the Schottky-barrier FET structure. The interplay of quantum mechanical effects and the highly anisotropic band structure plays a critical role in phosphorene transistors with a sub-10nm channel length. The optimum choice of the crystalline transport direction is there completely different from that in phosphorene FETs with a longer channel length. With a sub-10nm channel length, and considerable quantum tunnelling effects, the anisotropic band structure still provides better device performance in terms of on-current when compared with other two-dimensional semiconductors with nearly isotropic band structures, such as MoS_2.[1032] Enhanced electron and hole transport is found in few-layer phosphorene field effect transistors when using titanium as the source/drain contact electrode and 20nm of SiO_2 as the back gate dielectric. The field effect mobility values are about $38cm^2/Vs$ for electrons and some $172cm^2/Vs$ for holes. Contact resistances arising due to the Schottky barriers at the source and drain end ultimately limit the on-state performance. The transport gap and the Schottky barrier height at the metal/phosphorene contact interface can be deduced from the ambipolar transfer characteristics of phosphorene FETs. This is true of any ultra-thin body semiconductor which exhibits symmetrical ambipolar conduction. A high-gain high noise margin chemical dopant free fully complementary logic inverter can be based upon ambipolar phosphorene FETs.[1033]

It is thought to be possible to convert a normal insulator into a topological one by applying an external electric field that shifts various bands by different energies and induces a band inversion, leading to a topological state. Unlike graphene and silicene, single-layer phosphorene has a band-gap of 1.5 to 2.2eV and is not expected to exhibit band inversion and the resultant topological insulator behavior. First-principles calculations involving the application of a perpendicular electric field, $F_{\perp}$, to few-layer phosphorene predict a continuous transition from a normal insulator to a topological insulator - and eventually to a metal - as a function of $F_{\perp}$. The tuning of the topological behavior by means of an electric field would lead to spin-separated gap-less edge states: a quantum spin Hall effect. This offers the possibility of converting normal insulating materials into topological ones via an electric field, thus creating a multifunctional field effect topological transistor that could simultaneously control both spin and charge carriers.[1034]

Density functional theory calculations and non-equilibrium Green's function technique investigations, of the electronic transport properties of zig-zag nanoribbon homostructured negative differential resistance devices, show that the proposed devices have a simple construction but exhibit a good negative differential resistance as well as supporting quite large currents. This offers great potential for the building of nano-electronic devices. Analysis of the electronic structures and transmission spectra gives a clear physical picture of the negative differential resistance mechanism, in which the current-voltage behavior originates from bias-dependent interaction between the discrete density-of-states peaks of the electrodes and the narrow states around the Fermi level in the scattering region.[1035]

The transport properties of a Schottky contact structure based upon zig-zag phosphorene nanoribbon has been investigated using the non-equilibrium Green's function formalism and density functional theory. The calculated band structures show that the unpassivated nanoribbon is metallic, while H-passivated material is semiconducting. The in-plane contact structure of unpassivated and H-passivated nanoribbons leads to the formation of a Schottky barrier. The rectification ratio can attain 10^3 in the bias region from 0.3 to 0.5V.[1036] Upon extending the length of the H-passivated material in the Schottky contact structure, the currents in the Schottky contact structure are decreased, but the rectification ratio can be increased: when the length of the H-passivated nanoribbon is 4 times that of the unpassivated nanoribbon, the average rectification ratio increases to 10^6 in the bias region from 0.3 to 0.5V while the maximum rectification ratio attains 10^7 at 0.45V.

Ultra-small phosphorene quantum dots with an average size of 2.6±0.9nm have been prepared using a liquid exfoliation method involving ultrasound probe sonification followed by bath sonification. Upon coupling the as-prepared quantum dots to a microfiber light field, the quantum-dot based saturable absorber device exhibits ultra-fast non-linear saturable absorption properties with an optical modulation depth of 8.1% in the telecommunication band. When integrated with an all-fiber quantum-dot based saturable absorber, a continuous-wave passively mode-locked erbium-doped laser cavity delivers stable self-starting pulses having a pulse duration of 0.88ps and a cavity repetition rate of 5.47MHz.[1037] These results contribute to the growing body of work on the non-linear optical properties of ultra-small phosphorene quantum dots, and illustrate the new opportunities for this 2-dimensional nanomaterial in ultra-fast photonic technologies. A tight-binding Hamiltonian investigation of the energy spectra of triangular zig-zag phosphorene quantum dots and rings shows that, in comparison with the outer edge states in quantum dots, new extra inner edge states can be produced in quantum rings by the internal hole. A transition from uncoupled to coupled edge states can be induced by decreasing the width between the outer and inner edges of the quantum rings. The edge states in dots or rings are all anisotropically localized to one side, rather than on three sides as in triangular graphene quantum dots and rings.[1038] The dot and ring energy levels can be anisotropically manipulated by applying external electric fields or strain. This clearly demonstrates their potential for application in field effect transistors or electromechanical devices.

Variations in the electronic properties of zigzag-edge nanoribbons under a perpendicular electric field have been studied using the tight-binding Hamiltonian, combined with the surface lattice Green's function approach. This shows that the response of edge states to a electric field applied perpendicularly to an N-nanoribbon, where N is the number of zig-zag chains, depends upon the parity of N. The field opens up a gap between two edge bands near to the Fermi energy in the case of even N but, in the case of odd N the two edge bands are nearly always degenerate. This difference arises from the fact that the Stark-effect induced energies at the upper and lower edges for even and odd N are different due to the phosphorene lattice structure. The electronic densities are therefore more localized at the edges driven by the field, for even N, and not for odd N.[1039] This even-odd effect is also reflected by the conductance, so that odd-N nanoribbons may be more suitable for use in field-effect transistors.

7.14.8 Biological

The villin headpiece is the smallest known example of a cooperatively-folded naturally occurring protein. Large-scale molecular dynamics simulations have used it as a model protein for investigating the disruption, of protein structure and function, by phosphorene. Graphene has also been studied in order to compare its biological effect. It is found that graphene's disruptive influence is more severe, indicating that phosphorene's interaction with the villin headpiece is weaker than graphene's. The use of other model materials, pseudo-phosphorene and pseudo-graphene, reveals that the puckered surface can have a marked effect in weakening the ability of the material to disrupt proteins. These findings shed some light on understanding protein-nanomaterial interactions and may further bio-applications of phosphorene.[1040]

Chapter 8

Bismuthene

A double cantilever beam fracture technique has been used to transfer epitaxial bismuth thin films, grown on silicon (111), to silicon strips coated with epoxy. The transferred films retain electrical, optical and structural properties which are comparable to those of as-grown epitaxial films. The adhesion energy at the bismuth/silicon interface has been measured to be about $1J/m^2$; comparable to that of exfoliated and wet-transferred graphene.[1041] This low adhesion energy and unexpected ease-of-transfer for an epitaxially grown film may permit study of the material's properties on arbitrary substrates. This then suggests a means for integrating other group-V epitaxial films such as phosphorene with arbitrary substrates, or for isolating bismuthene.

First-principles phonon and finite-temperature molecular dynamics calculations, incorporating spin-orbit coupling, show that free-standing single-layer buckled-honeycomb and asymmetrical-washboard bismuthene structures are stable at high temperatures. Spin-orbit coupling is found to be essential to determining lattice constants, phonon frequencies, band-gaps and cohesion. The phonons of 3-dimensional hexagonal crystals, as well as those of single-layer bismuthene, are softened by spin orbit coupling. In going from the 3-dimensional hexagonal crystal to free-standing single-layer structures, the 2-dimensional hexagonal lattice is compressed and the semimetal is transformed into a semiconductor by the confinement effect.[1042] On the other hand, by going from single-layer to bilayer bismuthene, the lattice is slightly expanded and the fundamental band-gaps are narrowed. Thus interlayer coupling in multilayer and 3-dimensional bismuth crystals is essential to trivial to non-trivial topological transitions and semimetal-to-semiconductor transitions.

Bismuth exhibits the highest spin-orbit coupling strength among the main group elements. The quantum anomalous Hall state is predicted to occur in half-hydrogenated bismuth honeycomb monolayers, Bi_2H, but experimental proof is lacking. Halogenated films, Bi_2X (X = F, Cl, Br), have a metallic nature which blocks quantum anomalous Hall effects. First-principle calculations show that fully-cyanided bismuthene, $Bi_2(CN)_2$, a 2-dimensional topological insulator which supports a quantum spin Hall state with a gap of up to 1.10eV, and half-saturated, $Bi_2(CN)$, a Chern insulator which supports a valley-polarized quantum anomalous Hall state with a Curie temperature of 164K and a gap of up to 0.348eV, might be further tuned by means of bi-axial strain and the spin-orbit coupling strength.[1043]

Chapter 9

Borophene

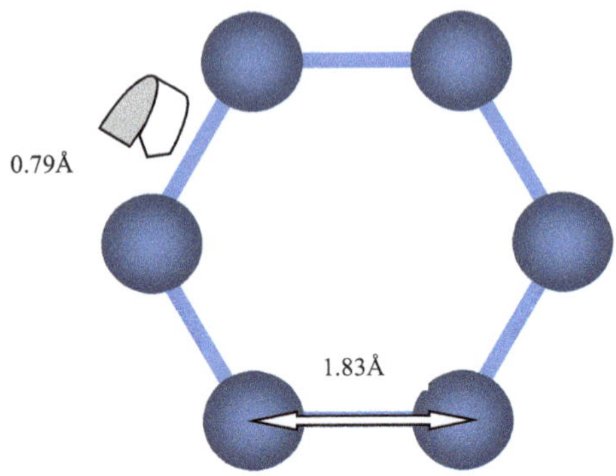

Pure boron is quite similar to carbon, and forms simple planar molecules and cage-like fullerenes. Theory predicts that two-dimensional boron sheets will take up an atomic configuration which is similar to that of boron atomic clusters. Atomically thin crystalline two-dimensional boron sheets, known as borophene, can be prepared on silver surfaces under ultra-high vacuum conditions. Detailed characterization reveals structures which recall fused boron clusters, with multiple scales of anisotropic out-of-plane buckling. Unlike bulk boron, borophene exhibits metallic characteristics which are consistent with a highly anisotropic two-dimensional metal.[1044] Elemental boron, being electron-deficient, cannot form graphene-like structures but triangular lattices containing hexagonal vacancies have been predicted to be stable.

9.1 Structure

Experimental and computational studies show that the B_{36} cluster has a planar C_{6v} structure with a central hexagonal hole, thus providing experimental evidence for the viability of atomically-thin boron sheets. Neutral B_{36} is the smallest boron cluster to have six-fold symmetry, plus a perfect hexagonal vacancy.[1045] A boron cluster with a double-hexagonal vacancy is a more flexible structural motif. The photo-electron spectrum of B_{35} has a simple pattern, having some similarity to that of B_{36}. Global-minimum searches indicate that B_{35} possesses a planar hexagonal structure which is similar to that of B_{36}, apart from a missing interior boron atom that creates the double-hexagonal vacancy.

Closed-shell B_{35} exhibits triple π aromaticity, with 11 delocalized π-bonds; analogous to benzo(g,h,i)perylene ($C_{22}H_{12}$). The B_{35} cluster can be used to build atomically-thin boron sheets having various hexagonal hole densities.[1046] Combined photo-electron spectroscopic and theoretical studies also indicate the existence of inherently chiral boron clusters, $[B_{30}]^-$. The most stable structure of $[B_{30}]^-$ is quasi-planar, with an hexagonal hole. Two enantiomers, arising from differing positions of the hexagonal hole, are degenerate. Both of them should co-exist experimentally because they have identical electronic structures and give rise to identical simulated photo-electron spectra.[1047]

First-principles calculations of monolayer 8-Pmmn borophene attribute amazing electronic properties to the 2-dimensional material. On the basis of Born effective charge analysis, 8-Pmmn borophene is deduced to be the first single-element based monolayered material to exhibit 2 sub-lattices having substantial ionic features. A tight-binding model confirms that the crystal hosts massless Dirac fermions and that the Dirac points are protected by symmetry.[1048] When strain is introduced into the model, it introduces a pseudomagnetic field vector potential and a scalar potential. The observed Dirac cones are formed by the p_z orbitals of one of the inequivalent sub-lattices, which are composed of just 4 atoms, yielding an underlying hexagonal network which is topologically equivalent to that of distorted graphene. This offers the possibility of converting metallic 8-Pmmn borophene into an indirect band gap semiconductor by applying an external shear stress.[1049] Stability of the strained structures is supported by phonon frequency analysis. The Dirac cones are sensitive to the formation of vacancies only in the inequivalent sub-lattice which is active at the Fermi level.

9.2 Defects

There is some experimental evidence that borophene which is formed on a Ag(111) surface consists of a buckled triangular lattice without vacancies. Calculation suggests a nucleation mechanism for boron clusters and emphasizes boron–silver interactions during the growth of borophene, including the structural evolution of triangular fragments having various profiles and vacancy distributions. When compared with a triangular lattice without vacancies, it is confirmed that the sheet which is energetically favored during nucleation and growth is that which contains 1/6 vacancies in a stripe pattern.[1050] The associated scanning tunnelling microscopic image is in better agreement with experimental observations.

9.3 Mechanical Properties

As usual, the celebrated mechanical properties of graphene lead one to look firstly at the analogous properties of borophene. Molecular dynamics results show that the Young's moduli of armchair and zig-zag borophene nanoribbons are 1.093 and 0.978TPa, respectively. The corresponding elastic strains are about 15.30 and 22.03%, with distinct ductile and brittle fractures, respectively. When the nanoribbons are grafted with amine groups (-NH_2), the moduli of the armchair and zig-zag nanoribbons increase to 1.125 and 1.016TPa, respectively. The ultimate elastic strain for the former increases to 18.23%, but that of the armchair nanoribbon is slightly reduced, to 21.12%. The fracture modes remain the same. The structural deformation following exposure to a temperature of 1500K shows that there is little difference between the armchair and zig-zag nanoribbons, but the structural deformation of grafted nanoribbons is clearly less than that for non grafted nanoribbons. The overall results indicate that the grafting of amine groups can increase the Young's modulus, lengthen the elastic strain range, reduce in-plane elastic anisotropy, increase crack resistance, increase the capacity to resist deformation at high temperatures, reduce thermal expansion anisotropy and improve structural stability.[1051] Two-dimensional sheets having rectangular structures have been successfully grown onto monocrystalline Ag(111) substrates. First-principles calculations predict that monolayer borophene can withstand stresses of up to 20.26 and 12.98N/m in the a and b directions, respectively. The critical strain is small however. In the a-direction, the critical value is 8%, and this still seems to be lowest among all of the studied 2-dimensional materials.[1052] A tensile strain, applied in the b-direction, enhances the bucking height and results in a negative out-of-plane Poisson ratio; leading boron sheet to exhibit a superior mechanical flexibility along the b direction. Molecular dynamics simulations of the properties of 5 types of boron film, having vacancy ratios ranging from 0.1 to 0.15, show that the Young's modulus and tensile strength decrease with increasing temperature. Boron sheets exhibit an anisotropic mechanical response, due to the differing arrangements of atoms along the armchair and zig-zag directions. At room temperature, the 2-dimensional Young's modulus and fracture stress range from 63 to 136N/m and from 12 to 19N/m, respectively. The strains at the ultimate tensile strength range from 9 to 14%, from 11 to 19% and from 10 to 16% at 1, 300 and 600K, respectively.[1053] These results reflect the marked stiffness of 2-dimensional boron.

Single-layer borophene has a buckled structure, with atomic ridges. First-principles density functional theory calculations show that the 2-dimensional borophene is very stretchable, with a marked anisotropy The strain-to-failure in the direction along the atomic ridges is nearly twice as high as that across the atomic ridges. Straining-induced flattening, and subsequent stretching of the flattened structure, account for the large strain-to-failure in tension along the atomic ridges. Biaxial tension increases the strength across the atomic ridges, but decreases the failure strain along the atomic ridges.[1054] When bilayer borophene is stretched along the cross-plane direction, the strength and failure strain are much higher than those of bilayer graphene. This is due to the very strong interlayer chemical bonding.

First-principles calculations, based upon density functional theory, show that monolayer 8-Pmmn borophene can withstand stresses of up to 21.59 and 27.79N/m in the a- and b-directions, respectively. The ultimate tensile strain along the a-direction is about 0.155 and that along the b-direction is 0.16; higher than that of stripe borophene.[1055] The phonon dispersion under applied strain indicates that mechanical failure is likely to originate from elastic instability. Overall, 8-Pmmn borophene is a hard but flexible material, with potential uses in high-strain engineering applications.

Two-dimensional materials tend to be mechanically flexible yet planar, especially when adhering to metal substrates. First-principles calculations show that periodic nanoscale one-dimensional undulations can be preferred in borophene located on reconstructed Ag(111). Such a wavy configuration is more stable, than its planar form on flat Ag(111), due to the high anisotropic bending flexibility of borophene. Atomic-scale ultra-high vacuum scanning tunnelling microscopy characterization of borophene, grown onto Ag(111), reveals just such undulations, which agree with theory in terms of topography, wavelength, moiré pattern and prevalence of vacancy defects.[1056] Although the lattice is coherent within a borophene island, undulations which have nucleated from different sides of the island form a distinctive domain boundary when they are laterally misaligned. This structural model suggests that transfer of undulated borophene to an elastomeric substrate would permit high levels of stretchability and compressibility; with potential application in stretchable and foldable devices.

9.4 Composites

A bilayer structure can markedly lower the anisotropy of elastic properties. Molecular dynamics studies of bilayer complex nanoribbons indicate that the optimum distance between layers is about 0.359nm for bilayer borophene, 0.340nm for bilayer graphene and 0.348nm for a hybrid borophene/graphene bilayer. Analysis of the results for the binding energy, interaction energy and van der Waals energies between nanosheets reveals that the structural stability of the bilayer complexes, from high to low, is borophene/graphene > borophene/borophene > graphene/graphene. The predominant interaction which maintains stability is van der Waals. Structural analysis of these complexes at 1500K indicates that the hybrid borophene/graphene bilayer nanosheet has the highest temperature tolerance, followed by graphene/graphene and then borophene/borophene.[1057] The excitation of a greater number of local electrons in borophene results in the high-temperature deformation resistance of borophene/borophene being inferior to that of graphene/graphene.

9.5 Superconductivity

Borophene has been successfully grown onto Ag(111) by direct evaporation of a pure boron source. This revealed two types of structure, β_{12} and χ_3. First-principles density-functional calculations of electron-phonon coupling and superconductivity in these structures show that both borophenes exhibit inherent metallicity, and that the electron-phonon coupling constants are greater than that in MgB_2. Using the McMillian-Allen-Dynes formula, the superconducting transition temperatures are predicted to be 18.7 and 24.7K for β_{12} and χ_3, respectively. These temperatures are much higher than the theoretically predicted 8.1K, and experimentally observed 7.4K, transition temperatures of graphene.[1058] First principles calculations predict that borophene is a pristine 2-dimensional monolayer superconductor in which the superconductivity can be markedly enhanced by straining and by charge-carrier doping. The intrinsic metallic ground-state, with a high density of states at the Fermi energy and strong Fermi surface nesting, leads to high electron-phonon coupling. This makes free-standing borophene a superconductor having a critical temperature which is close to 19.0K. Tensile straining can increase the critical temperature to 27.4K, while hole-doping can increase it to 34.8K.[1059] The results indicate that borophene which is grown onto substrates having large lattice parameters, or is subjected to photo-excitation, can exhibit superconductivity with a critical temperature that is far above the liquid hydrogen temperature (20.3K).

9.6 Magnetism

First-principles calculations of the electronic and magnetic properties of borophene, functionalized by 3d transition-metal adsorption, show that the transition-metal atoms are adsorbed with binding energies ranging from 5.9 to 8.3eV.[1060] The originally non-magnetic borophene tends to become ferromagnetic when titanium, vanadium, chromium, manganese and iron atoms are adsorbed and the magnetic moments are dominated by the transition-metal atoms. The origin of the ferromagnetism is explained in terms of the Stoner criterion. The results indicate that the magnetic properties of borophene can be effectively tuned by the adsorption of 3d transition-metal atoms.

9.7 Applications

9.7.1 Energy Storage

For the moment, interest in borophene is focussed on its possible energy-storage, hydrogen-storage and surface catalysis applications. First-principles calculations and simulations show that it has an adsorption energy, for lithium atoms, of -1.12eV. This is sufficiently high to ensure good lithium-borophene stability during lithiation. The fully lithiated phase, $Li_{0.75}B$, corresponds to a theoretical specific capacity of 1860mAh/g. This is some 4 times larger than that (372mAh/g) of commercial graphite anodes. The energy barrier along the furrows of corrugated borophene is only 2.6meV. This is much lower than that of anode materials such as phosphorene (80meV) and Ti_3C_2 (70meV). The results suggest that lithium diffusion on borophene can be extremely rapid, but with a strong directional anisotropy for lithium diffusion, with the barrier perpendicular to the furrows of borophene being 325.1meV. *Ab initio* molecular dynamics simulations for a temperature of 300K show that the lithium atom can drift freely along the furrows, but rarely jumps into neighboring furrows. Borophene exhibits metallic characteristics during the entire lithiation process, thus indicating an excellent electronic conductivity.[1061] Based upon density functional theory calculations, borophene is predicted to be a promising anode material for high-performance lithium-ion batteries. The lithium atom can strongly and easily combine with the borophene surface and exist in a pure Li^+ state. The small energy barrier of 0.007eV to lithium diffusion leads to ultra-high diffusivity along uncorrugated directions of borophene. This diffusivity is estimated to be 10^4 times faster than that on MoS_2 at room temperature and 10^5 times faster than that on graphene at room temperature; both of which are rival materials in this context. A lithium storage capacity of 1239mAh/g is achieved when the lithium content is 0.5. The low average operating voltage of 0.466V, and metallic properties, mean that borophene can be used as an anode material. The effect of an Ag(111) substrate is very weak. The lithium atom can still bind to borophene with a high binding energy of 2.648eV.[1062] A small energy barrier of 0.033eV still subsists for lithium diffusion along the uncorrugated direction, giving rise to high lithium diffusivity. The results imply that an extremely high-rate capability can be expected of borophene-based lithium-ion batteries. First-principles density functional theory calculations of the interaction of calcium, magnesium, sodium and lithium atoms with single-layer and free-standing borophene have identified the most stable binding sites and their corresponding binding energies. The calculations predict binding energies of about 4.03, 2.09, 2.92 and 3.28eV between a borophene substrate and calcium, magnesium, sodium and lithium ions, respectively. The binding energy generally decreases with increasing ion content. Borophene is expected to have a capacity of about 800, 1960, 1380 and 1720mAh/g for calcium, magnesium, sodium and lithium ion storage, respectively.[1063]

Using first-principles methods, it is predicted that 2-dimensional borophene sheets can serve as an ideal electrode material, with a high electrochemical performance for both lithium-ion and sodium-ion batteries. Calculations have been performed for the experimentally stable borophene structures, β_{12} and χ_3. The optimized lithium and sodium adsorption sites are identified, and the host materials are found to maintain good electrical conductivity before and after adsorption. The advantages include low diffusion barriers and low average open-circuit voltages. The storage capacity can be as high as 1984mAh/g in β_{12} borophene and 1240mAh/g in χ_3 borophene, for both lithium and sodium.[1064] These values are several times higher than those for commercial graphite electrodes and are the highest among all of the known 2-dimensional materials. These results confirm that borophenes will be valuable anode materials for both lithium-ion and sodium-ion batteries of extremely high power density.

Recent density functional theory calculations and first-principles molecular dynamics simulations of borophene anode materials have confirmed the exciting combination of ultra-high energy storage and ultra-fast ion diffusion in lithium, sodium, potassium, magnesium or aluminium ion batteries, such as a specific density of 3306mAh/g and a high charging-voltage of 1.46V at room temperature.[1065]

9.7.2 Hydrogen Storage

Due to the hexagonal holes in the structure, the tuning of the properties of B_{36} by chemical modification is not yet fully understood. Van der Waals corrected density functional theory shows that lithium atoms, adsorbed on B_{36} clusters, can serve as reversible hydrogen storage media. The curvature and ionization of substrates can enhance the bond strength of lithium, due to energetically favorable B 2p – Li 2p orbital hybridization. The polarization mechanism and orbital hybridization between H-s orbitals and Li-2s2p orbitals contribute to the adsorption of H_2 molecules. The resultant adsorption energy lies between the physisorbed and chemisorbed states. The number of H_2 molecules in the hydrogen storage medium can be measured by the appearance of a negative differential resistance behavior in various bias-voltage regions. Cluster-assembled hydrogen storage materials made from metalized B_{36} clusters do not cause a decrease in the number of adsorbed hydrogen molecules per lithium atom. The present system is favorable for reversible hydrogen adsorption/desorption under ambient conditions.[1066]

9.7.3 Sensors

Pristine carbon nanotubes and graphene are very sensitive to several lethal gases, but cannot identify formaldehyde. Density functional theory calculations of B_{36} borophene sheet predict that its electrical conductivity increases significantly in the presence of formaldehyde molecules, thus generating an electrical signal. The latter is amplified by increasing the number of adsorbed formaldehyde molecules, thus indicating that this sensor is sensitive to the concentration or pressure of formaldehyde gas.[1067] These results suggest that pristine borophene might be used as a formaldehyde detector.

9.7.4 Catalysis

Investigation of the dissociative adsorption of ammonia onto the surface of B_{36} borophene, by means of density functional theory calculations, shows that the N–H bond can be broken at the B_{36} edges, even at room temperature, in view of the small energy barrier (14.1 to 19.3kcal/mol) and more negative Gibbs free energy change. The type of exchange correlation functional significantly affects the electronic properties. By increasing the percentage of Hartree-Fock exchange of density functionals, the activation and adsorption energies are decreased. A linear relationship is predicted to exist between the highest occupied molecular orbital, or lowest unoccupied molecular orbital, of B_{36} borophene and the exchange of functionals. Pure whole boron nanosheets are thought to be promising metal-free materials for N–H bond-cleavage.[1068]

9.7.5 Biological

Density functional theory calculations of the energetic and electronic sensitivities of borophene sheet, B_{36}, towards the nucleobases, adenine, guanine, thymine and cytosine, show that attaching them to borophene will affect the electrical conductivity differently, and generate distinguishable electrical signals. The order of reactivity towards borophene is: adenine > guanine > cytosine > thymine. Borophene nanosheet can thus be used in devices which sequence DNA. Cytosine has the largest effect upon the conductivity of borophene and will generate the largest electrical signal.[1069] The second largest peak belongs to thymine, while adenine and guanine give the smallest peaks.

References

[1] The Einstein Approximation, 2010

[2] N.M.Pugno, Acta Astronautica, 82[2] 2013 221
https://doi.org/10.1016/j.actaastro.2012.01.008

[3] A.C.Clarke, The Fountains of Paradise, Gollancz, 1979. But, like many of Clarke's ideas, purloined from earlier writers such as Konstantin Tsiolkovsky and Herman Potočnik

[4] Tang, Q., Zhou, Z., Chen, Z., Wiley Interdisciplinary Reviews - Computational Molecular Science, 5[5] 2015, 360-379
https://doi.org/10.1002/wcms.1224

[5] Ezawa, M., Journal of the Physical Society of Japan, 84[12] 2015, 7566
https://doi.org/10.7566/JPSJ.84.121003

[6] Zhang, S., Xie, M., Li, F., Yan, Z., Li, Y., Kan, E., Liu, W., Chen, Z., Zeng, H., Angewandte Chemie – International, 55[5] 2016, 1666-1669

[7] Kara, A., Léandri, C., Dávila, M.E., De Padova, P., Ealet, B., Oughaddou, H., Aufray, B., Le Lay, G., Journal of Superconductivity and Novel Magnetism, 22[3] 2009, 259-263
https://doi.org/10.1007/s10948-008-0427-8

[8] O'Hare, A., Kusmartsev, F.V., Kugel, K.I., Nano Letters, 12[2] 2012, 1045-1052
https://doi.org/10.1021/nl204283q

[9] Chávez-Castillo, M.R., Rodríguez-Meza, M.A., Meza-Montes, L., Revista Mexicana de Fisica, 58[2] 2012, 139-143

[10] Jose, D., Nijamudheen, A., Datta, A., Physical Chemistry Chemical Physics, 15[22] 2013, 8700-8704
https://doi.org/10.1039/c3cp51028j

[11] Wang, Y.Y., Quhe, R.G., Yu, D.P., Lü, J., Chinese Physics B, 24[8] 2015, 087201
https://doi.org/10.1088/1674-1056/24/8/087201

[12] Soto, J.R., Molina, B., Castro, J.J., Physical Chemistry Chemical Physics, 17[12] 2015, 7624-7628
https://doi.org/10.1039/C4CP05912C

[13] Song, Y.L., Zhang, J.M., Lu, D.B., Xu, K.W., Physica E, 53 2013, 173-177
https://doi.org/10.1016/j.physe.2013.05.002

[14] Song, Y.L., Zhang, Y., Zhang, J.M., Lu, D.B., Applied Surface Science, 256[21] 2010, 6313-6317
https://doi.org/10.1016/j.apsusc.2010.04.009

[15] Sun, M., Ren, Q., Wang, S., Yu, J., Tang, W., Journal of Physics D, 49[44], 2016, 445305
https://doi.org/10.1088/0022-3727/49/44/445305

[16] Pei, Q.X., Sha, Z.D., Zhang, Y.Y., Zhang, Y.W., Journal of Applied Physics, 115[2] 2014, 023519
https://doi.org/10.1063/1.4861736

[17] Liu, N., Hong, J., Pidaparti, R., Wang, X., 2D Materials, 3[3], 2016, 035008

[18] Rouhi, S., Computational Materials Science, 131, 2017, 275-285
https://doi.org/10.1016/j.commatsci.2017.02.007

[19] Fan, Y.C., Fang, T.H., Chen, T.H., Nanomaterials, 6[7], 2016, 120
https://doi.org/10.3390/nano6070120

[20] Ding, N., Wang, H., Chen, X., Lawrence Wu, C.M., RSC Advances, 7[17], 2017, 10306-10315
https://doi.org/10.1039/C6RA27291F

[21] Davydov, S.Y., Physics of the Solid State, 52[1] 2010, 184-187
https://doi.org/10.1134/S1063783410010312

[22] Roman, R.E., Cranford, S.W., Computational Materials Science, 82, 2014, 50-55
https://doi.org/10.1016/j.commatsci.2013.09.030

[23] Botari, T., Perim, E., Autreto, P.A.S., Van Duin, A.C.T., Paupitz, R., Galvao, D.S., Physical Chemistry Chemical Physics, 16[36] 2014, 19417-19423
https://doi.org/10.1039/C4CP02902J

[24] Yang, C., Yu, Z., Lu, P., Liu, Y., Ye, H., Gao, T., Computational Materials Science, 95, 2014, 420-428
https://doi.org/10.1016/j.commatsci.2014.07.046

[25] Peng, Q., Wen, X., De, S., RSC Advances, 3[33] 2013, 13772-13781
https://doi.org/10.1039/c3ra41347k

[26] Qin, R., Wang, C.H., Zhu, W., Zhang, Y., AIP Advances, 2[2] 2012, 022159
https://doi.org/10.1063/1.4732134

[27] John, R., Merlin, B., Crystal Structure Theory and Applications, 5 (2016) 43–55
https://doi.org/10.4236/csta.2016.5304

[28] Mortazavi, B., Rahaman, O., Makaremi, M., Dianat, A., Cuniberti, G., Rabczuk, T., Physica E, 87, 2017, 228-232
https://doi.org/10.1016/j.physe.2016.10.047

[29] Lian, C., Ni, J., AIP Advances, 3[5] 2013, 052102
https://doi.org/10.1063/1.4804246

[30] Kamal, C., Chakrabarti, A., Banerjee, A., Deb, S.K., Journal of Physics - Condensed Matter, 25[8] 2013, 085508
https://doi.org/10.1088/0953-8984/25/8/085508

[31] Ince, A., Erkoç, S., Physica Status Solidi B, 249[1] 2012, 74-81
https://doi.org/10.1002/pssb.201147267

[32] Ansari, R., Rouhi, S., Ajori, S., Superlattices and Microstructures, 65, 2014, 64-70
https://doi.org/10.1016/j.spmi.2013.10.039

[33] Chávez-Castillo, M.R., Rodríguez-Meza, M.A., Meza-Montes, L., RSC Advances, 5[116] 2015, 96052-96061
https://doi.org/10.1039/C5RA15312C

[34] Jing, Y., Sun, Y., Niu, H., Shen, J., Physica Status Solidi B, 250[8] 2013, 1505-1509
https://doi.org/10.1002/pssb.201349023

[35] Kaloni, T.P., Cheng, Y.C., Schwingenschlögl, U., Journal of Applied Physics, 113[10] 2013, 104305
https://doi.org/10.1063/1.4794812

[36] Wang, B., Wu, J., Gu, X., Yin, H., Wei, Y., Yang, R., Dresselhaus, M., Applied Physics Letters, 104[8] 2014, 81902
https://doi.org/10.1063/1.4866415

[37] Huang, L.F., Gong, P.L., Zeng, Z., Physical Review B, 91[20] 2015, 205433
https://doi.org/10.1103/PhysRevB.91.205433

[38] Yuan, X., Lin, G., Wang, Y., Molecular Simulation, 42[14], 2016, 1157-1164
https://doi.org/10.1080/08927022.2016.1148266

[39] Lee, C.C., Fleurence, A., Friedlein, R., Yamada-Takamura, Y., Ozaki, T., Physical Review B, 88[16] 2013, 165404
https://doi.org/10.1103/PhysRevB.88.165404

[40] Fleurence, A., Yamada-Takamura, Y., Applied Physics Letters, 110[4], 2017, 041601
https://doi.org/10.1063/1.4974467

[41] Li, X.L., Liu, G., Du, T.Y., Zhao, J., Wu, M.S., Ouyang, C.Y., Xu, B., Acta Physica Sinica, 63[21] 2014, 217101

[42] Wang, Y., Ding, Y., Physica Status Solidi R, 7[6] 2013, 410-413
https://doi.org/10.1002/pssr.201307110

[43] Zhang, W.B., Song, Z.B., Dou, L.M., Journal of Materials Chemistry C, 3[13] 2015, 3087
https://doi.org/10.1039/C4TC02758B

[44] Davydov, S.Y., Physics of the Solid State, 54[3] 2012, 652-657
https://doi.org/10.1134/S1063783412030080

[45] Xu, P., Yu, Z., Yang, C., Lu, P., Liu, Y., Ye, H., Gao, T., Superlattices and Microstructures, 75, 2014, 647-656
https://doi.org/10.1016/j.spmi.2014.08.022

[46] Topsakal, M., Ciraci, S., Physical Review B, 81[2] 2010, 024107
https://doi.org/10.1103/PhysRevB.81.024107

[47] De Crescenzi, M., Berbezier, I., Scarselli, M., Castrucci, P., Abbarchi, M., Ronda, A., Jardali, F., Park, J., Vach, H., ACS Nano, 10[12], 2016, 11163-11171
https://doi.org/10.1021/acsnano.6b06198

[48] Liu, H.S., Han, N.N., Zhao, J.J., Chinese Physics B, 24[8] 2015, 087303
https://doi.org/10.1088/1674-1056/24/8/087303

[49] Pan, Y., Zhang, L., Huang, L., Li, L., Meng, L., Gao, M., Huan, Q., Lin, X., Wang, Y., Du, S., Freund, H.J., Gao, H.J., Small, 10[11] 2014, 2215-2225
https://doi.org/10.1002/smll.201303698

[50] Feng, B., Ding, Z., Meng, S., Yao, Y., He, X., Cheng, P., Chen, L., Wu, K., Nano Letters, 12[7] 2012, 3507-3511
https://doi.org/10.1021/nl301047g

[51] Shu, H., Cao, D., Liang, P., Wang, X., Chen, X., Lu, W., Physical Chemistry Chemical Physics, 16[1] 2014, 304-310
https://doi.org/10.1039/C3CP53933D

[52] Gao, J., Zhao, J., Scientific Reports, 2, 2012, 861

[53] Cinquanta, E., Scalise, E., Chiappe, D., Grazianetti, C., Van Den Broek, B., Houssa, M., Fanciulli, M., Molle, A., Journal of Physical Chemistry C, 117[32] 2013, 16719-16724
https://doi.org/10.1021/jp405642g

[54] Liu, G., Wu, M.S., Ouyang, C.Y., Xu, B., International Journal of Modern Physics B. 27[32] 2013, 1350188
https://doi.org/10.1142/S0217979213501889

[55] Van Hoang, V., Mi, H.T.C., Journal of Physics D, 47[49] 2014, 495303
https://doi.org/10.1088/0022-3727/47/49/495303

[56] Li, R., Liu, Z.L., Gu, Y., Zhang, W., Tan, Y., Physica E, 79, 2016, 152-159
https://doi.org/10.1016/j.physe.2015.12.033

[57] De Padova, P., Kubo, O., Olivieri, B., Quaresima, C., Nakayama, T., Aono, M., Le Lay, G., Nano Letters, 12[11] 2012, 5500-5503
https://doi.org/10.1021/nl302598x

[58] Kawahara, K., Shirasawa, T., Lin, C.L., Nagao, R., Tsukahara, N., Takahashi, T., Arafune, R., Kawai, M., Takagi, N., Surface Science, 651, 2016, 70-75
https://doi.org/10.1016/j.susc.2016.03.029

[59] Hu, W., Li, Z., Yang, J., Journal of Chemical Physics, 139[15] 2013, 154704
https://doi.org/10.1063/1.4824887

[60] Yu, S., Li, X.D., Wu, S.Q., Wen, Y.H., Zhou, S., Zhu, Z.Z., Materials Research Bulletin, 50, 2014, 268-272
https://doi.org/10.1016/j.materresbull.2013.11.033

[61] Berdiyorov, G.R., Bahlouli, H., Peeters, F.M., Journal of Applied Physics, 117[22] 2015, 225101
https://doi.org/10.1063/1.4921877

[62] Peymanirad, F., Neek-Amal, M., Beheshtian, J., Peeters, F.M., Physical Review B, 92[15] 2015, 155113
https://doi.org/10.1103/PhysRevB.92.155113

[63] Nigam, S., Majumder, C., Pandey, R., RSC Advances, 6[26] 2016, 21948-21953
https://doi.org/10.1039/C6RA00225K

[64] Jiang, Q.G., Zhang, J.F., Ao, Z.M., Wu, Y.P., Journal of Materials Chemistry C, 3[16] 2015, 3954-3959
https://doi.org/10.1039/C4TC02829E

[65] Cai, Y., Pei, Q.X., Zhang, G., Zhang, Y.W., Journal of Applied Physics, 119[6] 2016, 065102
https://doi.org/10.1063/1.4941534

[66] Noguchi, E., Sugawara, K., Yaokawa, R., Hitosugi, T., Nakano, H., Takahashi, T., Advanced Materials, 27[5] 2015, 856-860
https://doi.org/10.1002/adma.201403077

[67] Debernardi, A., Marchetti, L., Physical Review B, 93[24], 2016, 245426
https://doi.org/10.1103/PhysRevB.93.245426

[68] Yamada-Takamura, Y., Friedlein, R., Science and Technology of Advanced Materials, 15[6] 2014, 064404
https://doi.org/10.1088/1468-6996/15/6/064404

[69] Grazianetti, C., Chiappe, D., Cinquanta, E., Tallarida, G., Fanciulli, M., Molle, A., Applied Surface Science, 291, 2014, 109-112
https://doi.org/10.1016/j.apsusc.2013.08.121

[70] Sone, J., Yamagami, T., Aoki, Y., Nakatsuji, K., Hirayama, H., New Journal of Physics, 16, 2014, 095004
https://doi.org/10.1088/1367-2630/16/9/095004

[71] Feng, B.J., Li, W.B., Qiu, J.L., Cheng, P., Chen, L., Wu, K.H., Chinese Physics Letters, 32[3] 2015, 037302
https://doi.org/10.1088/0256-307X/32/3/037302

[72] Jamgotchian, H., Ealet, B., Maradj, H., Hoarau, J.Y., Bibérian, J.P., Aufray, B., Journal of Physics - Condensed Matter, 28[19], 2016, 195002
https://doi.org/10.1088/0953-8984/28/19/195002

[73] Díaz Álvarez, A., Zhu, T., Nys, J.P., Berthe, M., Empis, M., Schreiber, J., Grandidier, B., Xu, T., Surface Science, 653, 2016, 92-96
https://doi.org/10.1016/j.susc.2016.06.005

[74] Yuan, Y., Quhe, R., Zheng, J., Wang, Y., Ni, Z., Shi, J., Lu, J., Physica E, 58, 2014, 38-42
https://doi.org/10.1016/j.physe.2013.11.016

[75] Curcella, A., Bernard, R., Borensztein, Y., Resta, A., Lazzeri, M., Prévot, G., Physical Review B, 94[16], 2016, 165438
https://doi.org/10.1103/PhysRevB.94.165438

[76] Pflugradt, P., Matthes, L., Bechstedt, F., Physical Review B, 89[3], 2014, 035403
https://doi.org/10.1103/PhysRevB.89.035403

[77] Johnson, N.W., Muir, D., Kurmaev, E.Z., Moewes, A., Advanced Functional Materials, 25[26] 2015, 4083-4090
https://doi.org/10.1002/adfm.201501029

[78] Shirai, T., Shirasawa, T., Hirahara, T., Fukui, N., Takahashi, T., Hasegawa, S., Physical Review B, 89[24] 2014, 241403
https://doi.org/10.1103/PhysRevB.89.241403

[79] Chen, L., Li, H., Feng, B., Ding, Z., Qiu, J., Cheng, P., Wu, K., Meng, S., Physical Review Letters, 110[8] 2013, 085504
https://doi.org/10.1103/PhysRevLett.110.085504

[80] De Padova, P., Vogt, P., Resta, A., Avila, J., Razado-Colambo, I., Quaresima, C., Ottaviani, C., Olivieri, B., Bruhn, T., Hirahara, T., Shirai, T., Hasegawa, S., Carmen Asensio, M., Le Lay, G., Applied Physics Letters, 102[16] 2013, 163106
https://doi.org/10.1063/1.4802782

[81] Vogt, P., De Padova, P., Quaresima, C., Avila, J., Frantzeskakis, E., Asensio, M.C., Resta, A., Ealet, B., Le Lay, G., Physical Review Letters, 108[15] 2012, 155501
https://doi.org/10.1103/PhysRevLett.108.155501

[82] De Padova, P., Quaresima, C., Olivieri, B., Perfetti, P., Le Lay, G., Applied Physics Letters, 98[8] 2011, 081909
https://doi.org/10.1063/1.3557073

[83] Houssa, M., Pourtois, G., Heyns, M.M., Afanas'Ev, V.V., Stesmans, A., Journal of the Electrochemical Society, 158[2] 2011, H107-H110
https://doi.org/10.1149/1.3522774

[84] Zhang, Y.Y., Tsai, W.F., Chang, K., An, X.T., Zhang, G.P., Xie, X.C., Li, S.S., Physical Review B, 88[12] 2013, 125431
https://doi.org/10.1103/PhysRevB.88.125431

[85] De Padova, P., Avila, J., Resta, A., Razado-Colambo, I., Quaresima, C., Ottaviani, C., Olivieri, B., Bruhn, T., Vogt, P., Asensio, M.C., Le Lay, G., Journal of Physics - Condensed Matter, 25[38] 2013, 382202
https://doi.org/10.1088/0953-8984/25/38/382202

[86] Mahatha, S.K., Moras, P., Bellini, V., Sheverdyaeva, P.M., Struzzi, C., Petaccia, L., Carbone, C., Physical Review B, 89[20] 2014, 201416
https://doi.org/10.1103/PhysRevB.89.201416

[87] Zhuang, J., Xu, X., Du, Y., Wu, K., Chen, L., Hao, W., Wang, J., Yeoh, W.K., Wang, X., Dou, S.X., Physical Review B, 91[16] 2015, 161409
https://doi.org/10.1103/PhysRevB.91.161409

[88] Moras, P., Mentes, T.O., Sheverdyaeva, P.M., Locatelli, A., Carbone, C., Journal of Physics - Condensed Matter, 26[18] 2014, 185001
https://doi.org/10.1088/0953-8984/26/18/185001

[89] Scalise, E., Cinquanta, E., Houssa, M., Van Den Broek, B., Chiappe, D., Grazianetti, C., Pourtois, G., Ealet, B., Molle, A., Fanciulli, M., Afanasev, V.V., Stesmans, A., Applied Surface Science, 291, 2014, 113-117
https://doi.org/10.1016/j.apsusc.2013.08.113

[90] Lin, C.L., Arafune, R., Kawahara, K., Tsukahara, N., Minamitani, E., Kim, Y., Takagi, N., Kawai, M., Applied Physics Express, 5[4] 2012, 45802
https://doi.org/10.1143/APEX.5.045802

[91] Takagi, N., Lin, C.L., Kawahara, K., Minamitani, E., Tsukahara, N., Kawai, M., Arafune, R., Progress in Surface Science, 90[1] 2015, 1-20
https://doi.org/10.1016/j.progsurf.2014.10.001

[92] Arafune, R., Lin, C.L., Takagi, N., Journal of the Vacuum Society of Japan, 57[11] 2014, 428-433
https://doi.org/10.3131/jvsj2.57.428

[93] Liu, Z.L., Wang, M.X., Xu, J.P., Ge, J.F., Lay, G.L., Vogt, P., Qian, D., Gao, C.L., Liu, C., Jia, J.F., New Journal of Physics, 16, 2014, 075006
https://doi.org/10.1088/1367-2630/16/7/075006

[94] Lalmi, B., Oughaddou, H., Enriquez, H., Kara, A., Vizzini, S., Ealet, B., Aufray, B., Applied Physics Letters, 97[22] 2010, 223109
https://doi.org/10.1063/1.3524215

[95] Le Lay, G., De Padova, P., Resta, A., Bruhn, T., Vogt, P., Journal of Physics D, 45[39] 2012, 392001
https://doi.org/10.1088/0022-3727/45/39/392001

[96] Acun, A., Poelsema, B., Zandvliet, H.J.W., Van Gastel, R., Applied Physics Letters, 103[26] 2013, 263119
https://doi.org/10.1063/1.4860964

[97] Arafune, R., Lin, C.L., Kawahara, K., Tsukahara, N., Minamitani, E., Kim, Y., Takagi, N., Kawai, M., Surface Science, 608, 2013, 297-300
https://doi.org/10.1016/j.susc.2012.10.022

[98] Chiappe, D., Grazianetti, C., Tallarida, G., Fanciulli, M., Molle, A., Advanced Materials, 24[37] 2012, 5088-5093
https://doi.org/10.1002/adma.201202100

[99] Aufray, B., Kara, A., Vizzini, S., Oughaddou, H., Ĺandri, C., Ealet, B., Le Lay, G., Applied Physics Letters, 96[18] 2010, 183102
https://doi.org/10.1063/1.3419932

[100] De Padova, P., Quaresima, C., Ottaviani, C., Sheverdyaeva, P.M., Moras, P., Carbone, C., Topwal, D., Olivieri, B., Kara, A., Oughaddou, H., Aufray, B., Le Lay, G., Applied Physics Letters, 96[26] 2010, 261905 *https://doi.org/10.1063/1.3459143*

[101] Bernard, R., Borensztein, Y., Cruguel, H., Lazzeri, M., Prévot, G., Physical Review B, 92[4] 2015, 045415 *https://doi.org/10.1103/PhysRevB.92.045415*

[102] Prévot, G., Bernard, R., Cruguel, H., Borensztein, Y., Applied Physics Letters, 105[21] 2014, 213106 *https://doi.org/10.1063/1.4902811*

[103] Tchalala, M.R., Enriquez, H., Yildirim, H., Kara, A., Mayne, A.J., Dujardin, G., Ali, M.A., Oughaddou, H., Applied Surface Science, 303, 2014, 61-66 *https://doi.org/10.1016/j.apsusc.2014.02.064*

[104] Kawahara, K., Shirasawa, T., Arafune, R., Lin, C.L., Takahashi, T., Kawai, M., Takagi, N., Surface Science, 623, 2014, 25-28 *https://doi.org/10.1016/j.susc.2013.12.013*

[105] Fukaya, Y., Mochizuki, I., Maekawa, M., Wada, K., Hyodo, T., Matsuda, I., Kawasuso, A., Physical Review B, 88[20] 2013, 205413 *https://doi.org/10.1103/PhysRevB.88.205413*

[106] Enriquez, H., Vizzini, S., Kara, A., Lalmi, B., Oughaddou, H., Journal of Physics - Condensed Matter, 24[31] 2012, 314211 *https://doi.org/10.1088/0953-8984/24/31/314211*

[107] Majzik, Z., Rachid Tchalala, M., Švec, M., Hapala, P., Enriquez, H., Kara, A., Mayne, A.J., Dujardin, G., Jelínek, P., Oughaddou, H., Journal of Physics - Condensed Matter, 25[22] 2013, 225301 *https://doi.org/10.1088/0953-8984/25/22/225301*

[108] Guo, Z.X., Oshiyama, A., New Journal of Physics, 17, 2015, 045028, Page 045028

[109] Salomon, E., El Ajjouri, R., Lay, G.L., Angot, T., Journal of Physics - Condensed Matter, 26[18] 2014, 185003 *https://doi.org/10.1088/0953-8984/26/18/185003*

[110] Pflugradt, P., Matthes, L., Bechstedt, F., Physical Review B, 89[20] 2014, 205428 *https://doi.org/10.1103/PhysRevB.89.205428*

[111] Borensztein, Y., Curcella, A., Royer, S., Prévot, G., Physical Review B, 92[15] 2015, 155407 *https://doi.org/10.1103/PhysRevB.92.155407*

[112] Grazianetti, C., Chiappe, D., Cinquanta, E., Fanciulli, M., Molle, A., Journal of Physics - Condensed Matter, 27[25] 2015, 255005
https://doi.org/10.1088/0953-8984/27/25/255005

[113] Jamgotchian, H., Colignon, Y., Hamzaoui, N., Ealet, B., Hoarau, J.Y., Aufray, B., Bibérian, J.P., Journal of Physics - Condensed Matter, 24[17] 2012, 172001
https://doi.org/10.1088/0953-8984/24/17/172001

[114] Wang, W., Olovsson, W., Uhrberg, R.I.G., Physical Review B, 92[20] 2015, 205427
https://doi.org/10.1103/PhysRevB.92.205427

[115] Jamgotchian, H., Ealet, B., Colignon, Y., Maradj, H., Hoarau, J.Y., Biberian, J.P., Aufray, B., Journal of Physics - Condensed Matter, 27[39] 2015, 395002
https://doi.org/10.1088/0953-8984/27/39/395002

[116] Resta, A., Leoni, T., Barth, C., Ranguis, A., Becker, C., Bruhn, T., Vogt, P., Le Lay, G., Scientific Reports, 3, 2013, 2399

[117] Fu, H., Chen, L., Chen, J., Qiu, J., Ding, Z., Zhang, J., Wu, K., Li, H., Meng, S., Nanoscale, 7[38] 2015, 15880-15885
https://doi.org/10.1039/C5NR04548G

[118] Cahangirov, S., Audiffred, M., Tang, P., Iacomino, A., Duan, W., Merino, G., Rubio, A., Physical Review B, 88[3] 2013, 035432
https://doi.org/10.1103/PhysRevB.88.035432

[119] Stephan, R., Hanf, M.C., Sonnet, P., Journal of Physics - Condensed Matter, 27[1] 2014, 015002
https://doi.org/10.1088/0953-8984/27/1/015002

[120] Guo, Z.X., Oshiyama, A., Physical Review B, 89[15] 2014, 155418
https://doi.org/10.1103/PhysRevB.89.155418

[121] Cahangirov, S., Özçelik, V.O., Xian, L., Avila, J., Cho, S., Asensio, M.C., Ciraci, S., Rubio, A., Physical Review B, 90[3] 2014, 035448
https://doi.org/10.1103/PhysRevB.90.035448

[122] Lin, C.L., Arafune, R., Kawahara, K., Kanno, M., Tsukahara, N., Minamitani, E., Kim, Y., Kawai, M., Takagi, N., Physical Review Letters, 110[7] 2013, 076801
https://doi.org/10.1103/PhysRevLett.110.076801

[123] Guo, Z.X., Furuya, S., Iwata, J.I., Oshiyama, A., Journal of the Physical Society of Japan, 82[6] 2013, 063714
https://doi.org/10.7566/JPSJ.82.063714

[124] Guo, Z.X., Furuya, S., Iwata, J.I., Oshiyama, A., Physical Review B, 87[23] 2013, 235435
https://doi.org/10.1103/PhysRevB.87.235435

[125] Feng, B., Li, H., Liu, C.C., Shao, T.N., Cheng, P., Yao, Y., Meng, S., Chen, L., Wu, K., ACS Nano, 7[10] 2013, 9049-9054
https://doi.org/10.1021/nn403661h

[126] Wang, Y.P., Cheng, H.P., Physical Review B, 87[24] 2013, 245430
https://doi.org/10.1103/PhysRevB.87.245430

[127] Johnson, N.W., Vogt, P., Resta, A., De Padova, P., Perez, I., Muir, D., Kurmaev, E.Z., Le Lay, G., Moewes, A., Advanced Functional Materials, 24[33] 2014, 5253-5259
https://doi.org/10.1002/adfm.201400769

[128] Wu, K.H., Chinese Physics B, 24[8] 2015, 086802
https://doi.org/10.1088/1674-1056/24/8/086802

[129] Prévot, G., Bernard, R., Cruguel, H., Curcella, A., Lazzeri, M., Leoni, T., Masson, L., Ranguis, A., Borensztein, Y., Physica Status Solidi B, 253[2] 2016, 206-217
https://doi.org/10.1002/pssb.201552524

[130] Lagarde, P., Chorro, M., Roy, D., Trcera, N., Journal of Physics - Condensed Matter, 28[7] 2016, 075002
https://doi.org/10.1088/0953-8984/28/7/075002

[131] Feng, B., Li, H., Meng, S., Chen, L., Wu, K., Surface Science, 645, 2016, 74-79
https://doi.org/10.1016/j.susc.2015.10.037

[132] Feng, Y., Liu, D., Feng, B., Liu, X., Zhao, L., Xie, Z., Liu, Y., Liang, A., Hu, C., Hu, Y., He, S., Liu, G., Zhang, J., Chen, C., Xu, Z., Chen, L., Wu, K., Liu, Y.T., Lin, H., Huang, Z.Q., Hsu, C.H., Chuang, F.C., Bansil, A., Zhou, X.J., Proceedings of the National Academy of Sciences of the United States of America, 113[51], 2016, 14656-14661
https://doi.org/10.1073/pnas.1613434114

[133] Liu, H., Gao, J., Zhao, J., Journal of Physical Chemistry C, 117[20] 2013, 10353-10359
https://doi.org/10.1021/jp311836m

[134] Li, L., Wang, X., Zhao, X., Zhao, M., Physics Letters A, 377[38] 2013, 2628-2632
https://doi.org/10.1016/j.physleta.2013.07.037

[135] Vargiamidis, V., Vasilopoulos, P., Applied Physics Letters, 105[22] 2014, 223105
https://doi.org/10.1063/1.4903248

[136] Kokott, S., Matthes, L., Bechstedt, F., Physica Status Solidi R, 7[8] 2013, 538-541
https://doi.org/10.1002/pssr.201307215

[137] Ding, Y., Wang, Y., Applied Physics Letters, 103[4] 2013, 043114
https://doi.org/10.1063/1.4816753

[138] Zhang, X., Guo, Z.X., Cao, J.X., Xiao, S.G., Ding, J.W., Acta Physica Sinica, 64[18] 2015, 186101

[139] Kaloni, T.P., Tahir, M., Schwingenschlögl, U., Scientific Reports, 3, 2013, 3192
https://doi.org/10.1038/srep03192

[140] Drummond, N.D., Zólyomi, V., Falko, V.I., Physical Review B, 85[7] 2012, 075423
https://doi.org/10.1103/PhysRevB.85.075423

[141] Kikutake, K., Ezawa, M., Nagaosa, N., Physical Review B, 88[11] 2013, 115432
https://doi.org/10.1103/PhysRevB.88.115432

[142] Schüler, M., Rösner, M., Wehling, T.O., Lichtenstein, A.I., Katsnelson, M.I., Physical Review Letters, 111[3] 2013, 036601
https://doi.org/10.1103/PhysRevLett.111.036601

[143] Liu, F., Liu, C.C., Yao, Y.G., Chinese Physics B, 24[8] 2015, 087503
https://doi.org/10.1088/1674-1056/24/8/087503

[144] Kokott, S., Pflugradt, P., Matthes, L., Bechstedt, F., Journal of Physics - Condensed Matter, 26[18] 2014, 185002
https://doi.org/10.1088/0953-8984/26/18/185002

[145] Li, S.S., Zhang, C.W., Ji, W.X., Materials Chemistry and Physics, 164, 2015, 150-156
https://doi.org/10.1016/j.matchemphys.2015.08.036

[146] Cheng, J.Y., Chan, M.K.Y., Lilley, C.M., Applied Physics Letters, 109[13], 2016, 133111
https://doi.org/10.1063/1.4963653

[147] Filippone, F., Journal of Physics - Condensed Matter, 26[39] 2014, 395009
https://doi.org/10.1088/0953-8984/26/39/395009

[148] Houssa, M., Van Den Broek, B., Scalise, E., Pourtois, G., Afanasev, V.V., Stesmans, A., Physical Chemistry Chemical Physics, 15[11] 2013, 3702-3705
https://doi.org/10.1039/c3cp50391g

[149] Friedlein, R., Yamada-Takamura, Y., Journal of Physics - Condensed Matter, 27[20] 2015, 203201
https://doi.org/10.1088/0953-8984/27/20/203201

[150] Friedlein, R., Fleurence, A., Aoyagi, K., De Jong, M.P., Van Bui, H., Wiggers, F.B., Yoshimoto, S., Koitaya, T., Shimizu, S., Noritake, H., Mukai, K., Yoshinobu, J., Yamada-Takamura, Y., Journal of Chemical Physics, 140[18] 2014, 184704
https://doi.org/10.1063/1.4875075

[151] Lee, C.C., Fleurence, A., Friedlein, R., Yamada-Takamura, Y., Ozaki, T., Physical Review B, 90[24] 2014, 241402
https://doi.org/10.1103/PhysRevB.90.241402

[152] Aizawa, T., Suehara, S., Otani, S., Journal of Physical Chemistry C, 118[40] 2014, 23049-23057
https://doi.org/10.1021/jp505602c

[153] Aizawa, T., Suehara, S., Otani, S., Journal of Physics - Condensed Matter, 27[30] 2015, 305002
https://doi.org/10.1088/0953-8984/27/30/305002

[154] Lee, C.C., Fleurence, A., Yamada-Takamura, Y., Ozaki, T., Friedlein, R., Physical Review B, 90[7] 2014, 075422
https://doi.org/10.1103/PhysRevB.90.075422

[155] Fleurence, A., Yoshida, Y., Lee, C.C., Ozaki, T., Yamada-Takamura, Y., Hasegawa, Y., Applied Physics Letters, 104[2] 2014, 021605
https://doi.org/10.1063/1.4862261

[156] Friedlein, R., Fleurence, A., Sadowski, J.T., Yamada-Takamura, Y., Applied Physics Letters, 102[22] 2013, 221603
https://doi.org/10.1063/1.4808214

[157] Zhu, J., Schwingenschlögl, U., 2D Materials, 2[4] 2015, 045004

[158] Kaltsas, D., Tsetseris, L., Dimoulas, A., Applied Surface Science, 291, 2014, 93-97
https://doi.org/10.1016/j.apsusc.2013.09.115

[159] Quhe, R., Yuan, Y., Zheng, J., Wang, Y., Ni, Z., Shi, J., Yu, D., Yang, J., Lu, J., Scientific Reports, 4, 2014, 5476
https://doi.org/10.1038/srep05476

[160] Pflugradt, P., Matthes, L., Bechstedt, F., New Journal of Physics, 16, 2014, 075004
https://doi.org/10.1088/1367-2630/16/7/075004

[161] Oughaddou, H., Enriquez, H., Tchalala, M.R., Yildirim, H., Mayne, A.J., Bendounan, A., Dujardin, G., Ait Ali, M., Kara, A., Progress in Surface Science, 90[1] 2015, 46-83
https://doi.org/10.1016/j.progsurf.2014.12.003

[162] Morishita, T., Spencer, M.J.S., Kawamoto, S., Snook, I.K., Journal of Physical Chemistry C, 117[42] 2013, 22142-22148
https://doi.org/10.1021/jp4080898

[163] Meng, L., Wang, Y., Zhang, L., Du, S., Wu, R., Li, L., Zhang, Y., Li, G., Zhou, H., Hofer, W.A., Gao, H.J., Nano Letters, 13[2] 2013, 685-690
https://doi.org/10.1021/nl304347w

[164] Meng, L., Wang, Y.L., Zhang, L.Z., Du, S.X., Gao, H.J., Chinese Physics B, 24[8] 2015, 086803
https://doi.org/10.1088/1674-1056/24/8/086803

[165] Wei, W., Dai, Y., Huang, B., Whangbo, M.H., Jacob, T., Journal of Physical Chemistry Letters, 6[6] 2015, 1065-1070
https://doi.org/10.1021/acs.jpclett.5b00106

[166] Podsiadły-Paszkowska, A., Krawiec, M., Physical Review B, 92[16] 2015, 165411
https://doi.org/10.1103/PhysRevB.92.165411

[167] Podsiadły-Paszkowska, A., Krawiec, M., Physical Chemistry Chemical Physics, 17[3] 2015, 2246-2251
https://doi.org/10.1039/C4CP05104A

[168] Kanno, M., Arafune, R., Liang Lin, C., Minamitani, E., Kawai, M., Takagi, N., New Journal of Physics, 16, 2014, 105019
https://doi.org/10.1088/1367-2630/16/10/105019

[169] Qin, Z., Xu, Z., Buehler, M.J., Journal of Applied Mechanics, 82[10] 2015, 101003
https://doi.org/10.1115/1.4030888

[170] Berdiyorov, G.R., Neek-Amal, M., Peeters, F.M., Van Duin, A.C.T., Physical Review B, 89[2] 2014, 024107
https://doi.org/10.1103/PhysRevB.89.024107

[171] Neek-Amal, M., Sadeghi, A., Berdiyorov, G.R., Peeters, F.M., Applied Physics Letters, 103[26] 2013, 261904
https://doi.org/10.1063/1.4852636

[172] Stephan, R., Hanf, M.C., Sonnet, P., Physical Chemistry Chemical Physics, 17[22] 2015, 14495-14501
https://doi.org/10.1039/C5CP00613A

[173] Švec, M., Hapala, P., OndráAek, M., Merino, P., Blanco-Rey, M., Mutombo, P., VondráAek, M., Polyak, Y., Cháb, V., Martín Gago, J.A., Jelínek, P., Physical Review B, 89[20] 2014, 201412
https://doi.org/10.1103/PhysRevB.89.201412

[174] Spencer, M.J.S., Bassett, M.R., Morishita, T., Snook, I.K., Nakano, H., New Journal of Physics, 15, 2013, 125018
https://doi.org/10.1088/1367-2630/15/12/125018

[175] Chen, M.X., Zhong, Z., Weinert, M., Physical Review B, 94[7], 2016, 075409
https://doi.org/10.1103/PhysRevB.94.075409

[176] Yang, K., Cahangirov, S., Cantarero, A., Rubio, A., D'Agosta, R., Physical Review B, 89[12] 2014, 125403
https://doi.org/10.1103/PhysRevB.89.125403

[177] Zhao, W., Guo, Z.X., Zhang, Y., Ding, J.W., Zheng, X.J., Solid State Communications, 227, 2016, 1-8
https://doi.org/10.1016/j.ssc.2015.11.012

[178] Yang, X.F., Liu, Y.S., Feng, J.F., Wang, X.F., AIP Advances, 4[8] 2014, 087116
https://doi.org/10.1063/1.4892956

[179] An, R.L., Wang, X.F., Vasilopoulos, P., Liu, Y.S., Chen, A.B., Dong, Y.J., Zhai, M.X., Journal of Physical Chemistry C, 118[37] 2014, 21339-21346
https://doi.org/10.1021/jp506111a

[180] Wierzbicki, M., Barnaš, J., Swirkowicz, R., Journal of Physics - Condensed Matter, 27[48] 2015, 485301
https://doi.org/10.1088/0953-8984/27/48/485301

[181] Niu, Z.P., Zhang, Y.M., Dong, S., New Journal of Physics, 17[7] 2015, 073026
https://doi.org/10.1088/1367-2630/17/7/073026

[182] Zberecki, K., Swirkowicz, R., Wierzbicki, M., Barnaś, J., Physical Chemistry Chemical Physics, 16[25] 2014, 12900-12908
https://doi.org/10.1039/c4cp01039f

[183] Ping Niu, Z., Dong, S., Applied Physics Letters, 104[20] 2014, 202401

[184] Zberecki, K., Swirkowicz, R., Barnaś, J., Physical Review B, 89[16] 2014, 165419
https://doi.org/10.1103/PhysRevB.89.165419

[185] Zberecki, K., Swirkowicz, R., Barnaš, J., Journal of Magnetism and Magnetic Materials, 393 2015, 305-309
https://doi.org/10.1016/j.jmmm.2015.05.077

[186] Pan, L., Liu, H.J., Tan, X.J., Lv, H.Y., Shi, J., Tang, X.F., Zheng, G., Physical Chemistry Chemical Physics, 14[39] 2012, 13588-13593
https://doi.org/10.1039/c2cp42645e

[187] Fu, H.H., Wu, D.D., Zhang, Z.Q., Gu, L., Scientific Reports, 5, 2015, 10547
https://doi.org/10.1038/srep10547

[188] Sadeghi, H., Sangtarash, S., Lambert, C.J., Scientific Reports, 5, 2015, 9514
https://doi.org/10.1038/srep09514

[189] Wierzbicki, M., Barnaš, J., Swirkowicz, R., Physical Review B, 91[16] 2015, 165417
https://doi.org/10.1103/PhysRevB.91.165417

[190] Zhou, B., Zhou, B., Zeng, Y., Zhou, G., Duan, M., Physics Letters A, 380[16] 2016, 1469-1474
https://doi.org/10.1016/j.physleta.2016.02.014

[191] Li, H.P., Zhang, R.Q., EPL, 99[3] 2012, 36001
https://doi.org/10.1209/0295-5075/99/36001

[192] Xie, H., Hu, M., Bao, H., Applied Physics Letters, 104[13] 2014, 131906
https://doi.org/10.1063/1.4870586

[193] Peng, B., Zhang, H., Shao, H., Xu, Y., Zhang, R., Lu, H., Zhang, D.W., Zhu, H., ACS Applied Materials and Interfaces, 8[32], 2016, 20977-20985
https://doi.org/10.1021/acsami.6b04211

[194] Srinivasan, S., Ray, U., Balasubramanian, G., Chemical Physics Letters, 650, 2016, 88-93
https://doi.org/10.1016/j.cplett.2016.02.073

[195] Gu, X., Yang, R., Journal of Applied Physics, 117[2] 2015, 025102
https://doi.org/10.1063/1.4905540

[196] Li, X., Mullen, J.T., Jin, Z., Borysenko, K.M., Buongiorno Nardelli, M., Kim, K.W., Physical Review B, 87[11] 2013, 115418
https://doi.org/10.1103/PhysRevB.87.115418

[197] Yeo, J.J., Ng, T.Y., Liu, Z.S., Journal of Computational and Theoretical Nanoscience, 11[8] 2014, 1790-1796
https://doi.org/10.1166/jctn.2014.3568

[198] Ng, T.Y., Yeo, J., Liu, Z., International Journal of Mechanics and Materials in Design, 9[2] 2013, 105-114
https://doi.org/10.1007/s10999-013-9215-0

[199] Kuang, Y.D., Lindsay, L., Shi, S.Q., Zheng, G.P., Nanoscale, 8[6] 2016, 3760-3767
https://doi.org/10.1039/C5NR08231E

[200] Xie, H., Ouyang, T., Germaneau, É., Qin, G., Hu, M., Bao, H., Physical Review B, 93[7] 2016, 075404
https://doi.org/10.1103/PhysRevB.93.075404

[201] Zhang, X., Xie, H., Hu, M., Bao, H., Yue, S., Qin, G., Su, G., Physical Review B, 89[5] 2014, 054310
https://doi.org/10.1103/PhysRevB.89.054310

[202] Guo, Y., Zhou, S., Bai, Y., Zhao, J., Journal of Superconductivity and Novel Magnetism, 29[3] 2016, 717-720
https://doi.org/10.1007/s10948-015-3305-1

[203] Liu, B., Reddy, C.D., Jiang, J., Zhu, H., Baimova, J.A., Dmitriev, S.V., Zhou, K., Journal of Physics D, 47[16] 2014, 165301
https://doi.org/10.1088/0022-3727/47/16/165301

[204] Pei, Q.X., Zhang, Y.W., Sha, Z.D., Shenoy, V.B., Journal of Applied Physics, 114[3] 2013, 033526
https://doi.org/10.1063/1.4815960

[205] Hu, M., Zhang, X., Poulikakos, D., Physical Review B, 87[19] 2013, 195417
https://doi.org/10.1103/PhysRevB.87.195417

[206] Wang, Z., Feng, T., Ruan, X., Journal of Applied Physics, 117[8] 2015, 084317
https://doi.org/10.1063/1.4913600

[207] Liu, B., Baimova, J.A., Reddy, C.D., Dmitriev, S.V., Law, W.K., Feng, X.Q., Zhou, K., Carbon, 79[1] 2014, 236-244
https://doi.org/10.1016/j.carbon.2014.07.064

[208] Liu, B., Baimova, J.A., Reddy, C.D., Law, A.W.K., Dmitriev, S.V., Wu, H., Zhou, K., ACS Applied Materials and Interfaces, 6[20] 2014, 18180-18188
https://doi.org/10.1021/am505173s

[209] Zhang, J., Hong, Y., Tong, Z., Xiao, Z., Bao, H., Yue, Y., Physical Chemistry Chemical Physics, 17[37] 2015, 23704-23710
https://doi.org/10.1039/C5CP03323C

[210] Xu, C., Luo, G., Liu, Q., Zheng, J., Zhang, Z., Nagase, S., Gao, Z., Lu, J., Nanoscale, 4[10] 2012, 3111-3117
https://doi.org/10.1039/c2nr00037g

[211] Kang, J., Wu, F., Li, J., Applied Physics Letters, 100[23] 2012, 233122
https://doi.org/10.1063/1.4726276

[212] Xu, L., Wang, X.F., Zhou, L., Yang, Z.Y., Journal of Physics D, 48[21] 2015, 215306
https://doi.org/10.1088/0022-3727/48/21/215306

[213] Yang, X.F., Zhang, X., Hong, X.K., Liu, Y.S., Feng, J.F., Wang, X.F., Zhang, C.W., RSC Advances, 4[89] 2014, 48539-48546
https://doi.org/10.1039/C4RA07791A

[214] Zhang, D., Long, M., Zhang, X., Cao, C., Xu, H., Li, M., Chan, K., Chemical Physics Letters, 616-617, 2014, 178-183
https://doi.org/10.1016/j.cplett.2014.10.041

[215] Saxena, R., Saha, A., Rao, S., Physical Review B, 92[24] 2015, 245412
https://doi.org/10.1103/PhysRevB.92.245412

[216] Tahir, M., Schwingenschlögl, U., European Physical Journal B, 88[11] 2015, 285
https://doi.org/10.1140/epjb/e2015-60719-7

[217] Ping Niu, Z., Dong, S., EPL, 111[3] 2015, 37007

[218] Rachel, S., Ezawa, M., Physical Review B, 89[19] 2014, 195303
https://doi.org/10.1103/PhysRevB.89.195303

[219] Chen, A.B., Wang, X.F., Vasilopoulos, P., Zhai, M.X., Liu, Y.S., Physical Chemistry Chemical Physics, 16[11] 2014, 5113-5118
https://doi.org/10.1039/c3cp55447c

[220] Wang, Y., Lou, Y., Journal of Applied Physics, 114[18] 2013, 183712
https://doi.org/10.1063/1.4830020

[221] Ezawa, M., Physical Review B, 87[15] 2013, 155415
https://doi.org/10.1103/PhysRevB.87.155415

[222] Ezawa, M., European Physical Journal B, 86[4] 2013, 139
https://doi.org/10.1140/epjb/e2013-31029-1

[223] Liu, C.C., Jiang, H., Yao, Y., Physical Review B, 84[19] 2011, 195430
https://doi.org/10.1103/PhysRevB.84.195430

[224] Wang, S.K., Wang, J., Chan, K.S., New Journal of Physics, 16, 2014, 045015
https://doi.org/10.1088/1367-2630/16/4/045015

[225] Tahir, M., Manchon, A., Sabeeh, K., Schwingenschlögl, U., Applied Physics Letters, 102[16] 2013, 162412
https://doi.org/10.1063/1.4803084

[226] Tahir, M., Schwingenschlögl, U., Scientific Reports, 3, 2013, 1075

[227] Li, H., Wang, X., Manchon, A., Physical Review B, 93[3] 2016, 035417
https://doi.org/10.1103/PhysRevB.93.035417

[228] Tabert, C.J., Nicol, E.J., Physical Review B, 87[23] 2013, 235426
https://doi.org/10.1103/PhysRevB.87.235426

[229] Goswami, P., Indian Journal of Physics, 89[7] 2015, 677-686
https://doi.org/10.1007/s12648-014-0637-9

[230] Ezawa, M., Physical Review Letters, 109[5] 2012, 055502
https://doi.org/10.1103/PhysRevLett.109.055502

[231] An, X.T., Zhang, Y.Y., Liu, J.J., Li, S.S., Applied Physics Letters, 102[21] 2013, 213115
https://doi.org/10.1063/1.4808367

[232] Ezawa, M., Physical Review B, 86[16] 2012, 161407
https://doi.org/10.1103/PhysRevB.86.161407

[233] Stille, L., Tabert, C.J., Nicol, E.J., Physical Review B, 86[19] 2012, 195405
https://doi.org/10.1103/PhysRevB.86.195405

[234] Zhao, B., Zhang, J., Wang, Y., Yang, Z., Journal of Chemical Physics, 141[24] 2014, 244701
https://doi.org/10.1063/1.4904285

[235] Shirkani, H., Golshan, M.M., Physica E, 63, 2014, 81-86
https://doi.org/10.1016/j.physe.2014.04.022

[236] Wang, S.K., Tian, H.Y., Yang, Y.H., Wang, J., Chinese Physics B, 23[1] 2014, 017203
https://doi.org/10.1088/1674-1056/23/1/017203

[237] Kim, Y., Choi, K., Ihm, J., Jin, H., Physical Review B, 89[8] 2014, 085429
https://doi.org/10.1103/PhysRevB.89.085429

[238] Pan, H., Li, Z., Liu, C.C., Zhu, G., Qiao, Z., Yao, Y., Physical Review Letters, 112[10] 2014, 106802
https://doi.org/10.1103/PhysRevLett.112.106802

[239] Shakouri, K., Vasilopoulos, P., Vargiamidis, V., Hai, G.Q., Peeters, F.M., Applied Physics Letters, 104[21] 2014, 213109
https://doi.org/10.1063/1.4878509

[240] Shakouri, K., Vasilopoulos, P., Vargiamidis, V., Peeters, F.M., Physical Review B, 90[12] 2014, 125444
https://doi.org/10.1103/PhysRevB.90.125444

[241] Ji, Q.S., Hao, H.Y., Zhang, C.X., Wang, R., Acta Physica Sinica, 64[8] 2015, 087302

[242] Wu, J.Y., Lin, C.Y., Gumbs, G., Lin, M.F., RSC Advances, 5[64] 2015, 51912-51918
https://doi.org/10.1039/C5RA07721D

[243] Tabert, C.J., Nicol, E.J., Physical Review B, 89[19] 2014, 195410
https://doi.org/10.1103/PhysRevB.89.195410

[244] Wu, J.Y., Chen, S.C., Lin, M.F., New Journal of Physics, 16, 2014, 125002
https://doi.org/10.1088/1367-2630/16/12/125002

[245] Van Duppen, B., Vasilopoulos, P., Peeters, F.M., Physical Review B, 90[3] 2014, 035142
https://doi.org/10.1103/PhysRevB.90.035142

[246] Shen, M., Zhang, Y.Y., An, X.T., Liu, J.J., Li, S.S., Journal of Applied Physics, 115[23] 2014, 233702
https://doi.org/10.1063/1.4883193

[247] Qiu, X., Cheng, Y., Cao, Z., Lei, J., Current Applied Physics, 15[6] 2015, 722-726
https://doi.org/10.1016/j.cap.2015.03.014

[248] Yesilyurt, C., Tan, S.G., Liang, G., Jalil, M.B.A., Applied Physics Express, 8[10] 2015, 105201
https://doi.org/10.7567/APEX.8.105201

[249] Wang, S.K., Wang, J., Chinese Physics B, 24[3] 2015, 037202
https://doi.org/10.1088/1674-1056/24/3/037202

[250] Yang, M., Song, X.L., Chen, D.H., Bai, Y.K., Physics Letters A, 379[16-17] 2015, 1149-1152
https://doi.org/10.1016/j.physleta.2015.02.021

[251] Wang, P., Zhou, M., Liu, G., Liu, Y., Long, M.Q., Zhou, G., European Physical Journal B, 88[9] 2015, 243
https://doi.org/10.1140/epjb/e2015-60316-x

[252] Linder, J., Yokoyama, T., Physical Review B, 89[2] 2014, 020504
https://doi.org/10.1103/PhysRevB.89.020504

[253] Yu, Z., Pan, H., Yao, Y., Physical Review B, 92[15] 2015, 155419
https://doi.org/10.1103/PhysRevB.92.155419

[254] Lu, Y., Zhao, S., Zhang, Y., Liu, H., Lu, W., Liang, W., Materials Research Express, 1[4] 2015, 045009
https://doi.org/10.1088/2053-1591/1/4/045009

[255] Zhang, Q., Chan, K.S., Long, M., Journal of Physics - Condensed Matter, 28[5] 2016, 055301
https://doi.org/10.1088/0953-8984/28/5/055301

[256] Miwa, R.H., Kagimura, R., Lima, M.P., Fazzio, A., Physical Review B, 91[20] 2015, 205442
https://doi.org/10.1103/PhysRevB.91.205442

[257] Mohammadi, Y., Nia, B.A., Superlattices and Microstructures, 88, 2015, 442-449
https://doi.org/10.1016/j.spmi.2015.10.006

[258] Yao, Y., Liu, S.Y., Lei, X.L., Physical Review B, 91[11] 2015, 115411
https://doi.org/10.1103/PhysRevB.91.115411

[259] Hajati, Y., Rashidian, Z., AIP Advances, 6[2] 2016, 025307
https://doi.org/10.1063/1.4942043

[260] Hajati, Y., Rashidian, Z., Superlattices and Microstructures, 92, 2016, 264-277
https://doi.org/10.1016/j.spmi.2016.02.032

[261] Missault, N., Vasilopoulos, P., Vargiamidis, V., Peeters, F.M., Van Duppen, B., Physical Review B, 92[19] 2015, 195423
https://doi.org/10.1103/PhysRevB.92.195423

[262] Vargiamidis, V., Vasilopoulos, P., Journal of Applied Physics, 117[9] 2015, 094305
https://doi.org/10.1063/1.4913934

[263] Qiu, X.J., Cheng, Y.F., Cao, Z.Z., Lei, J.M., Journal of Physics D, 48[46] 2015, 465105
https://doi.org/10.1088/0022-3727/48/46/465105

[264] Wu, X.Q., Meng, H., Journal of Applied Physics, 117[20] 2015, 203903
https://doi.org/10.1063/1.4921799

[265] An, X.T., Physics Letters A, 379[7] 2015, 723-727
https://doi.org/10.1016/j.physleta.2014.11.059

[266] Soodchomshom, B., Journal of Applied Physics, 115[2] 2014, 023706
https://doi.org/10.1063/1.4861644

[267] Tabert, C.J., Nicol, E.J., Physical Review Letters, 110[19] 2013, 197402
https://doi.org/10.1103/PhysRevLett.110.197402

[268] Wang, Y., Applied Physics Letters, 104[3] 2014, 032105
https://doi.org/10.1063/1.4863091

[269] Yokoyama, T., Physical Review B, 87[24] 2013, 241409
https://doi.org/10.1103/PhysRevB.87.241409

[270] Zhang, L.D., Yang, F., Yao, Y., Physical Review B, 92[10] 2015, 104504
https://doi.org/10.1103/PhysRevB.92.104504

[271] Chen, C., Zhu, Z., Zha, D., Qi, M., Wu, J., Chemical Physics Letters, 646, 2016, 148-152
https://doi.org/10.1016/j.cplett.2016.01.027

[272] Zheng, R., Chen, Y., Ni, J., Applied Physics Letters, 107[26] 2015, 263104
https://doi.org/10.1063/1.4938755

[273] Zheng, R., Lin, X., Ni, J., Applied Physics Letters, 105[9] 2014, 092410
https://doi.org/10.1063/1.4895036

[274] Sun, X., Wang, L., Lin, H., Hou, T., Li, Y., Applied Physics Letters, 106[22] 2015, 222401
https://doi.org/10.1063/1.4921699

[275] Dzade, N.Y., Obodo, K.O., Adjokatse, S.K., Ashu, A.C., Amankwah, E., Atiso, C.D., Bello, A.A., Igumbor, E., Nzabarinda, S.B., Obodo, J.T., Ogbuu, A.O., Emmanuel Femi, O., Udeigwe, J.O., Waghmare, U.V., Journal of Physics - Condensed Matter, 22[37] 2010, 375502
https://doi.org/10.1088/0953-8984/22/37/375502

[276] Lee, Y., Yun, K.H., Cho, S.B., Chung, Y.C., ChemPhysChem, 15[18] 2014, 4095-4099
https://doi.org/10.1002/cphc.201402613

[277] Zhao, T., Zhang, S., Wang, Q., Kawazoe, Y., Jena, P., Physical Chemistry Chemical Physics, 16[42] 2014, 22979-22986
https://doi.org/10.1039/C4CP02758B

[278] Xiao, X., Liu, Y., Wen, W., Journal of Physics - Condensed Matter, 26[26] 2014, 266001
https://doi.org/10.1088/0953-8984/26/26/266001

[279] Luan, H.X., Zhang, C.W., Li, F., Wang, P.J., Physics Letters A, 377[39] 2013, 2792-2795
https://doi.org/10.1016/j.physleta.2013.08.016

[280] Zhang, R.W., Zhang, C.W., Li, S.S., Ji, W.X., Wang, P.J., Li, F., Li, P., Ren, M.J., Yuan, M., Solid State Communications, 191, 2014, 49-53
https://doi.org/10.1016/j.ssc.2014.04.015

[281] Zheng, F.B., Zhang, C.W., Nanoscale Research Letters, 7, 2012

[282] Fang, D.Q., Zhang, S.L., Xu, H., RSC Advances, 3[46] 2013, 24075-24080
https://doi.org/10.1039/c3ra42720j

[283] Ezawa, M., European Physical Journal B, 85[11] 2012, 363
https://doi.org/10.1140/epjb/e2012-30577-0

[284] Song, Y.L., Zhang, Y., Zhang, J.M., Lu, D.B., Xu, K.W., European Physical Journal B, 79[2] 2011, 197-202
https://doi.org/10.1140/epjb/e2010-10627-5

[285] Song, Y.L., Zhang, J.M., Lu, D.B., Xu, K.W., Physica E, 56, 2014, 205-210
https://doi.org/10.1016/j.physe.2013.08.028

[286] Zheng, F.B., Zhang, C.W., Yan, S.S., Li, F., Journal of Materials Chemistry C, 1[15] 2013, 2735-2743
https://doi.org/10.1039/c3tc30097h

[287] Zhang, C.W., Yan, S.S., Journal of Physical Chemistry C, 116[6] 2012, 4163-4166
https://doi.org/10.1021/jp2104177

[288] Wang, X.Q., Li, H.D., Wang, J.T., Physical Chemistry Chemical Physics, 14[9] 2012, 3031-3036
https://doi.org/10.1039/c2cp23385a

[289] Tahir, M., Schwingenschlögl, U., Applied Physics Letters, 101[13] 2012, 132412
https://doi.org/10.1063/1.4754711

[290] Islam, S.K.F., Ghosh, T.K., Journal of Physics - Condensed Matter, 26[33] 2014, 335303
https://doi.org/10.1088/0953-8984/26/33/335303

[291] Drissi, L.B., Ramadan, F.Z., Saidi, E.H., Bousmina, M., Fassi-Fehri, O., Journal of the Physical Society of Japan, 82[10] 2013, 104711
https://doi.org/10.7566/JPSJ.82.104711

[292] Aghaei, S.M., Monshi, M.M., Torres, I., Calizo, I., RSC Advances, 6[21] 2016, 17046-17058
https://doi.org/10.1039/C5RA26107D

[293] Matthes, L., Pulci, O., Bechstedt, F., New Journal of Physics, 16, 2014, 105007
https://doi.org/10.1088/1367-2630/16/10/105007

[294] Tabert, C.J., Nicol, E.J., Physical Review B, 88[8] 2013, 085434
https://doi.org/10.1103/PhysRevB.88.085434

[295] Bao, H., Liao, W., Guo, J., Zhao, H., Zhou, G., Laser Physics Letters, 12[9] 2015, 095902
https://doi.org/10.1088/1612-2011/12/9/095902

[296] Bechstedt, F., Matthes, L., Gori, P., Pulci, O., Applied Physics Letters, 100[26] 2012, 261906
https://doi.org/10.1063/1.4731626

[297] Guo, Z.X., Zhang, Y.Y., Xiang, H., Gong, X.G., Oshiyama, A., Physical Review B, 92[20] 2015, 201413
https://doi.org/10.1103/PhysRevB.92.201413

[298] Cinquanta, E., Fratesi, G., Dal Conte, S., Grazianetti, C., Scotognella, F., Stagira, S., Vozzi, C., Onida, G., Molle, A., Physical Review B, 92[16] 2015, 165427
https://doi.org/10.1103/PhysRevB.92.165427

[299] Das, R., Chowdhury, S., Majumdar, A., Jana, D., RSC Advances, 5[1] 2015, 41-50
https://doi.org/10.1039/C4RA07976K

[300] Bao, H., Guo, J., Liao, W., Zhao, H., Applied Physics A, 118[2] 2014, 431-435
https://doi.org/10.1007/s00339-014-8837-x

[301] Ye, X.S., Shao, Z.G., Zhao, H., Yang, L., Wang, C.L., RSC Advances, 4[72], 2014, 37998-38003
https://doi.org/10.1039/C4RA03942D

[302] Konobeeva, N.N., Belonenko, M.B., Technical Physics Letters, 39[6] 2013, 579-581
https://doi.org/10.1134/S1063785013060205

[303] Vogt, P., Capiod, P., Berthe, M., Resta, A., De Padova, P., Bruhn, T., Le Lay, G., Grandidier, B., Applied Physics Letters, 104[2] 2014, 021602
https://doi.org/10.1063/1.4861857

[304] Yamacli, S., Journal of Nanoparticle Research, 16[8] 2014, 2576
https://doi.org/10.1007/s11051-014-2576-y

[305] Suwanvarangkoon, A., Soodchomshom, B., Journal of Magnetism and Magnetic Materials, 374, 2015, 479-483
https://doi.org/10.1016/j.jmmm.2014.08.095

[306] Esmailpour, A., Abdolmaleki, M., Saadat, M., Physica E, 77, 2016, 144-148
https://doi.org/10.1016/j.physe.2015.11.006

[307] Yang, T., Lin, Q., Wang, C.M., EPL, 112[1] 2015, 17009
https://doi.org/10.1209/0295-5075/112/17009

[308] Romera, E., Calixto, M., EPL, 111[3] 2015, 37006
https://doi.org/10.1209/0295-5075/111/37006

[309] Rakyta, P., Vigh, M., Csordás, A., Cserti, J., Physical Review B, 91[12] 2015, 125412
https://doi.org/10.1103/PhysRevB.91.125412

[310] Yin, H.F., Xiang, G.Z., Yue, L., Zhang, H., Acta Physico - Chimica Sinica, 31[1] 2015, 67-72

[311] Wang, X.S., Shen, M., An, X.T., Liu, J.J., Physics Letters A, 380[18-19] 2016 1663-1667
https://doi.org/10.1016/j.physleta.2016.02.046

[312] Deng, X.Q., Zhang, Z.H., Fan, Z.Q., Tang, G.P., Sun, L., Li, C.X., Organic Electronics, 32, 2016, 41-46
https://doi.org/10.1016/j.orgel.2016.01.036

[313] Deng, X.Q., Zhang, Z.H., Tang, G.P., Fan, Z.Q., Yang, C.H., RSC Advances, 4[103] 2014, 58941-58948
https://doi.org/10.1039/C4RA09566A

[314] Zha, D., Chen, C., Wu, J., Solid State Communications, 219, 2015, 12715, 21-24

[315] Ezawa, M., New Journal of Physics, 14, 2012, 033003
https://doi.org/10.1088/1367-2630/14/3/033003

[316] An, X.T., Diao, S.M., Acta Physica Sinica, 63[18] 2014, 187304

[317] Wang, Y.P., Fry, J.N., Cheng, H.P., Physical Review B, 88[12] 2013, 125428
https://doi.org/10.1103/PhysRevB.88.125428

[318] Sardroodi, J.J., Afshari, S., Ebrahimzadeh, A.R., Ghavami, B., Notash, M.Y., Journal of Theoretical and Computational Chemistry, 13[6] 2014, 1450046
https://doi.org/10.1142/S0219633614500461

[319] Zhang, L.D., Yang, F., Yao, Y., Scientific Reports, 5, 2015, 8203
https://doi.org/10.1038/srep08203

[320] Liu, F., Liu, C.C., Wu, K., Yang, F., Yao, Y., Physical Review Letters, 111[6] 2013, 066804
https://doi.org/10.1103/PhysRevLett.111.066804

[321] Wan, W., Ge, Y., Yang, F., Yao, Y., EPL, 104[3] 2013, 36001
https://doi.org/10.1209/0295-5075/104/36001

[322] Chen, L., Feng, B., Wu, K., Applied Physics Letters, 102[8] 2013, 081602
https://doi.org/10.1063/1.4793998

[323] Ezawa, M., Journal of Superconductivity and Novel Magnetism, 28[4] 2015, 1249-1253
https://doi.org/10.1007/s10948-014-2900-x

[324] Durajski, A.P., Szczešniak, D., Szczešniak, R., Solid State Communications, 200, 2014, 17-21
https://doi.org/10.1016/j.ssc.2014.09.007

[325] Dyrdał, A., Barnaś, J., Physica Status Solidi R, 6[8] 2012, 340-342
https://doi.org/10.1002/pssr.201206202

[326] Ezawa, M., Journal of the Physical Society of Japan, 81[6] 2012, 064705
https://doi.org/10.1143/JPSJ.81.064705

[327] Liu, C.C., Feng, W., Yao, Y., Physical Review Letters, 107[7] 2011, 076802
https://doi.org/10.1103/PhysRevLett.107.076802

[328] Cao, G., Zhang, Y., Cao, J., Physics Letters A, 379[22-23] 2015, 1475-1479
https://doi.org/10.1016/j.physleta.2015.03.021

[329] Zhang, X.L., Liu, L.F., Liu, W.M., Scientific Reports, 3, 2013, 2908

[330] An, X.T., Zhang, Y.Y., Liu, J.J., Li, S.S., Applied Physics Letters, 102[4] 2013, 043113
https://doi.org/10.1063/1.4790147

[331] Shakouri, K., Vasilopoulos, P., Vargiamidis, V., Peeters, F.M., Physical Review B, 90[23] 2014, 235423
https://doi.org/10.1103/PhysRevB.90.235423

[332] Geissler, F., Budich, J.C., Trauzettel, B., New Journal of Physics, 15, 2013, 085030
https://doi.org/10.1088/1367-2630/15/8/085030

[333] Singh, N., Schwingenschlögl, U., Physica Status Solidi R, 8[4] 2014, 353-356
https://doi.org/10.1002/pssr.201409025

[334] Lew Yan Voon, L.C., Chinese Physics B, 24[8] 2015, 087309
https://doi.org/10.1088/1674-1056/24/8/087309

[335] Sakai, Y., Oshiyama, A., Physical Review B, 91[20] 2015, 201405
https://doi.org/10.1103/PhysRevB.91.201405

[336] Shiraishi, K., Hatsugai, Y., Journal of the Vacuum Society of Japan, 57[11] 2014, 423-427
https://doi.org/10.3131/jvsj2.57.423

[337] Huang, B., Xiang, H.J., Wei, S.H., Physical Review Letters, 111[14] 2013, 145502
https://doi.org/10.1103/PhysRevLett.111.145502

[338] Ezawa, M., Journal of the Physical Society of Japan, 81[10] 2012, 104713
https://doi.org/10.1143/JPSJ.81.104713

[339] Ezawa, M., Physical Review Letters, 110[2] 2013, 026603
https://doi.org/10.1103/PhysRevLett.110.026603

[340] Luan, H.X., Zhang, C.W., Zheng, F.B., Wang, P.J., Journal of Physical Chemistry C, 117[26] 2013, 13620-13626
https://doi.org/10.1021/jp4005357

[341] Ma, L., Zhang, J.M., Xu, K.W., Ji, V., Physica B, 425, 2013, 66-71
https://doi.org/10.1016/j.physb.2013.05.022

[342] Ma, L., Zhang, J.M., Xu, K.W., Ji, V., Physica E, 60, 2014, 112-117
https://doi.org/10.1016/j.physe.2014.02.013

[343] Liang, Y., Wang, V., Mizuseki, H., Kawazoe, Y., Journal of Physics - Condensed Matter, 24[45] 2012, 455302
https://doi.org/10.1088/0953-8984/24/45/455302

[344] Jose, D., Datta, A., Journal of Physical Chemistry C, 116[46] 2012, 24639-24648
https://doi.org/10.1021/jp3084716

[345] Osborn, T.H., Farajian, A.A., Journal of Physical Chemistry C, 116[43] 2012, 22916-22920
https://doi.org/10.1021/jp306889x

[346] Liu, J., Zhang, W., RSC Advances, 3[44] 2013, 21943-21948
https://doi.org/10.1039/c3ra44392b

[347] Mohan, B., Kumar, A., Ahluwalia, P.K., Physica E, 53, 2013, 233-239
https://doi.org/10.1016/j.physe.2013.05.014

[348] Chen, K., Wan, X., Xu, J.B., Journal of Materials Chemistry C, 1[32] 2013, 4869-4878
https://doi.org/10.1039/c3tc30567h

[349] Zhou, B., Zhou, B., Zeng, Y., Zhou, G., Duan, M., Physics Letters A, 380[1-2] 2016, 282-287
https://doi.org/10.1016/j.physleta.2015.09.054

[350] Mohan, B., Kumar, A., Ahluwalia, P.K., Physica E, 61, 2014, 40-47
https://doi.org/10.1016/j.physe.2014.03.013

[351] Chen, M.X., Weinert, M., Nano Letters, 14[9] 2014, 5189-5193
https://doi.org/10.1021/nl502107v

[352] Avila, J., De Padova, P., Cho, S., Colambo, I., Lorcy, S., Quaresima, C., Vogt, P., Resta, A., Le Lay, G., Asensio, M.C., Journal of Physics - Condensed Matter, 25[26] 2013, 262001
https://doi.org/10.1088/0953-8984/25/26/262001

[353] Tsoutsou, D., Xenogiannopoulou, E., Golias, E., Tsipas, P., Dimoulas, A., Applied Physics Letters, 103[23] 2013, 231604
https://doi.org/10.1063/1.4841335

[354] Rastgoo, S., Shirkani, H., Golshan, M.M., Physics Letters A, 379[14-15] 2015, 1048-1052
https://doi.org/10.1016/j.physleta.2015.01.039

[355] Wei, W., Jacob, T., Physical Review B, 88[4] 2013, 045203
https://doi.org/10.1103/PhysRevB.88.045203

[356] Yamakage, A., Ezawa, M., Tanaka, Y., Nagaosa, N., Physical Review B, 88[8] 2013, 085322
https://doi.org/10.1103/PhysRevB.88.085322

[357] Yan, J.A., Stein, R., Schaefer, D.M., Wang, X.Q., Chou, M.Y., Physical Review B, 88[12] 2013, 121403
https://doi.org/10.1103/PhysRevB.88.121403

[358] Liu, Y., Shu, H., Liang, P., Cao, D., Chen, X., Lu, W., Journal of Applied Physics, 114[9] 2013, 094308
https://doi.org/10.1063/1.4820566

[359] Liu, H., Han, N., Zhao, J., Journal of Physics - Condensed Matter, 26[47] 2014, 475303
https://doi.org/10.1088/0953-8984/26/47/475303

[360] Wei, W., Dai, Y., Huang, B., Jacob, T., Physical Chemistry Chemical Physics, 15[22] 2013, 8789-8794
https://doi.org/10.1039/c3cp51078f

[361] Matthes, L., Pulci, O., Bechstedt, F., Journal of Physics - Condensed Matter, 25[39] 2013, 395305
https://doi.org/10.1088/0953-8984/25/39/395305

[362] Zakerian, F., Berahman, M., Optical and Quantum Electronics, 48[7], 2016, 370
https://doi.org/10.1007/s11082-016-0637-4

[363] Roome, N.J., Carey, J.D., ACS Applied Materials and Interfaces, 6[10] 2014, 7743-7750
https://doi.org/10.1021/am501022x

[364] Wang, G., EPL, 101[2] 2013, 27005
https://doi.org/10.1209/0295-5075/101/27005

[365] Shao, Z.G., Ye, X.S., Yang, L., Wang, C.L., Journal of Applied Physics, 114[9] 2013, 093712
https://doi.org/10.1063/1.4820526

[366] Yu-Zhong, Z., Rowlands, D.A., Zhang, Y.Z., Chinese Physics B, 23[3] 2014, 037101
https://doi.org/10.1088/1674-1056/23/3/037101

[367] Song, Y.L., Zhang, S., Lu, D.B., Xu, H.R., Wang, Z., Zhang, Y., Lu, Z.W., European Physical Journal B, 86[12] 2013, 488
https://doi.org/10.1140/epjb/e2013-31078-4

[368] Yan, Y., Wu, H., Jiang, F., Zhao, H., European Physical Journal B, 86[11] 2013, 457
https://doi.org/10.1140/epjb/e2013-40818-3

[369] Zhang, J.M., Song, W.T., Xu, K.W., Ji, V., Computational Materials Science, 95, 2014, 429-434
https://doi.org/10.1016/j.commatsci.2014.08.019

[370] Pournaghavi, N., Esmaeilzadeh, M., Ahmadi, S., Farokhnezhad, M., Solid State Communications, 226, 2016, 33-38
https://doi.org/10.1016/j.ssc.2015.11.005

[371] Kaloni, T.P., Schwingenschlögl, U., Physica Status Solidi – R, 8[8] 2014, 685-687
https://doi.org/10.1002/pssr.201409245

[372] Li, X., Wu, S., Zhou, S., Zhu, Z., Nanoscale Research Letters, 9[1] 2014, 1-9
https://doi.org/10.1186/1556-276X-9-1

[373] Gao, N., Li, J.C., Jiang, Q., Physical Chemistry Chemical Physics, 16[23] 2014, 11673-11678
https://doi.org/10.1039/c4cp00089g

[374] Pan, F., Wang, Y., Jiang, K., Ni, Z., Ma, J., Zheng, J., Quhe, R., Shi, J., Yang, J., Chen, C., Lu, J., Scientific Reports, 5, 2015, 9075
https://doi.org/10.1038/srep09075

[375] Gao, N., Li, J.C., Jiang, Q., Chemical Physics Letters, 592, 2014, 222-226
https://doi.org/10.1016/j.cplett.2013.12.036

[376] Du, Y., Zhuang, J., Liu, H., Xu, X., Eilers, S., Wu, K., Cheng, P., Zhao, J., Pi, X., See, K.W., Peleckis, G., Wang, X., Dou, S.X., ACS Nano, 8[10] 2014, 10019-10025
https://doi.org/10.1021/nn504451t

[377] Zhu, J., Schwingenschlögl, U., ACS Applied Materials and Interfaces, 6[21] 2014, 19242-19246
https://doi.org/10.1021/am5052697

[378] Fleurence, A., Friedlein, R., Ozaki, T., Kawai, H., Wang, Y., Yamada-Takamura, Y., Physical Review Letters, 108[24] 2012, 245501
https://doi.org/10.1103/PhysRevLett.108.245501

[379] Kaloni, T.P., Singh, N., Schwingenschlögl, U., Physical Review B, 89[3] 2014, 035409
https://doi.org/10.1103/PhysRevB.89.035409

[380] Mohan, B., Kumar, A., Ahluwalia, P.K., RSC Advances, 4[60] 2014, 31700-31705
https://doi.org/10.1039/C4RA02711F

[381] Zhang, R.W., Zhang, C.W., Ji, W.X., Hu, S.J., Yan, S.S., Li, S.S., Li, P., Wang, P.J., Liu, Y.S., Journal of Physical Chemistry C, 118[43] 2014, 25278-25283
https://doi.org/10.1021/jp508253x

[382] Kaloni, T.P., Gangopadhyay, S., Singh, N., Jones, B., Schwingenschlögl, U., Physical Review B, 88[23] 2013, 235418
https://doi.org/10.1103/PhysRevB.88.235418

[383] Van Den Broek, B., Houssa, M., Scalise, E., Pourtois, G., Afanasev, V.V., Stesmans, A., Applied Surface Science, 291, 2014, 104-108
https://doi.org/10.1016/j.apsusc.2013.09.032

[384] Trivedi, S., Srivastava, A., Kurchania, R., Journal of Computational and Theoretical Nanoscience, 11[3] 2014, 789-794
https://doi.org/10.1166/jctn.2014.3429

[385] Chang, H.R., Zhou, J., Zhang, H., Yao, Y., Physical Review B, 89[20] 2014, 201411
https://doi.org/10.1103/PhysRevB.89.201411

[386] Priede, L.M., Meza-Montes, L., Journal of Nano Research, 28, 2014, 1-7
https://doi.org/10.4028/www.scientific.net/JNanoR.28.1

[387] Rui, W., Shaofeng, W., Xiaozhi, W., Journal of Applied Physics, 116[2] 2014, 024303
https://doi.org/10.1063/1.4887353

[388] Lopez-Bezanilla, A., Journal of Physical Chemistry C, 118[32] 2014, 18788-18792
https://doi.org/10.1021/jp5060809

[389] Vargiamidis, V., Vasilopoulos, P., Hai, G.Q., Journal of Physics - Condensed Matter, 26[34] 2014, 345303
https://doi.org/10.1088/0953-8984/26/34/345303

[390] Chen, J., Wang, X.F., Vasilopoulos, P., Chen, A.B., Wu, J.C., ChemPhysChem, 15[13] 2014, 2701-2706
https://doi.org/10.1002/cphc.201402171

[391] Gong, L., Xiu, S.L., Zheng, M.M., Zhao, P., Zhang, Z., Liang, Y.Y., Chen, G., Kawazoe, Y., Journal of Materials Chemistry C, 2[41] 2014, 8773-8779
https://doi.org/10.1039/C4TC01665C

[392] Feng, Y., Feng, B.J., Xie, Z.J., Li, W.B., Liu, X., Liu, D.F., Zhao, L., Chen, L., Zhou, X.J., Wu, K.H., Chinese Physics Letters, 31[12] 2014, 127303
https://doi.org/10.1088/0256-307X/31/12/127303

[393] Huang, Y.P., Yuan, J.M., Guo, G., Mao, Y.L., Acta Physica Sinica, 64[1] 2015, 013101

[394] Ding, Y., Wang, Y., Nanoscale Research Letters, 10[1] 2015, 1-11
https://doi.org/10.1186/1556-276X-10-1

[395] Drissi, L.B., Ramadan, F.Z., Physica E, 68, 2015, 38-41
https://doi.org/10.1016/j.physe.2014.12.009

[396] Padilha, J.E., Pontes, R.B., Journal of Physical Chemistry C, 119[7] 2015, 3818-3825
https://doi.org/10.1021/jp512489m

[397] Das, D.K., Sahoo, S., Advances in Intelligent Systems and Computing, 343, 2015, 97-103
https://doi.org/10.1007/978-81-322-2268-2_11

[398] Zhang, R.W., Zhang, C.W., Ji, W.X., Ren, M.J., Li, F., Yuan, M., Materials Chemistry and Physics, 156, 2015, 89-94
https://doi.org/10.1016/j.matchemphys.2015.02.029

[399] Gao, T.H., Acta Physica Sinica, 64[7] 2015, 076801

[400] Wang, R., Pi, X., Ni, Z., Liu, Y., Yang, D., RSC Advances, 5[43] 2015, 33831-33837
https://doi.org/10.1039/C5RA05751E

[401] Calixto, M., Romera, E., EPL, 109[4] 2015, 40003
https://doi.org/10.1209/0295-5075/109/40003

[402] Hatsugai, Y., Shiraishi, K., Aoki, H., New Journal of Physics, 17, 2015, 025009
https://doi.org/10.1088/1367-2630/17/2/025009

[403] Lew Yan Voon, L.C., Lopez-Bezanilla, A., Wang, J., Zhang, Y., Willatzen, M., New Journal of Physics, 17, 2015, 025004
https://doi.org/10.1088/1367-2630/17/2/025004

[404] López, A., Scholz, A., Santos, B., Schliemann, J., Physical Review B, 91[12] 2015, 125105
https://doi.org/10.1103/PhysRevB.91.125105

[405] Lin, X., Ni, J., Journal of Applied Physics, 117[16] 2015, 164305
https://doi.org/10.1063/1.4919223

[406] Zhu, J., Schwingenschlögl, U., Journal of Materials Chemistry C, 3[16] 2015, 3946-3953
https://doi.org/10.1039/C5TC00435G

[407] Nigam, S., Gupta, S.K., Majumder, C., Pandey, R., Physical Chemistry Chemical Physics, 17[17] 2015, 11324-11328
https://doi.org/10.1039/C4CP05462H

[408] Noor-A-Alam, M., Kim, H.J., Shin, Y.H., Journal of Applied Physics, 117[22] 2015, 224304
https://doi.org/10.1063/1.4922404

[409] Hu, B., Journal of Physics - Condensed Matter, 27[24] 2015, 245301
https://doi.org/10.1088/0953-8984/27/24/245301

[410] Shakouri, K., Simchi, H., Esmaeilzadeh, M., Mazidabadi, H., Peeters, F.M., Physical Review B, 92[3] 2015, 035413
https://doi.org/10.1103/PhysRevB.92.035413

[411] Dai, J., Zeng, X.C., Physical Chemistry Chemical Physics, 17[27] 2015, 17957-17961
https://doi.org/10.1039/C4CP04953E

[412] Ma, T., Wen, S., Wu, C., Yan, L., Zhang, M., Kan, Y., Su, Z., Journal of Materials Chemistry C, 3[39] 2015, 10085-10090 *https://doi.org/10.1039/C5TC00792E*

[413] Jia, T.T., Zheng, M.M., Fan, X.Y., Su, Y., Li, S.J., Liu, H.Y., Chen, G., Kawazoe, Y., Journal of Physical Chemistry C, 119[35] 2015, 20747-20754 *https://doi.org/10.1021/acs.jpcc.5b06626*

[414] Kou, L., Ma, Y., Yan, B., Tan, X., Chen, C., Smith, S.C., ACS Applied Materials and Interfaces, 7[34] 2015, 19226-19233 *https://doi.org/10.1021/acsami.5b05063*

[415] Yang, J.Y., Liu, L.H., Applied Physics Letters, 107[9] 2015, 091902 *https://doi.org/10.1063/1.4930025*

[416] Wang, N., Guo, H., Liu, Y.J., Zhao, J.X., Cai, Q.H., Wang, X.Z., Physica E, 73, 2015, 11963, 21-26

[417] Mehdi Aghaei, S., Calizo, I., Journal of Applied Physics, 118[10] 2015, 104304 *https://doi.org/10.1063/1.4930139*

[418] Bao, H., Liao, W., Guo, J., Yang, X., Zhao, H., Zhou, G., Journal of Physics D, 48[45] 2015, 455306 *https://doi.org/10.1088/0022-3727/48/45/455306*

[419] Zhang, M.M., Xu, L., Zhang, J., Journal of Physics - Condensed Matter, 27[44] 2015, 445301 *https://doi.org/10.1088/0953-8984/27/44/445301*

[420] Pogorelov, Y.G., Loktev, V.M., Physical Review B, 93[4] 2016, 045117 *https://doi.org/10.1103/PhysRevB.93.045117*

[421] Zhu, B.S., Wang, Y., Lou, Y.Y., Physics Letters A, 380[3] 2016, 502-508 *https://doi.org/10.1016/j.physleta.2015.11.001*

[422] Nawaz, S., Tahir, M., Physica E, 76, 2016, 169-172 *https://doi.org/10.1016/j.physe.2015.10.023*

[423] Jia, T.T., Fan, X.Y., Zheng, M.M., Chen, G., Scientific Reports, 6, 2016, 20971 *https://doi.org/10.1038/srep20971*

[424] Fu, B., Shi, Q., Li, Q., Yang, J., Physical Review B, 93[8] 2016, 085430 *https://doi.org/10.1103/PhysRevB.93.085430*

[425] Le, H.M., Pham, T.T., Dinh, T.S., Kawazoe, Y., Nguyen-Manh, D., Journal of Physics - Condensed Matter, 28[13] 2016, 135301 *https://doi.org/10.1088/0953-8984/28/13/135301*

[426] Liao, W., Bao, H., Zhang, X., Yang, X., Zhang, Z., Zhao, H., Applied Physics A, 122[4] 2016, 464
https://doi.org/10.1007/s00339-016-9988-8

[427] Ishida, H., Hamamoto, Y., Morikawa, Y., Minamitani, E., Arafune, R., Takagi, N., New Journal of Physics, 17, 2015, 015013
https://doi.org/10.1088/1367-2630/17/1/015013

[428] Gert, A.V., Nestoklon, M.O., Yassievich, I.N., Journal of Experimental and Theoretical Physics, 121[1] 2015, 115-121
https://doi.org/10.1134/S1063776115060072

[429] Lian, C., Ni, J., Physical Chemistry Chemical Physics, 17[20] 2015, 13366-13373
https://doi.org/10.1039/C5CP01557J

[430] Yan, J.A., Gao, S.P., Stein, R., Coard, G., Physical Review B, 91[24] 2015, 245403
https://doi.org/10.1103/PhysRevB.91.245403

[431] Wang, W., Olovsson, W., Uhrberg, R.I.G., Physical Review B, 93[8] 2016, 081406
https://doi.org/10.1103/PhysRevB.93.081406

[432] Ding, Y., Wang, Y., Journal of Materials Chemistry C, 3[43] 2015, 11341-11348
https://doi.org/10.1039/C5TC02504D

[433] Huang, Z.Q., Chou, B.H., Hsu, C.H., Chuang, F.C., Lin, H., Bansil, A., Physical Review B, 90[24] 2014, 245433
https://doi.org/10.1103/PhysRevB.90.245433

[434] Li, S.S., Zhang, C.W., Yan, S.S., Hu, S.J., Ji, W.X., Wang, P.J., Li, P., Journal of Physics - Condensed Matter, 26[39] 2014, 395003
https://doi.org/10.1088/0953-8984/26/39/395003

[435] Ersan, F., Arslanalp, Ö., Gökoğlu, G., Aktürk, E., Applied Surface Science, 311, 2014, 9-13
https://doi.org/10.1016/j.apsusc.2014.04.176

[436] Li, L., Zhao, M., Journal of Physical Chemistry C, 118[33] 2014, 19129-19138
https://doi.org/10.1021/jp5043359

[437] Li, S., Wu, Y., Liu, W., Zhao, Y., Chemical Physics Letters, 609, 2014, 161-166
https://doi.org/10.1016/j.cplett.2014.06.047

[438] Zhu, J., Schwingenschlögl, U., ACS Applied Materials and Interfaces, 6[14] 2014, 11675-11681
https://doi.org/10.1021/am502469m

[439] Wu, Y., Zhang, K., Huang, Y., Wu, S., Zhu, H., Cheng, P., Ni, J., European Physical Journal B, 87[4] 2014, 94, 1-6

[440] Kamal, C., Chakrabarti, A., Banerjee, A., Physics Letters A, 378[16-17] 2014, 1162-1169
https://doi.org/10.1016/j.physleta.2014.02.011

[441] Fu, H., Zhang, J., Ding, Z., Li, H., Meng, S., Applied Physics Letters, 104[13] 2014, 131904
https://doi.org/10.1063/1.4870534

[442] Li, G., Tan, J., Liu, X., Wang, X., Li, F., Zhao, M., Chemical Physics Letters, 595-596, 2014, 20-24
https://doi.org/10.1016/j.cplett.2014.01.042

[443] Rahman, G., EPL, 105[3] 2014, 37012
https://doi.org/10.1209/0295-5075/105/37012

[444] Zberecki, K., Wierzbicki, M., Barnaå, J., Swirkowicz, R., Physical Review B, 88[11] 2013, 115404
https://doi.org/10.1103/PhysRevB.88.115404

[445] Cai, Y., Chuu, C.P., Wei, C.M., Chou, M.Y., Physical Review B, 88[24] 2013, 245408
https://doi.org/10.1103/PhysRevB.88.245408

[446] Qin, R., Zhu, W., Zhang, Y., Deng, X., Nanoscale Research Letters, 9[1] 2014
https://doi.org/10.1186/1556-276X-9-521

[447] Zhao, H., Physics Letters A, 376[46] 2012, 3546-3550
https://doi.org/10.1016/j.physleta.2012.10.024

[448] Wang, Y., Ding, Y., Solid State Communications, 155, 2013, 6-11
https://doi.org/10.1016/j.ssc.2012.10.044

[449] Liu, G., Wu, M.S., Ouyang, C.Y., Xu, B., EPL, 99[1] 2012, 17010
https://doi.org/10.1209/0295-5075/99/17010

[450] Pan, F., Quhe, R., Ge, Q., Zheng, J., Ni, Z., Wang, Y., Gao, Z., Wang, L., Lu, J., Physica E, 56, 2014, 43-47
https://doi.org/10.1016/j.physe.2013.08.011

[451] Huang, S., Kang, W., Yang, L., Applied Physics Letters, 102[13] 2013, 133106
https://doi.org/10.1063/1.4801309

[452] Ding, Y., Wang, Y., Applied Physics Letters, 102[14] 2013, 143115
https://doi.org/10.1063/1.4801948

[453] Zheng, F.B., Zhang, C.W., Wang, P.J., Li, S.S., Journal of Applied Physics, 113[15] 2013, 154302
https://doi.org/10.1063/1.4801882

[454] Bhattacharya, A., Bhattacharya, S., Das, G.P., Applied Physics Letters, 103[12] 2013, 123113
https://doi.org/10.1063/1.4821993

[455] Zhao, Y.C., Ni, J., Physical Chemistry Chemical Physics, 16[29] 2014, 15477-15482
https://doi.org/10.1039/c4cp01549e

[456] Sivek, J., Sahin, H., Partoens, B., Peeters, F.M., Physical Review B, 87[8] 2013, 085444
https://doi.org/10.1103/PhysRevB.87.085444

[457] Yang, X.F., Zou, X.L., Kuang, Y.W., Shao, Z.G., Zhang, J., Hong, X.K., Zhang, D.B., Feng, J.F., Chen, X.S., Liu, Y.S., Chemical Physics Letters, 667, 2017, 113-119
https://doi.org/10.1016/j.cplett.2016.11.049

[458] Zaminpayma, E., Nayebi, P., Physica E, 84, 2016, 555-563
https://doi.org/10.1016/j.physe.2016.06.016

[459] Ongun Özçelik, V., Ciraci, S., Journal of Physical Chemistry C, 117[49] 2013, 26305-26315
https://doi.org/10.1021/jp408647t

[460] Ongun Özçelik, V., Kecik, D., Durgun, E., Ciraci, S., Journal of Physical Chemistry C, 119[1] 2015, 845-853
https://doi.org/10.1021/jp5106554

[461] Wang, R., Xu, M.S., Pi, X.D., Chinese Physics B, 24[8] 2015, 086807
https://doi.org/10.1088/1674-1056/24/8/086807

[462] Wang, X., Liu, H., Tu, S.T., RSC Advances, 5[9] 2015, 6238-6245
https://doi.org/10.1039/C4RA12257G

[463] Gao, N., Zheng, W.T., Jiang, Q., Physical Chemistry Chemical Physics, 14[1] 2012, 257-261
https://doi.org/10.1039/C1CP22719J

[464] Chandiramouli, R., Srivastava, A., Nagarajan, V., Applied Surface Science, 351, 2015, 662-672
https://doi.org/10.1016/j.apsusc.2015.05.166

[465] Liu, G., Lei, X.L., Wu, M.S., Xu, B., Ouyang, C.Y., Journal of Physics - Condensed Matter, 26[35] 2014, 355007
https://doi.org/10.1088/0953-8984/26/35/355007

[466] Nguyen, M.T., Phong, P.N., Tuyen, N.D., ChemPhysChem, 16[8] 2015, 1733-1738
https://doi.org/10.1002/cphc.201402902

[467] Lin, S.Y., Chang, S.L., Thuy Tran, N.T., Yang, P.H., Lin, M.F., Physical Chemistry Chemical Physics, 17[39] 2015, 26443-26450
https://doi.org/10.1039/C5CP04841A

[468] Houssa, M., Scalise, E., Sankaran, K., Pourtois, G., Afanasev, V.V., Stesmans, A., Applied Physics Letters, 98[22] 2011, 223107
https://doi.org/10.1063/1.3595682

[469] Zhang, P., Li, X.D., Hu, C.H., Wu, S.Q., Zhu, Z.Z., Physics Letters A, 376[14] 2012, 1230-1233
https://doi.org/10.1016/j.physleta.2012.02.030

[470] Osborn, T.H., Farajian, A.A., Pupysheva, O.V., Aga, R.S., Lew Yan Voon, L.C., Chemical Physics Letters, 511[1-3] 2011, 101-105
https://doi.org/10.1016/j.cplett.2011.06.009

[471] Huang, B., Deng, H.X., Lee, H., Yoon, M., Sumpter, B.G., Liu, F., Smith, S.C., Wei, S.H., Physical Review X, 4[2] 2014, 021029
https://doi.org/10.1103/PhysRevX.4.021029

[472] Le Lay, G., Aufray, B., Léandri, C., Oughaddou, H., Biberian, J.P., De Padova, P., Dávila, M.E., Ealet, B., Kara, A., Applied Surface Science, 256[2] 2009, 524-529
https://doi.org/10.1016/j.apsusc.2009.07.114

[473] Wu, W.C., Ao, Z.M., Yang, C.H., Li, S., Wang, G.X., Li, C.M., Li, S., Journal of Materials Chemistry C, 3[11] 2015, 2593-2602
https://doi.org/10.1039/C4TC02095B

[474] Wu, W., Ao, Z., Wang, T., Li, C., Li, S., Physical Chemistry Chemical Physics, 16[31] 2014, 16588-16594
https://doi.org/10.1039/C4CP01416B

[475] Qiu, J., Fu, H., Xu, Y., Oreshkin, A.I., Shao, T., Li, H., Meng, S., Chen, L., Wu, K., Physical Review Letters, 114[12] 2015, 126101
https://doi.org/10.1103/PhysRevLett.114.126101

[476] Qiu, J., Fu, H., Xu, Y., Zhou, Q., Meng, S., Li, H., Chen, L., Wu, K., ACS Nano, 9[11] 2015, 11192-11199
https://doi.org/10.1021/acsnano.5b04722

[477] Trivedi, S., Srivastava, A., Kurchania, R., Journal of Computational and Theoretical Nanoscience, 11[3] 2014, 781-788
https://doi.org/10.1166/jctn.2014.3428

[478] Yang, X.F., Liu, Y.S., Feng, J.F., Wang, X.F., Zhang, C.W., Chi, F., Journal of Applied Physics, 116[12] 2014, 124312
https://doi.org/10.1063/1.4896630

[479] Ding, Y., Wang, Y., Applied Physics Letters, 100[8] 2012, 083102
https://doi.org/10.1063/1.3688035

[480] Drissi, L.B., Saidi, E.H., Bousmina, M., Fassi-Fehri, O., Journal of Physics - Condensed Matter, 24[48] 2012, 485502
https://doi.org/10.1088/0953-8984/24/48/485502

[481] Ding, Y., Wang, Y., Applied Physics Letters, 104[8] 2014, 83111
https://doi.org/10.1063/1.4866786

[482] Rupp, C.J., Chakraborty, S., Ahuja, R., Baierle, R.J., Physical Chemistry Chemical Physics, 17[34] 2015, 22210-22216
https://doi.org/10.1039/C5CP03489B

[483] Johnson, N.W., Muir, D.I., Moewes, A., Scientific Reports, 6, 2016, 22510
https://doi.org/10.1038/srep22510

[484] Wang, R., Pi, X., Ni, Z., Liu, Y., Lin, S., Xu, M., Yang, D., Scientific Reports, 3, 2013, 3507

[485] Liu, G., Lei, X.L., Wu, M.S., Xu, B., Ouyang, C.Y., EPL, 106[4] 2014, 47001
https://doi.org/10.1209/0295-5075/106/47001

[486] Wang, Y., Ding, Y., Physica Status Solidi R, 7[6] 2013, 410-413
https://doi.org/10.1002/pssr.201307110

[487] De Padova, P., Quaresima, C., Olivieri, B., Perfetti, P., Le Lay, G., Journal of Physics D, 44[31] 2011, 312001
https://doi.org/10.1088/0022-3727/44/31/312001

[488] De Padova, P., Ottaviani, C., Quaresima, C., Olivieri, B., Imperatori, P., Salomon, E., Angot, T., Quagliano, L., Romano, C., Vona, A., Muniz-Miranda, M., Generosi, A., Paci, B., Lay, G.L., 2D Materials, 1[2] 2014, A3

[489] Xu, X., Zhuang, J., Du, Y., Feng, H., Zhang, N., Liu, C., Lei, T., Wang, J., Spencer, M., Morishita, T., Wang, X., Dou, S.X., Scientific Reports, 4, 2014, 7543 *https://doi.org/10.1038/srep07543*

[490] Molle, A., Grazianetti, C., Chiappe, D., Cinquanta, E., Cianci, E., Tallarida, G., Fanciulli, M., Advanced Functional Materials, 23[35] 2013, 4340-4344 *https://doi.org/10.1002/adfm.201300354*

[491] Friedlein, R., Van Bui, H., Wiggers, F.B., Yamada-Takamura, Y., Kovalgin, A.Y., De Jong, M.P., Journal of Chemical Physics, 140[20] 2014, 204705 *https://doi.org/10.1063/1.4878375*

[492] Morishita, T., Spencer, M.J.S., Scientific Reports, 5, 2015, 17570 *https://doi.org/10.1038/srep17570*

[493] Zhong, S., Ning, F., Rao, F., Lei, X., Wu, M., Zhou, L., International Journal of Modern Physics B, 30[25], 2016, 1650176 *https://doi.org/10.1142/S0217979216501769*

[494] Qian, Y., Wu, H., Kan, E., Lu, R., Deng, K., Journal of Applied Physics, 120[23], 2016, 234303 *https://doi.org/10.1063/1.4972337*

[495] Lin, X., Ni, J., Physical Review B, 86[7] 2012, 075440 *https://doi.org/10.1103/PhysRevB.86.075440*

[496] Gürel, H.H., Özçelik, V.O., Ciraci, S., Journal of Physics - Condensed Matter, 25[30] 2013, 305007 *https://doi.org/10.1088/0953-8984/25/30/305007*

[497] Li, Z., Feng, H., Zhuang, J., Pu, N., Wang, L., Xu, X., Hao, W., Du, Y., Journal of Physics - Condensed Matter, 28[3] 2015, 034002 *https://doi.org/10.1088/0953-8984/28/3/034002*

[498] Johll, H., Lee, M.D.K., Ng, S.P.N., Kang, H.C., Tok, E.S., Scientific Reports, 4, 2014, 7594 *https://doi.org/10.1038/srep07594*

[499] Bui, V.Q., Pham, T.T., Nguyen, H.V.S., Le, H.M., Journal of Physical Chemistry C, 117[44] 2013, 23364-23371 *https://doi.org/10.1021/jp407601d*

[500] Zhang, J., Zhao, B., Yang, Z., Physical Review B, 88[16] 2013, 165422
https://doi.org/10.1103/PhysRevB.88.165422

[501] Sahin, H., Peeters, F.M., Physical Review B, 87[8] 2013, 085423
https://doi.org/10.1103/PhysRevB.87.085423

[502] Liu, S., Li, H., He, Y., Li, X., Li, Y., Wang, X., Materials and Design, 85, 15, 2015, 60-66

[503] Ince, A., Erkoc, S., Computational Materials Science, 50[3] 2011, 865-870
https://doi.org/10.1016/j.commatsci.2010.10.023

[504] Van Bui, H., Wiggers, F.B., Friedlein, R., Yamada-Takamura, Y., Kovalgin, A.Y., De Jong, M.P., Journal of Chemical Physics, 142[6] 2015, 064702
https://doi.org/10.1063/1.4907375

[505] Brennan, M.D., Morishita, T., Spencer, M.J.S., Journal of Chemical Physics, 144[11] 2016, 114704
https://doi.org/10.1063/1.4943880

[506] Stephan, R., Hanf, M.C., Sonnet, P., Journal of Chemical Physics, 143[15] 2015, 154706
https://doi.org/10.1063/1.4933369

[507] García, G., Atilhan, M., Aparicio, S., Physical Chemistry Chemical Physics, 17[25] 2015, 16315-16326
https://doi.org/10.1039/C5CP02432C

[508] Rubio-Pereda, P., Takeuchi, N., Journal of Chemical Physics, 138[19] 2013, 194702
https://doi.org/10.1063/1.4804545

[509] Kaloni, T.P., Schreckenbach, G., Freund, M.S., Journal of Physical Chemistry C, 118[40] 2014, 23361-23367
https://doi.org/10.1021/jp505814v

[510] Haldar, S., Amorim, R.G., Sanyal, B., Scheicher, R.H., Rocha, A.R., RSC Advances, 6[8] 2016, 6702-6708
https://doi.org/10.1039/C5RA23052G

[511] Lima, M.P., Fazzio, A., Da Silva, A.J.R., Physical Review B, 88[23] 2013, 235413
https://doi.org/10.1103/PhysRevB.88.235413

[512] Yang, M., Chen, D.H., Wang, R.Q., Bai, Y.K., Physics Letters A, 379[4] 2015, 396-400
https://doi.org/10.1016/j.physleta.2014.11.055

[513] Ghosh, D., Parida, P., Pati, S.K., Physical Review B, 92[19] 2015, 195136
https://doi.org/10.1103/PhysRevB.92.195136

[514] Li, S., Wu, Y., Tu, Y., Wang, Y., Jiang, T., Liu, W., Zhao, Y., Scientific Reports, 5, 2015, 7881
https://doi.org/10.1038/srep07881

[515] Li, R., Liu, Z., Ma, W., Tan, Y., AIP Advances, 6[5], 2016, 055204
https://doi.org/10.1063/1.4948980

[516] Liu, H., Feng, H., Du, Y., Chen, J., Wu, K., Zhao, J., 2D Materials, 3[2], 2016, 025034

[517] Galashev, A.E., Rakhmanova, O.R., Zaikov, Y.P., Physics of the Solid State, 58[9], 2016, 1850-1857
https://doi.org/10.1134/S1063783416090146

[518] Berdiyorov, G.R., Peeters, F.M., RSC Advances, 4[3] 2014, 1133-1137
https://doi.org/10.1039/C3RA43487G

[519] Song, Y.L., Zhang, Y., Zhang, J.M., Lu, D.B., Xu, K.W., Journal of Molecular Structure, 990[1-3] 2011, 75-78
https://doi.org/10.1016/j.molstruc.2011.01.020

[520] Saxena, N., Srivastava, A., Verma, S., Advanced Science Letters, 20[7-9] 2014, 1605-1607
https://doi.org/10.1166/asl.2014.5552

[521] Lan, M., Xiang, G., Zhang, C., Zhang, X., Journal of Applied Physics, 114[16] 2013, 163711
https://doi.org/10.1063/1.4828482

[522] Majumdar, A., Chowdhury, S., Nath, P., Jana, D., RSC Advances, 4[61] 2014, 32221-32227
https://doi.org/10.1039/C4RA04174G

[523] Li, R., Han, Y., Hu, T., Dong, J., Kawazoe, Y., Physical Review B, 90[4] 2014 045425
https://doi.org/10.1103/PhysRevB.90.045425

[524] Li, R., Han, Y., Dong, J., Physical Chemistry Chemical Physics, 17[35] 2015, 22969-22976
https://doi.org/10.1039/C5CP02538A

[525] Hakan Gürel, H., Özcelik, V.O., Ciraci, S., Journal of Physical Chemistry C, 118[47] 2014, 27574-27582
https://doi.org/10.1021/jp509260c

[526] Özçelik, V.O., Gurel, H.H., Ciraci, S., Physical Review B, 88[4] 2013, 045440
https://doi.org/10.1103/PhysRevB.88.045440

[527] Wang, X., Liu, H., Tu, S.T., RSC Advances, 5[80] 2015, 65255-65263
https://doi.org/10.1039/C5RA12096A

[528] Jamdagni, P., Kumar, A., Sharma, M., Thakur, A., Ahluwalia, P.K., Physica E, Nanostructures, 85, 2017, 65-73

[529] Chowdhury, S., Jana, D., Materials Chemistry and Physics, 183, 2016, 580-587
https://doi.org/10.1016/j.matchemphys.2016.09.018

[530] Sahin, H., Sivek, J., Li, S., Partoens, B., Peeters, F.M., Physical Review B, 88[4] 2013, 045434
https://doi.org/10.1103/PhysRevB.88.045434

[531] Ting, X., Rui, W., Shaofeng, W., Xiaozhi, W., Applied Surface Science, 383, 2016, 310-316
https://doi.org/10.1016/j.apsusc.2016.04.172

[532] Manjanath, A., Singh, A.K., Chemical Physics Letters, 592, 2014, 52-55
https://doi.org/10.1016/j.cplett.2013.12.010

[533] Zha, D., Chen, C., Wu, J., Wang, M., International Journal of Modern Physics B, 29[9] 2015, 1550061
https://doi.org/10.1142/S0217979215500617

[534] Iordanidou, K., Houssa, M., Van Den Broek, B., Pourtois, G., Afanasev, V.V., Stesmans, A., Journal of Physics - Condensed Matter, 28[3] 2016, 035302
https://doi.org/10.1088/0953-8984/28/3/035302

[535] Gao, J., Zhang, J., Liu, H., Zhang, Q., Zhao, J., Nanoscale, 5[20] 2013, 9785-9792
https://doi.org/10.1039/c3nr02826g

[536] Le, M.Q., Nguyen, D.T., Applied Physics A, 118[4] 2014, 1437-1445
https://doi.org/10.1007/s00339-014-8904-3

[537] Dong, H., Fang, D., Gong, B., Zhang, Y., Zhang, E., Zhang, S., Journal of Applied Physics, 117[6] 2015, 064307
https://doi.org/10.1063/1.4907582

[538] Le, N.B., Huan, T.D., Woods, L.M., Physical Review Applied, 1[5] 2014, 054002
https://doi.org/10.1103/PhysRevApplied.1.054002

[539] Scalise, E., Houssa, M., Pourtois, G., van den Broek, B., Afanas'ev, V., Stesmans, A., Nano Research, 6[1] 2013, 19-28
https://doi.org/10.1007/s12274-012-0277-3

[540] Guo, X.X., Guo, P., Zheng, J.M., Cao, L.K., Zhao, P.J., Applied Surface Science, 341, 2015, 69-74
https://doi.org/10.1016/j.apsusc.2015.03.002

[541] Jiang, Q.G., Zhang, J.F., Ao, Z.M., Wu, Y.P., Scientific Reports, 5, 2015, 15734
https://doi.org/10.1038/srep15734

[542] Huang, J., Chen, H.J., Wu, M.S., Liu, G., Ouyang, C.Y., Xu, B., Chinese Physics Letters, 30[1] 2013, 017103
https://doi.org/10.1088/0256-307X/30/1/017103

[543] Deng, J., Liu, J.Z., Medhekar, N.V., RSC Advances, 3[43] 2013, 20338-20344
https://doi.org/10.1039/c3ra43326a

[544] Guo, Y.H., Cao, J.X., Xu, B., Chinese Physics B, 25[1] 2015, 017101
https://doi.org/10.1088/1674-1056/25/1/017101

[545] Xiao, H.J., Zhang, L.Z., Du, S.X., Gao, H.J., Chinese Physics B, 24[8] 2015, 086806
https://doi.org/10.1088/1674-1056/24/8/086806

[546] Scalise, E., Houssa, M., Cinquanta, E., Grazianetti, C., Van Den Broek, B., Pourtois, G., Stesmans, A., Fanciulli, M., Molle, A., 2D Materials, 1[1] 2014, 011010

[547] Nguyen, N.T., Micro and Nanosystems, 6[4] 2014, 205-206

[548] Zandvliet, H.J.W., Nano Today, 9[6] 2014, 691-694
https://doi.org/10.1016/j.nantod.2014.09.007

[549] Kaneko, S., Tsuchiya, H., Kamakura, Y., Mori, N., Ogawa, M., Applied Physics Express, 7[3] 2014, 035102
https://doi.org/10.7567/APEX.7.035102

[550] Quhe, R.G., Wang, Y.Y., Lü, J., Chinese Physics B, 24[8] 2015, 088105
https://doi.org/10.1088/1674-1056/24/8/088105

[551] Li, Y., Chen, Z., Journal of Physical Chemistry Letters, 4[2] 2013, 269-275
https://doi.org/10.1021/jz301821n

[552] Chen, S.H., Journal of Magnetism and Magnetic Materials, 405, 2016, 317
https://doi.org/10.1016/j.jmmm.2015.12.088

[553] Mahmoudi, M., Ahangari, Z., Fathipour, M., Chinese Physics B, 25[1] 2015, 018501
https://doi.org/10.1088/1674-1056/25/1/018501

[554] Sadeghi, H., Journal of Nanoscience and Nanotechnology, 14[6] 2014, 4178-4184
https://doi.org/10.1166/jnn.2014.8914

[555] Ezawa, M., Applied Physics Letters, 102[17] 2013, 172103
https://doi.org/10.1063/1.4803010

[556] Ezawa, M., Nagaosa, N., Physical Review B, 88[12] 2013, 121401
https://doi.org/10.1103/PhysRevB.88.121401

[557] Zhao, H., Zhang, C., Li, S., Ji, W., Wang, P., Journal of Applied Physics, 117[8] 2015, 085306
https://doi.org/10.1063/1.4913480

[558] Jin, C., Dai, Y., Wei, W., Sun, Q., Li, X., Huang, B., RSC Advances, 7[10], 2017, 5827-5835
https://doi.org/10.1039/C6RA25614G

[559] Ren, M., Li, M., Zhang, C., Yuan, M., Li, P., Li, F., Ji, W., Chen, X., Solid State Communications, 239, 2016, 32-36
https://doi.org/10.1016/j.ssc.2016.04.013

[560] Vali, M., Dideban, D., Moezi, N., Journal of Computational Electronics, 15[1] 2016, 138
https://doi.org/10.1007/s10825-015-0758-1

[561] Ni, Z., Liu, Q., Tang, K., Zheng, J., Zhou, J., Qin, R., Gao, Z., Yu, D., Lu, J., Nano Letters, 12[1] 2012, 113-118
https://doi.org/10.1021/nl203065e

[562] Li, H., Wang, L., Liu, Q., Zheng, J., Mei, W.N., Gao, Z., Shi, J., Lu, J., European Physical Journal B, 85[8] 2012, 274
https://doi.org/10.1140/epjb/e2012-30220-2

[563] Ni, Z., Zhong, H., Jiang, X., Quhe, R., Luo, G., Wang, Y., Ye, M., Yang, J., Shi, J., Lu, J., Nanoscale, 6[13] 2014, 7609-7618
https://doi.org/10.1039/C4NR00028E

[564] Quhe, R., Fei, R., Liu, Q., Zheng, J., Li, H., Xu, C., Ni, Z., Wang, Y., Yu, D., Gao, Z., Lu, J., Scientific Reports, 2, 2012, 853
https://doi.org/10.1038/srep00853

[565] Li, G., Zhao, Y.C., Zheng, R., Ni, J., Wu, Y.N., Chinese Physics B, 24[8] 2015, 087302
https://doi.org/10.1088/1674-1056/24/8/087302

[566] Al-Dirini, F., Hossain, F.M., Mohammed, M.A., Nirmalathas, A., Skafidas, E., Scientific Reports, 5, 2015, 14815
https://doi.org/10.1038/srep14815

[567] Tsetseris, L., Kaltsas, D., Physical Chemistry Chemical Physics, 16[11] 2014, 5183-5187
https://doi.org/10.1039/c3cp55529a

[568] Cheng, Y.C., Zhu, Z.Y., Schwingenschlögl, U., EPL, 95[1] 2011, 17005
https://doi.org/10.1209/0295-5075/95/17005

[569] Wang, Y., Zheng, J., Ni, Z., Fei, R., Liu, Q., Quhe, R., Xu, C., Zhou, J., Gao, Z., Lu, J., Nano, 7[5] 2012, 1250037
https://doi.org/10.1142/S1793292012500373

[570] An, X.T., Zhang, Y.Y., Liu, J.J., Li, S.S., New Journal of Physics, 14, 2012, 083039
https://doi.org/10.1088/1367-2630/14/8/083039

[571] Farokhnezhad, M., Esmaeilzadeh, M., Ahmadi, S., Pournaghavi, N., Journal of Applied Physics, 117[17] 2015, 173913
https://doi.org/10.1063/1.4919659

[572] Bishnoi, B., Ghosh, B., Computational Materials Science, 85, 2014, 16-19
https://doi.org/10.1016/j.commatsci.2013.12.041

[573] Bishnoi, B., Ghosh, B., Journal of Computational Electronics, 13[1] 2014, 186-191
https://doi.org/10.1007/s10825-013-0498-z

[574] Bishnoi, B., Ghosh, B., Journal of Computational and Theoretical Nanoscience, 11[5] 2014, 1271-1274
https://doi.org/10.1166/jctn.2014.3493

[575] Bishnoi, B., Ghosh, B., RSC Advances, 3[48] 2013, 26153-26159
https://doi.org/10.1039/c3ra43491e

[576] Saari, T., Huang, C.Y., Nieminen, J., Tsai, W.F., Lin, H., Bansil, A., Applied Physics Letters, 104[17] 2014, 173104
https://doi.org/10.1063/1.4873716

[577] Wang, Y., Lou, Y., Physics Letters A, 378[35] 2014, 2627-2631
https://doi.org/10.1016/j.physleta.2014.05.056

[578] Dong, Y.J., Wang, X.F., Vasilopoulos, P., Zhai, M.X., Wu, X.M., Journal of Physics D, 47[10] 2014, 105304
https://doi.org/10.1088/0022-3727/47/10/105304

[579] Tsai, W.F., Huang, C.Y., Chang, T.R., Lin, H., Jeng, H.T., Bansil, A., Nature Communications, 4, 2013, 1500
https://doi.org/10.1038/ncomms2525

[580] Nú-ez, C., Domínguez-Adame, F., Orellana, P.A., Rosales, L., Römer, R.A., 2D Materials, 3[2], 2016, 025006

[581] Li, W., Lu, W.T., Li, Y.F., Han, H.H., Physica E, 88, 2017, 284-288
https://doi.org/10.1016/j.physe.2017.01.016

[582] Xu, X., Li, J., Zhang, X., Xu, H., Ke, Z.F., Zhao, C., RSC Advances, 5[28] 2015, 22135-22147
https://doi.org/10.1039/C4RA13754J

[583] Zhu, J., Chroneos, A., Schwingenschlögl, U., Physica Status Solidi R, 10[6], 458-461
https://doi.org/10.1002/pssr.201600102

[584] Osborn, T.H., Farajian, A.A., Nano Research, 7[7] 2014, 945-952
https://doi.org/10.1007/s12274-014-0454-7

[585] Ersan, F., Arslanalp, Ö., Gökoılu, G., Aktürk, E., Applied Surface Science, 371, 2016, 314-321
https://doi.org/10.1016/j.apsusc.2016.02.225

[586] Feng, J.W., Liu, Y.J., Wang, H.X., Zhao, J.X., Cai, Q.H., Wang, X.Z., Computational Materials Science, 87, 2014, 218-226
https://doi.org/10.1016/j.commatsci.2014.02.025

[587] Hu, W., Xia, N., Wu, X., Li, Z., Yang, J., Physical Chemistry Chemical Physics, 16[15] 2014, 6957-6962
https://doi.org/10.1039/c3cp55250k

[588] Prasongkit, J., Amorim, R.G., Chakraborty, S., Ahuja, R., Scheicher, R.H., Amornkitbamrung, V., Journal of Physical Chemistry C, 119[29] 2015, 16934-16940
https://doi.org/10.1021/acs.jpcc.5b03635

[589] Amorim, R.G., Scheicher, R.H., Nanotechnology, 26[15] 2015, 154002
https://doi.org/10.1088/0957-4484/26/15/154002

[590] Sadeghi, H., Bailey, S., Lambert, C.J., Applied Physics Letters, 104[10] 2014, 103104
https://doi.org/10.1063/1.4868123

[591] Hu, W., Wu, X., Li, Z., Yang, J., Physical Chemistry Chemical Physics, 15[16] 2013, 5753-5757
https://doi.org/10.1039/c3cp00066d

[592] Hu, W., Wu, X., Li, Z., Yang, J., Nanoscale, 5[19] 2013, 9062-9066
https://doi.org/10.1039/c3nr02326e

[593] Jose, D., Datta, A., Physical Chemistry Chemical Physics, 13[16] 2011, 7304-7311
https://doi.org/10.1039/c0cp02580a

[594] Li, F., Zhang, C.W., Luan, H.X., Wang, P.J., Journal of Nanoparticle Research, 15[10] 2013, 1972

[595] Hussain, T., Kaewmaraya, T., Chakraborty, S., Ahuja, R., Physical Chemistry Chemical Physics, 15[43] 2013, 18900-18905
https://doi.org/10.1039/c3cp52830h

[596] Hussain, T., Chakraborty, S., Ahuja, R., ChemPhysChem, 14[15] 2013, 3463-3466
https://doi.org/10.1002/cphc.201300548

[597] Hussain, T., Chakraborty, S., De Sarkar, A., Johansson, B., Ahuja, R., Applied Physics Letters, 105[12] 2014, 123903
https://doi.org/10.1063/1.4896503

[598] Song, E.H., Yoo, S.H., Kim, J.J., Lai, S.W., Jiang, Q., Cho, S.O., Physical Chemistry Chemical Physics, 16[43] 2014, 23985-23992
https://doi.org/10.1039/C4CP02638A

[599] Wang, J., Li, J., Li, S.S., Liu, Y., Journal of Applied Physics, 114[12] 2013, 124309
https://doi.org/10.1063/1.4823738

[600] Li, F., Zhang, C., Ji, W.X., Zhao, M., Physica Status Solidi B, 252[9] 2015, 2072-2078
https://doi.org/10.1002/pssb.201552151

[601] Hwang, Y., Yun, K.H., Chung, Y.C., Journal of Power Sources, 275, 2015, 32-37
https://doi.org/10.1016/j.jpowsour.2014.11.016

[602] Seyed-Talebi, S.M., Kazeminezhad, I., Beheshtian, J., Physical Chemistry Chemical Physics, 17[44] 2015, 29689-29696
https://doi.org/10.1039/C5CP04666A

[603] Zhu, J., Chroneos, A., Schwingenschlögl, U., Nanoscale, 8[13], 2016, 7272-7277
https://doi.org/10.1039/C6NR00913A

[604] Tan, X., Cabrera, C.R., Chen, Z., Journal of Physical Chemistry C, 118[45] 2014, 25836-25843
https://doi.org/10.1021/jp503597n

[605] Tan, X., Li, F., Chen, Z., Journal of Physical Chemistry C, 118[45] 2014, 25825-25835
https://doi.org/10.1021/jp507011p

[606] Zhu, J., Schwingenschlögl, U., 2D Materials, 3[3], 2016, 035012

[607] Li, C., Yang, S., Li, S.S., Xia, J.B., Li, J., Journal of Physical Chemistry C, 117[1] 2013, 483-488
https://doi.org/10.1021/jp310746m

[608] Mondal, K., Kamal, C., Banerjee, A., Chakrabarti, A., Ghanty, T.K., Journal of Physical Chemistry C, 119[6] 2015, 3192-3198
https://doi.org/10.1021/jp5119579

[609] Ding, Y., Wang, Y., Journal of Physical Chemistry C, 117[35] 2013, 18266-18278
https://doi.org/10.1021/jp407666m

[610] Balendhran, S., Walia, S., Nili, H., Sriram, S., Bhaskaran, M., Small, 11[6] 2015, 640-652
https://doi.org/10.1002/smll.201402041

[611] Dimoulas, A., Microelectronic Engineering, 131, 2015, 68-78
https://doi.org/10.1016/j.mee.2014.08.013

[612] Persichetti, L., Jardali, F., Vach, H., Sgarlata, A., Berbezier, I., Crescenzi, M.D., Balzarotti, A., Journal of Physical Chemistry Letters, 7[16], 2016, 3246-3251
https://doi.org/10.1021/acs.jpclett.6b01284

[613] Zhang, L., Bampoulis, P., Rudenko, A.N., Yao, Q., Van Houselt, A., Poelsema, B., Katsnelson, M.I., Zandvliet, H.J.W., Physical Review Letters, 116[25], 2016, 256804
https://doi.org/10.1103/PhysRevLett.116.256804

[614] Gou, J., Zhong, Q., Sheng, S., Li, W., Cheng, P., Li, H., Chen, L., Wu, K., 2D Materials, 3[4], 2016, 045005

[615] Dávila, M.E., Le Lay, G., Scientific Reports, 6, 2016, 20714
https://doi.org/10.1038/srep20714

[616] Dávila, M.E., Xian, L., Cahangirov, S., Rubio, A., Le Lay, G., New Journal of Physics, 16, 2014, 095002
https://doi.org/10.1088/1367-2630/16/9/095002

[617] Naji, S., Belhaj, A., Labrim, H., Benyoussef, A., El Kenz, A., Modern Physics Letters B, 27[29] 2013, 1350212
https://doi.org/10.1142/S0217984913502126

[618] Li, L., Lu, S.Z., Pan, J., Qin, Z., Wang, Y.Q., Wang, Y., Cao, G.Y., Du, S., Gao, H.J., Advanced Materials, 26[28] 2014, 4820-4824
https://doi.org/10.1002/adma.201400909

[619] Özçelik, V.O., Durgun, E., Ciraci, S., Journal of Physical Chemistry Letters, 5[15] 2014, 2694-2699
https://doi.org/10.1021/jz500977v

[620] Roome, N.J., Carey, J.D., ACS Applied Materials and Interfaces, 6[10] 2014, 7743-7750
https://doi.org/10.1021/am501022x

[621] Cai, Y., Chuu, C.P., Wei, C.M., Chou, M.Y., Physical Review B, 88[24] 2013, 245408
https://doi.org/10.1103/PhysRevB.88.245408

[622] Tsai, H.S., Chen, Y.Z., Medina, H., Su, T.Y., Chou, T.S., Chen, Y.H., Chueh, Y.L., Liang, J.H., Physical Chemistry Chemical Physics, 17[33] 2015, 21389-21393
https://doi.org/10.1039/C5CP02469B

[623] Li, P., Cao, J., Guo, Z.X., Journal of Materials Chemistry C, 4[8] 2016, 1736-1740
https://doi.org/10.1039/C5TC03442F

[624] Tomberli, B., Rahemtulla, A., Kim, E., Roorda, S., Kycia, S., Physical Review B, 92[6] 2015, 064204
https://doi.org/10.1103/PhysRevB.92.064204

[625] Nijamudheen, A., Bhattacharjee, R., Choudhury, S., Datta, A., Journal of Physical Chemistry C, 119[7] 2015, 3802-3809
https://doi.org/10.1021/jp511488m

[626] Trivedi, S., Srivastava, A., Kurchania, R., Journal of Computational and Theoretical Nanoscience, 11[3] 2014, 781-788
https://doi.org/10.1166/jctn.2014.3428

[627] Acun, A., Zhang, L., Bampoulis, P., Farmanbar, M., Van Houselt, A., Rudenko, A.N., Lingenfelder, M., Brocks, G., Poelsema, B., Katsnelson, M.I., Zandvliet, H.J.W., Journal of Physics - Condensed Matter, 27[44] 2015, 443002
https://doi.org/10.1088/0953-8984/27/44/443002

[628] Stephan, R., Hanf, M.C., Derivaz, M., Dentel, D., Asensio, M.C., Avila, J., Mehdaoui, A., Sonnet, P., Pirri, C., Journal of Physical Chemistry C, 120[3] 2016, 1580-1585
https://doi.org/10.1021/acs.jpcc.5b10307

[629] D'Acapito, F., Torrengo, S., Xenogiannopoulou, E., Tsipas, P., Marquez Velasco, J., Tsoutsou, D., Dimoulas, A., Journal of Physics - Condensed Matter, 28[4] 2016, 045002
https://doi.org/10.1088/0953-8984/28/4/045002

[630] Liang, P., Liu, Y., Xing, S., Shu, H., Tai, B., Solid State Communications, 226, 2016, 19-24
https://doi.org/10.1016/j.ssc.2015.10.021

[631] Yuan, J., Tang, C., Zhong, J., Mao, Y., Applied Surface Science, 360, 2016, 707-714
https://doi.org/10.1016/j.apsusc.2015.11.050

[632] Kaloni, T.P., Schwingenschlögl, U., Chemical Physics Letters, 583, 2013, 137-140
https://doi.org/10.1016/j.cplett.2013.08.001

[633] Huang, L.F., Gong, P.L., Zeng, Z., Physical Review B, 91[20] 2015, 205433
https://doi.org/10.1103/PhysRevB.91.205433

[634] Mortazavi, B., Rahaman, O., Makaremi, M., Dianat, A., Cuniberti, G., Rabczuk, T., Physica E, 87, 2017, 228-232
https://doi.org/10.1016/j.physe.2016.10.047

[635] John, R., Merlin, B., Crystal Structure Theory and Applications, 5 (2016) 43–55
https://doi.org/10.4236/csta.2016.5304

[636] Wang, Y., Ding, Y., Solid State Communications, 155, 2013, 6-11
https://doi.org/10.1016/j.ssc.2012.10.044

[637] Shi, W., Gu, W., Ma, X., Emerging Materials Research, 5[1], 2016, 114-118

[638] Tahir, M., Schwingenschlögl, U., European Physical Journal B, 88[11] 2015, 285
https://doi.org/10.1140/epjb/e2015-60719-7

[639] Li, S.J., Su, Y., Chen, G., Chemical Physics Letters, 638, 2015, 187-190
https://doi.org/10.1016/j.cplett.2015.08.044

[640] Liu, C.C., Feng, W., Yao, Y., Physical Review Letters, 107[7] 2011, 076802
https://doi.org/10.1103/PhysRevLett.107.076802

[641] Jomehpour Zaveh, S., Roknabadi, M.R., Morshedloo, T., Modarresi, M., Superlattices and Microstructures, 91, 2016, 383-390
https://doi.org/10.1016/j.spmi.2016.01.033

[642] Yan, J.A., Gao, S.P., Stein, R., Coard, G., Physical Review B, 91[24] 2015, 245403
https://doi.org/10.1103/PhysRevB.91.245403

[643] Ding, Y., Wang, Y., Nanoscale Research Letters, 10[1] 2015, 1-11
https://doi.org/10.1186/1556-276X-10-1

[644] Si, C., Liu, J., Xu, Y., Wu, J., Gu, B.L., Duan, W., Physical Review B, 89[11] 2014, 115429
https://doi.org/10.1103/PhysRevB.89.115429

[645] Bayani, A.H., Dideban, D., Moezi, N., Superlattices and Microstructures, 100, 2016, 198-208
https://doi.org/10.1016/j.spmi.2016.09.021

[646] Chen, X., Sun, X., Jiang, J., Liang, Q., Yang, Q., Meng, R., Journal of Physical Chemistry C, 120[36], 2016, 20350-20356
https://doi.org/10.1021/acs.jpcc.6b06161

[647] Li, L., Zhao, M., Physical Chemistry Chemical Physics, 15[39] 2013, 16853-16863
https://doi.org/10.1039/c3cp52619d

[648] Liu, G., Liu, S.B., Xu, B., Ouyang, C.Y., Song, H.Y., Guan, S., Yang, S.A., Journal of Physical Chemistry Letters, 6[24] 2015, 4936-4942
https://doi.org/10.1021/acs.jpclett.5b02413

[649] Fukaya, Y., Matsuda, I., Feng, B., Mochizuki, I., Hyodo, T., Shamoto, S.I., 2D Materials, 3[3], 2016, 035019

[650] Zhang, L., Bampoulis, P., Van Houselt, A., Zandvliet, H.J.W., Applied Physics Letters, 107[11] 2015, 111605
https://doi.org/10.1063/1.4931102

[651] Bampoulis, P., Zhang, L., Safaei, A., Van Gastel, R., Poelsema, B., Zandvliet, H.J.W., Journal of Physics - Condensed Matter, 26[44] 2014, 442001
https://doi.org/10.1088/0953-8984/26/44/442001

[652] Ni, Z., Minamitani, E., Ando, Y., Watanabe, S., Physical Chemistry Chemical Physics, 17[29] 2015, 19039-19044
https://doi.org/10.1039/C5CP02428E

[653] Zhang, X., Guo, Z.X., Cao, J.X., Xiao, S.G., Ding, J.W., Acta Physica Sinica, 64[18] 2015, 186101

[654] Kaloni, T.P., Schwingenschlögl, U., Journal of Applied Physics, 114[18] 2013, 184307
https://doi.org/10.1063/1.4830016

[655] Chen, M.X., Zhong, Z., Weinert, M., Physical Review B, 94[7], 2016, 075409
https://doi.org/10.1103/PhysRevB.94.075409

[656] Kaloni, T.P., Journal of Physical Chemistry C, 118[43] 2014, 25200-25208
https://doi.org/10.1021/jp5058644

[657] Pang, Q., Li, L., Zhang, C.L., Wei, X.M., Song, Y.L., Materials Chemistry and Physics, 160, 2015, 96-104
https://doi.org/10.1016/j.matchemphys.2015.04.011

[658] Pang, Q., Zhang, C.L., Li, L., Fu, Z.Q., Wei, X.M., Song, Y.L., Applied Surface Science, 314, 2014, 15-20
https://doi.org/10.1016/j.apsusc.2014.06.138

[659] Pang, Q., Li, L., Zhang, L.L., Zhang, C.L., Song, Y.L., Canadian Journal of Physics, 93[11] 2015, 1310-1318
https://doi.org/10.1139/cjp-2015-0206

[660] Li, S.S., Zhang, C.W., Ji, W.X., Li, F., Wang, P.J., Hu, S.J., Yan, S.S., Liu, Y.S., Physical Chemistry Chemical Physics, 16[30] 2014, 15968-15978
https://doi.org/10.1039/C4CP01211A

[661] Van Den Broek, B., Houssa, M., Scalise, E., Pourtois, G., Afanas'Ev, V.V., Stesmans, A., Applied Surface Science, 291, 2014, 104-108
https://doi.org/10.1016/j.apsusc.2013.09.032

[662] Gürel, H.H., Özçelik, V.O., Ciraci, S., Journal of Physics - Condensed Matter, 25[30] 2013, 305007
https://doi.org/10.1088/0953-8984/25/30/305007

[663] Ongun Özçelik, V., Kecik, D., Durgun, E., Ciraci, S., Journal of Physical Chemistry C, 119[1] 2015, 845-853
https://doi.org/10.1021/jp5106554

[664] Rubio-Pereda, P., Takeuchi, N., Journal of Physical Chemistry C, 119[50] 2015, 27995-28004
https://doi.org/10.1021/acs.jpcc.5b08370

[665] Houssa, M., Scalise, E., Sankaran, K., Pourtois, G., Afanas'Ev, V.V., Stesmans, A., Applied Physics Letters, 98[22] 2011, 223107
https://doi.org/10.1063/1.3595682

[666] Wella, S.A., Suprijadi, ARPN Journal of Engineering and Applied Sciences, 10[18] 2015, 8250-8253

[667] Jamdagni, P., Kumar, A., Sharma, M., Thakur, A., Ahluwalia, P.K., Physica E, 85, 2017, 65-73
https://doi.org/10.1016/j.physe.2016.08.015

[668] Xiao, P., Fan, X.L., Liu, L.M., Computational Materials Science, 92, 2014, 244-252
https://doi.org/10.1016/j.commatsci.2014.05.041

[669] Wang, X.Q., Li, H.D., Wang, J.T., Physical Chemistry Chemical Physics, 14[9] 2012, 3031-3036
https://doi.org/10.1039/c2cp23385a

[670] Rupp, C.J., Chakraborty, S., Ahuja, R., Baierle, R.J., Physical Chemistry Chemical Physics, 17[34] 2015, 22210-22216
https://doi.org/10.1039/C5CP03489B

[671] Liu, G., Liu, S.B., Xu, B., Ouyang, C.Y., Song, H.Y., Journal of Applied Physics, 118[12] 2015, 124303
https://doi.org/10.1063/1.4931057

[672] Xia, W., Hu, W., Li, Z., Yang, J., Physical Chemistry Chemical Physics, 16[41] 2014, 22495-22498
https://doi.org/10.1039/C4CP03292F

[673] Pang, Q., Li, L., Gao, D.L., Chai, R.P., Zhang, C.L., Song, Y.L., Physica E, 88, 2017, 237-242
https://doi.org/10.1016/j.physe.2017.01.018

[674] García, G., Atilhan, M., Aparicio, S., Physical Chemistry Chemical Physics, 17[25] 2015, 16315-16326
https://doi.org/10.1039/C5CP02432C

[675] Padilha, J.E., Pontes, R.B., Solid State Communications, 225, 2016, 38-43
https://doi.org/10.1016/j.ssc.2015.10.019

[676] Scalise, E., Houssa, M., Pourtois, G., van den Broek, B., Afanasev, V., Stesmans, A., Nano Research, 6[1] 2013, 19-28
https://doi.org/10.1007/s12274-012-0277-3

[677] Kuang, Y.D., Lindsay, L., Shi, S.Q., Zheng, G.P., Nanoscale, 8[6] 2016, 3760-3767
https://doi.org/10.1039/C5NR08231E

[678] Zheng, J., Chi, F., Guo, Y., Journal of Applied Physics, 116[24] 2014, 243907
https://doi.org/10.1063/1.4905114

[679] Zheng, J., Chi, F., Guo, Y., Journal of Physics - Condensed Matter, 27[29] 2015, 295302
https://doi.org/10.1088/0953-8984/27/29/295302

[680] Yang, K., Cahangirov, S., Cantarero, A., Rubio, A., D'Agosta, R., Physical Review B, 89[12] 2014, 125403
https://doi.org/10.1103/PhysRevB.89.125403

[681] Matthes, L., Pulci, O., Bechstedt, F., Journal of Physics - Condensed Matter, 25[39] 2013, 395305
https://doi.org/10.1088/0953-8984/25/39/395305

[682] Wei, W., Dai, Y., Huang, B., Jacob, T., Physical Chemistry Chemical Physics, 15[22] 2013, 8789-8794
https://doi.org/10.1039/c3cp51078f

[683] Bechstedt, F., Matthes, L., Gori, P., Pulci, O., Applied Physics Letters, 100[26] 2012, 261906
https://doi.org/10.1063/1.4731626

[684] Matthes, L., Pulci, O., Bechstedt, F., New Journal of Physics, 16, 2014, 105007
https://doi.org/10.1088/1367-2630/16/10/105007

[685] Dhar, N., Bandyopadhyay, A., Jana, D., Current Applied Physics, 17[4], 573-583
https://doi.org/10.1016/j.cap.2017.02.003

[686] Zhang, R.W., Zhang, C.W., Ji, W.X., Ren, M.J., Li, F., Yuan, M., Materials Chemistry and Physics, 156, 2015, 89-94
https://doi.org/10.1016/j.matchemphys.2015.02.029

[687] Yan, J.A., Stein, R., Schaefer, D.M., Wang, X.Q., Chou, M.Y., Physical Review B, 88[12] 2013, 121403
https://doi.org/10.1103/PhysRevB.88.121403

[688] Zhang, R.W., Zhang, C.W., Li, S.S., Ji, W.X., Wang, P.J., Li, F., Li, P., Ren, M.J., Yuan, M., Solid State Communications, 191, 2014, 49-53
https://doi.org/10.1016/j.ssc.2014.04.015

[689] Goswami, P., Indian Journal of Physics. 89[7] 2015, 677-686
https://doi.org/10.1007/s12648-014-0637-9

[690] Li, X., Wu, S., Zhou, S., Zhu, Z., Nanoscale Research Letters, 9[1] 2014, 1-9
https://doi.org/10.1186/1556-276X-9-1

[691] Seixas, L., Padilha, J.E., Fazzio, A., Physical Review B, 89[19] 2014, 195403
https://doi.org/10.1103/PhysRevB.89.195403

[692] Jing, Y., Zhang, X., Wu, D., Zhao, X., Zhou, Z., Journal of Physical Chemistry Letters, 6[21] 2015, 4252-4258
https://doi.org/10.1021/acs.jpclett.5b01848

[693] Zhu, J., Chroneos, A., Schwingenschlögl, U., Nanoscale, 8[13], 2016, 7272-7277
https://doi.org/10.1039/C6NR00913A

[694] Rupp, C.J., Chakraborty, S., Anversa, J., Baierle, R.J., Ahuja, R., ACS Applied Materials and Interfaces, 8[2] 2016, 1536-1544
https://doi.org/10.1021/acsami.5b11513

[695] Ye, M., Quhe, R., Zheng, J., Ni, Z., Wang, Y., Yuan, Y., Tse, G., Shi, J., Gao, Z., Lu, J., Physica E, 59, 2014, 60-65
https://doi.org/10.1016/j.physe.2013.12.016

[696] Ye, X.S., Shao, Z.G., Zhao, H., Yang, L., Wang, C.L., RSC Advances, 4[41] 2014, 21216-21220
https://doi.org/10.1039/C4RA01802H

[697] Zhang, R.W., Zhang, C.W., Ji, W.X., Li, F., Ren, M.J., Li, P., Yuan, M., Wang, P.J., Physical Chemistry Chemical Physics, 17[18] 2015, 12194-12198
https://doi.org/10.1039/C5CP00875A

[698] Ni, Z., Liu, Q., Tang, K., Zheng, J., Zhou, J., Qin, R., Gao, Z., Yu, D., Lu, J., Nano Letters, 12[1] 2012, 113-118
https://doi.org/10.1021/nl203065e

[699] Kaneko, S., Tsuchiya, H., Kamakura, Y., Mori, N., Ogawa, M., Applied Physics Express, 7[3] 2014, 035102
https://doi.org/10.7567/APEX.7.035102

[700] Pang, Q., Zhang, Y., Zhang, J.M., Ji, V., Xu, K.W., Nanoscale, 3[10] 2011, 4330-4338
https://doi.org/10.1039/c1nr10594a

[701] Zhang, R.W., Ji, W.X., Zhang, C.W., Li, S.S., Li, P., Wang, P.J., Journal of Materials Chemistry C, 4[10] 2016, 2088-2094
https://doi.org/10.1039/C6TC00160B

[702] Wang, Y.P., Ji, W.X., Zhang, C.W., Li, S.S., Li, F., Li, P., Ren, M.J., Chen, X.L., Yuan, M., Wang, P.J., Materials Chemistry and Physics, 173, 2016, 379-384 *https://doi.org/10.1016/j.matchemphys.2016.02.026*

[703] Rachel, S., Ezawa, M., Physical Review B, 89[19] 2014, 195303 *https://doi.org/10.1103/PhysRevB.89.195303*

[704] Cao, C., Long, M., Mao, X., Journal of Nanomaterials, 2015, 2015, 810659

[705] Matthes, L., Bechstedt, F., Physical Review B, 90[16] 2014, 165431 *https://doi.org/10.1103/PhysRevB.90.165431*

[706] Chang, H.R., Zhou, J., Zhang, H., Yao, Y., Physical Review B, 89[20] 2014, 201411 *https://doi.org/10.1103/PhysRevB.89.201411*

[707] Chaudhary, P., Salimath, A., Ghosh, B., Journal of Computational and Theoretical Nanoscience, 11[12] 2014, 2437-2439 *https://doi.org/10.1166/jctn.2014.3659*

[708] Bishnoi, B., Ghosh, B., RSC Advances, 3[48] 2013, 26153-26159 *https://doi.org/10.1039/c3ra43491e*

[709] Saxena, S., Chaudhary, R.P., Shukla, S., Scientific Reports, 6, 2016, 31073 *https://doi.org/10.1038/srep31073*

[710] Zhu, F.F., Chen, W.J., Xu, Y., Gao, C.L., Guan, D.D., Liu, C.H., Qian, D., Zhang, S.C., Jia, J.F., Nature Materials, 14[10] 2015, 1020-1025 *https://doi.org/10.1038/nmat4384*

[711] Van den Broek, B., Houssa, M., Pourtois, G., Afanasev, V.V., Stesmans, A., Physica Status Solidi R, 8[11] 2014, 931-934 *https://doi.org/10.1002/pssr.201400073*

[712] Matusalem, F., Marques, M., Teles, L.K., Bechstedt, F., Physical Review B, 92[4] 2015, 045436 *https://doi.org/10.1103/PhysRevB.92.045436*

[713] Mogulkoc, A., Modarresi, M., Kandemir, B.S., Roknabadi, M.R., Physica Status Solidi B, 253[2] 2016, 300-307 *https://doi.org/10.1002/pssb.201552341*

[714] Van Den Broek, B., Houssa, M., Scalise, E., Pourtois, G., Afanasev, V.V., Stesmans, A., 2D Materials, 1[2] 2014, 021004

[715] Cai, B., Zhang, S., Hu, Z., Hu, Y., Zou, Y., Zeng, H., Physical Chemistry Chemical Physics, 17[19] 2015, 12634-12638
https://doi.org/10.1039/C5CP00563A

[716] Gao, J., Zhang, G., Zhang, Y.W., Scientific Reports, 6, 2016, 29107
https://doi.org/10.1038/srep29107

[717] Mojumder, S., Amin, A.A., Islam, M.M., Journal of Applied Physics, 118[12] 2015, 124305
https://doi.org/10.1063/1.4931572

[718] Barraza-Lopez, S., Synthetic Metals, 210, 2015, 32-41
https://doi.org/10.1016/j.synthmet.2015.06.025

[719] Modarresi, M., Kakoee, A., Mogulkoc, Y., Roknabadi, M.R., Computational Materials Science, 101, 2015, 164-167
https://doi.org/10.1016/j.commatsci.2015.01.039

[720] Ribeiro-Soares, J., Almeida, R.M., Cançado, L.G., Dresselhaus, M.S., Jorio, A., Physical Review B, 91[20] 2015, 205421
https://doi.org/10.1103/PhysRevB.91.205421

[721] Mortazavi, B., Rahaman, O., Makaremi, M., Dianat, A., Cuniberti, G., Rabczuk, T., Physica E, 87, 2017, 228-232
https://doi.org/10.1016/j.physe.2016.10.047

[722] John, R., Merlin, B., Crystal Structure Theory and Applications, 5 (2016) 43–55
https://doi.org/10.4236/csta.2016.5304

[723] Tao, L., Yang, C., Wu, L., Han, L., Song, Y., Wang, S., Lu, P., Modern Physics Letters B, 30[12], 2016, 1650146
https://doi.org/10.1142/S0217984916501463

[724] Zhang, H., Zhou, T., Zhang, J., Zhao, B., Yao, Y., Yang, Z., Physical Review B, 94[23], 2016, 235409
https://doi.org/10.1103/PhysRevB.94.235409

[725] Guo, Y., Pan, F., Ye, M., Wang, Y., Pan, Y., Zhang, X., Li, J., Zhang, H., Lu, J., 2D Materials, 3[3], 2016, 035020

[726] Xing, D.X., Ren, C.C., Zhang, S.F., Feng, Y., Chen, X.L., Zhang, C.W., Wang, P.J., Superlattices and Microstructures, 103, 2017, 139-144
https://doi.org/10.1016/j.spmi.2017.01.033

[727] Kadioglu, Y., Ersan, F., Gökoğlu, G., Aktürk, O.Ü., Aktürk, E., Materials Chemistry and Physics, 180, 2016, 326-331
https://doi.org/10.1016/j.matchemphys.2016.06.014

[728] Mortazavi, B., Dianat, A., Cuniberti, G., Rabczuk, T., Electrochimica Acta, 213, 2016, 865-870
https://doi.org/10.1016/j.electacta.2016.08.027

[729] Shaidu, Y., Akin-Ojo, O., Computational Materials Science, 118, 2016, 11-15
https://doi.org/10.1016/j.commatsci.2016.02.029

[730] Ezawa, M., Journal of Superconductivity and Novel Magnetism, 28[4] 2015, 1249-1253
https://doi.org/10.1007/s10948-014-2900-x

[731] Shen, L., Lan, M., Zhang, X., Xiang, G., RSC Advances, 7[16], 2017, 9840-9846
https://doi.org/10.1039/C6RA28155A

[732] Zhu, Z., Sun, Q., Jia, Y., AIP Advances, 6[3] 2016, 035012
https://doi.org/10.1063/1.4944621

[733] Ongun Özçelik, V., Kecik, D., Durgun, E., Ciraci, S., Journal of Physical Chemistry C, 119[1] 2015, 845-853
https://doi.org/10.1021/jp5106554

[734] Bao, W.C., Zou, L.J., Chinese Physics Letters, 31[9] 2014, 097501
https://doi.org/10.1088/0256-307X/31/9/097501

[735] Lado, J.L., Fernández-Rossier, J., Physical Review Letters, 113[2] 2014, 027203
https://doi.org/10.1103/PhysRevLett.113.027203

[736] Rachel, S., Ezawa, M., Physical Review B, 89[19] 2014, 195303
https://doi.org/10.1103/PhysRevB.89.195303

[737] Ezawa, M., Journal of the Physical Society of Japan, 84[12] 2015, 7566
https://doi.org/10.7566/JPSJ.84.121003

[738] Lütken, C.A., Contemporary Physics, 56[3] 2015, 292-318
https://doi.org/10.1080/00107514.2014.949445

[739] Wang, Y.P., Ji, W.X., Zhang, C.W., Li, P., Li, F., Wang, P.J., Li, S.S., Yan, S.S., Applied Physics Letters, 108[7] 2016, 073104
https://doi.org/10.1063/1.4942380

[740] Zhang, R.W., Zhang, C.W., Ji, W.X., Li, S.S., Hu, S.J., Yan, S.S., Li, P., Wang, P.J., Li, F., New Journal of Physics, 17[8] 2015, 083036
https://doi.org/10.1088/1367-2630/17/8/083036

[741] Zhang, R.W., Zhang, C.W., Ji, W.X., Li, S.S., Yan, S.S., Hu, S.J., Li, P., Wang, P.J., Li, F., Scientific Reports, 6, 2016, 18879
https://doi.org/10.1038/srep18879

[742] Xu, Y., Tang, P., Zhang, S.C., Physical Review B, 92[8] 2015, 081112
https://doi.org/10.1103/PhysRevB.92.081112

[743] Xiong, W., Xia, C., Peng, Y., Du, J., Wang, T., Zhang, J., Jia, Y., Physical Chemistry Chemical Physics, 18[9] 2016, 6534-6540
https://doi.org/10.1039/C5CP07140B

[744] Ji, W.X., Zhang, C.W., Ding, M., Li, P., Li, F., Ren, M.J., Wang, P.J., Hu, S.J., Yan, S.S., Scientific Reports, 5, 2015, 18604
https://doi.org/10.1038/srep18604

[745] Fang, Y., Huang, Z.Q., Hsu, C.H., Li, X., Xu, Y., Zhou, Y., Wu, S., Chuang, F.C., Zhu, Z.Z., Scientific Reports, 5, 2015, 14196
https://doi.org/10.1038/srep14196

[746] Ding, Y., Wang, Y., Journal of Physical Chemistry C, 119[49] 2015, 27848-27854
https://doi.org/10.1021/acs.jpcc.5b08946

[747] Chen, X., Li, L., Zhao, M., Physical Chemistry Chemical Physics, 17[25] 2015, 16624-16629
https://doi.org/10.1039/C5CP00046G

[748] Tang, P., Chen, P., Cao, W., Huang, H., Cahangirov, S., Xian, L., Xu, Y., Zhang, S.C., Duan, W., Rubio, A., Physical Review B, 90[12] 2014, 121408
https://doi.org/10.1103/PhysRevB.90.121408

[749] Wang, D., Chen, L., Wang, X., Cui, G., Zhang, P., Physical Chemistry Chemical Physics, 17[40] 2015, 26979-26987
https://doi.org/10.1039/C5CP04322K

[750] Nigam, S., Gupta, S., Banyai, D., Pandey, R., Majumder, C., Physical Chemistry Chemical Physics, 17[10] 2015, 6705-6712
https://doi.org/10.1039/C4CP04861J

[751] Kuang, Y.D., Lindsay, L., Shi, S.Q., Zheng, G.P., Nanoscale, 8[6] 2016, 3760-3767
https://doi.org/10.1039/C5NR08231E

[752] Peng, B., Zhang, H., Shao, H., Xu, Y., Zhang, X., Zhu, H., Scientific Reports, 6, 2016, 20225
https://doi.org/10.1038/srep20225

[753] Jomehpour Zaveh, S., Roknabadi, M.R., Morshedloo, T., Modarresi, M., Superlattices and Microstructures, 91, 2016, 383-390
https://doi.org/10.1016/j.spmi.2016.01.033

[754] Cherukara, M.J., Narayanan, B., Kinaci, A., Sasikumar, K., Gray, S.K., Chan, M.K.Y., Sankaranarayanan, S.K.R.S., Journal of Physical Chemistry Letters, 7[19], 2016, 3752-3759
https://doi.org/10.1021/acs.jpclett.6b01562

[755] Bhimanapati, G.R., Lin, Z., Meunier, V., Jung, Y., Cha, J., Das, S., Xiao, D., Son, Y., Strano, M.S., Cooper, V.R., Liang, L., Louie, S.G., Ringe, E., Zhou, W., Kim, S.S., Naik, R.R., Sumpter, B.G., Terrones, H., Xia, F., Wang, Y., Zhu, J., Akinwande, D., Alem, N., Schuller, J.A., Schaak, R.E., Terrones, M., Robinson, J.A., ACS Nano, 9[12] 2015, 11509-11539
https://doi.org/10.1021/acsnano.5b05556

[756] Ezawa, M., Le Lay, G., New Journal of Physics, 17[9] 2015, 090201
https://doi.org/10.1088/1367-2630/17/9/090201

[757] Balendhran, S., Walia, S., Nili, H., Sriram, S., Bhaskaran, M., Small, 11[6] 2015, 640-652
https://doi.org/10.1002/smll.201402041

[758] Takahashi, L., Takahashi, K., Physical Chemistry Chemical Physics, 17[33] 2015, 21394-21396
https://doi.org/10.1039/C5CP03382A

[759] Wang, T., Zhao, R., Zhao, M., Zhao, X., An, Y., Dai, X., Xia, C., Journal of Materials Science, 52[9], 2017, 5083-5096
https://doi.org/10.1007/s10853-016-0745-3

[760] Chen, X., Tan, C., Yang, Q., Meng, R., Liang, Q., Cai, M., Zhang, S., Jiang, J., Journal of Physical Chemistry C, 120[26], 2016, 13987-13994
https://doi.org/10.1021/acs.jpcc.6b04481

[761] Houssa, M., van den Broek, B., Iordanidou, K., Lu, A.K.A., Pourtois, G., Locquet, J.P., Afanasev, V., Stesmans, A., Nano Research, 9[3] 2016, 774-778
https://doi.org/10.1007/s12274-015-0956-y

[762] Wang, H., Pi, S.T., Kim, J., Wang, Z., Fu, H.H., Wu, R.Q., Physical Review B, 94[3], 2016, 035112
https://doi.org/10.1103/PhysRevB.94.035112

[763] Aktürk, O.U., Özçelik, V.O., Ciraci, S., Physical Review B, 91[23] 2015, 235446
https://doi.org/10.1103/PhysRevB.91.235446

[764] Zhao, M., Zhang, X., Li, L., Scientific Reports, 5, 2015, 16108
https://doi.org/10.1038/srep16108

[765] Wang, G., Pandey, R., Karna, S.P., ACS Applied Materials and Interfaces, 7[21] 2015, 11490-11496
https://doi.org/10.1021/acsami.5b02441

[766] Yang, L.F., Song, Y., Mi, W.B., Wang, X.C., Applied Physics Letters, 109[2], 2016, 022103
https://doi.org/10.1063/1.4958702

[767] Ji, J., Song, X., Liu, J., Yan, Z., Huo, C., Zhang, S., Su, M., Liao, L., Wang, W., Ni, Z., Hao, Y., Zeng, H., Nature Communications, 7, 2016, 13352
https://doi.org/10.1038/ncomms13352

[768] Gibaja, C., Rodriguez-San-Miguel, D., Ares, P., Gómez-Herrero, J., Varela, M., Gillen, R., Maultzsch, J., Hauke, F., Hirsch, A., Abellán, G., Zamora, F., Angewandte Chemie, 55[46], 2016, 14345-14349
https://doi.org/10.1002/anie.201605298

[769] Sun, Q., Dai, Y., Ma, Y., Yin, N., Wei, W., Yu, L., Huang, B., 2D Materials, 3[3], 2016, 035017

[770] Lu, H., Gao, J., Hu, Z., Shao, X., RSC Advances, 6[104], 2016, 102724-102732
https://doi.org/10.1039/C6RA21781H

[771] Fang Zhang, Wei Li, Xianqi Dai, Superlattices and Microstructures, 100, 2016, 826-832
https://doi.org/10.1016/j.spmi.2016.10.040

[772] Chen, X., Yang, Q., Meng, R., Jiang, J., Liang, Q., Tan, C., Sun, X., Journal of Materials Chemistry C, 4[23], 2016, 5434-5441
https://doi.org/10.1039/C6TC01141A

[773] Yang, L., Mi, W., Wang, X., RSC Advances, 6[45], 2016, 38746-38752
https://doi.org/10.1039/C6RA03927H

[774] Üzengi Aktürk, O., Aktürk, E., Ciraci, S., Physical Review B, 93[3], 2016, 035450
https://doi.org/10.1103/PhysRevB.93.035450

[775] Yang, L., Song, Y., Mi, W., Wang, X., RSC Advances, 6[70], 2016, 66140-66146
https://doi.org/10.1039/C6RA13101H

[776] Hu, Y., Wu, Y., Zhang, S., Physica B, 503, 2016, 126-129
https://doi.org/10.1016/j.physb.2016.08.034

[777] Gupta, S.K., Sonvane, Y., Wang, G., Pandey, R., Chemical Physics Letters, 641, 2015, 169-172
https://doi.org/10.1016/j.cplett.2015.10.072

[778] Singh, D., Gupta, S.K., Sonvane, Y., Lukačević, I., Journal of Materials Chemistry C, 4[26], 2016, 6386-6390
https://doi.org/10.1039/C6TC01913G

[779] Xie, M., Zhang, S., Cai, B., Zou, Y., Zeng, H., RSC Advances, 6[18] 2016, 14620-14625
https://doi.org/10.1039/C5RA25572D

[780] Lei, T., Liu, C., Zhao, J.L., Li, J.M., Li, Y.P., Wang, J.O., Wu, R., Qian, H.J., Wang, H.Q., Ibrahim, K., Journal of Applied Physics, 119[1] 2016, 015302
https://doi.org/10.1063/1.4939281

[781] Zhang, S., Yan, Z., Li, Y., Chen, Z., Zeng, H., Angewandte Chemie – International, 54[10] 2015, 3112-3115

[782] Pizzi, G., Gibertini, M., Dib, E., Marzari, N., Iannaccone, G., Fiori, G., Nature Communications, 7, 2016, 12585
https://doi.org/10.1038/ncomms12585

[783] Singh, D., Gupta, S.K., Lukačević, I., Sonvane, Y., RSC Advances, 6[10] 2016, 8006-8014
https://doi.org/10.1039/C5RA25773E

[784] Kamal, C., Ezawa, M., Physical Review B, 91[8] 2015, 085423
https://doi.org/10.1103/PhysRevB.91.085423

[785] Wang, C., Xia, Q., Nie, Y., Rahman, M., Guo, G., AIP Advances, 6[3] 2016, 035204
https://doi.org/10.1063/1.4943548

[786] Chaves, A., Mayers, M.Z., Peeters, F.M., Reichman, D.R., Physical Review B, 93[11] 2016, 115314
https://doi.org/10.1103/PhysRevB.93.115314

[787] Cao, H., Yu, Z., Lu, P., Superlattices and Microstructures, 86, 2015, 501-507
https://doi.org/10.1016/j.spmi.2015.08.006

[788] Guo, D., Shao, B., Li, C., Ma, Y., Superlattices and Microstructures, 100, 2016, 324-334
https://doi.org/10.1016/j.spmi.2016.09.044

[789] Xu, W., Lu, P., Wu, L., Yang, C., Song, Y., Guan, P., Han, L., Wang, S., IEEE Journal on Selected Topics in Quantum Electronics, 23[1], 2017, 7562584
https://doi.org/10.1109/JSTQE.2016.2593106

[790] Wang, Y.R., Zhang, Z.Y., Cao, H.N., Wang, D., Wang, Y.H., Si, M.S., Materials Technology, 31, 2016, 4-6
https://doi.org/10.1080/10667857.2016.1199135

[791] Liu, M.Y., Huang, Y., Chen, Q.Y., Cao, C., He, Y., Scientific Reports, 6, 2016, 29114
https://doi.org/10.1038/srep29114

[792] Zhang, Z., Xie, J., Yang, D., Wang, Y., Si, M., Xue, D., Applied Physics Express, 8[5] 2015, 055201
https://doi.org/10.7567/APEX.8.055201

[793] Luo, K., Chen, S.Y., Duan, C.G., Science China - Physics, Mechanics and Astronomy, 58[8] 2015, A003

[794] Wang, Y.P., Zhang, C.W., Ji, W.X., Wang, P.J., Applied Physics Express, 8[6] 2015, 065202
https://doi.org/10.7567/APEX.8.065202

[795] Wang, Y., Ding, Y., Nanoscale Research Letters, 10[1] 2015, 254
https://doi.org/10.1186/s11671-015-0955-7

[796] Zhang, Z.Y., Cao, H.N., Zhang, J.C., Wang, Y.H., Xue, D.S., Si, M.S., AIP Advances, 5[6] 2015, 067117
https://doi.org/10.1063/1.4922329

[797] Zhang, H., Ma, Y., Chen, Z., Nanoscale, 7[45] 2015, 19152-19159
https://doi.org/10.1039/C5NR05006E

[798] Zhang, S., Yan, Z., Li, Y., Chen, Z., Zeng, H., Angewandte Chemie – International, 54[10] 2015, 3112-3115

[799] Li, Z., Xu, W., Yu, Y., Du, H., Zhen, K., Wang, J., Luo, L., Qiu, H., Yang, X., Journal of Materials Chemistry C, 4[2] 2016, 362-370
https://doi.org/10.1039/C5TC03001C

[800] Tsai, H.S., Wang, S.W., Hsiao, C.H., Chen, C.W., Ouyang, H., Chueh, Y.L., Kuo, H.C., Liang, J.H., Chemistry of Materials, 28[2] 2016, 425-429
https://doi.org/10.1021/acs.chemmater.5b04949

[801] Wang, Y., Ding, Y., Physical Chemistry Chemical Physics, 17[41] 2015, 27769-27776
https://doi.org/10.1039/C5CP04815J

[802] Xia, C., Xue, B., Wang, T., Peng, Y., Jia, Y., Applied Physics Letters, 107[19] 2015, 193107
https://doi.org/10.1063/1.4935602

[803] Li, W., Wang, T.X., Dai, X.Q., Wang, X.L., Ma, Y.Q., Chang, S.S., Tang, Y.N., Physica E, 88, 2017, 6-10
https://doi.org/10.1016/j.physe.2016.11.013

[804] Song, Y., Li, D., Mi, W., Wang, X., Cheng, Y., Journal of Physical Chemistry C, 120[10] 2016, 5613-5618
https://doi.org/10.1021/acs.jpcc.6b01062

[805] Li, Y., Xia, C., Wang, T., Tan, X., Zhao, X., Wei, S., Solid State Communications, 230, 2016, 6-10
https://doi.org/10.1016/j.ssc.2016.01.005

[806] Wang, D., Chen, L., Shi, C., Wang, X., Cui, G., Zhang, P., Chen, Y., Scientific Reports, 6, 2016, 28487
https://doi.org/10.1038/srep28487

[807] Liu, C., Liu, C.S., Yan, X., Physics Letters A, 381[12], 2017, 1092-1096
https://doi.org/10.1016/j.physleta.2017.01.048

[808] Ersan, F., Aktürk, E., Ciraci, S., Journal of Physical Chemistry C, 120[26], 2016, 14345-14355
https://doi.org/10.1021/acs.jpcc.6b02439

[809] Wang, Y.P., Ji, W.X., Zhang, C.W., Li, P., Li, F., Ren, M.J., Chen, X.L., Yuan, M., Wang, P.J., Scientific Reports, 6, 2016, 20342
https://doi.org/10.1038/srep20342

[810] Wang, Y.P., Zhang, C.W., Ji, W.X., Zhang, R.W., Li, P., Wang, P.J., Ren, M.J., Chen, X.L., Yuan, M., Journal of Physics D, 49[5] 2016, 055305
https://doi.org/10.1088/0022-3727/49/5/055305

[811] Wang, D., Chen, L., Shi, C., Wang, X., Cui, G., Zhang, P., Chen, Y., New Journal of Physics, 18[3], 2016, 033026
https://doi.org/10.1088/1367-2630/18/3/033026

[812] Li, G., Zhao, Y., Zeng, S., Ni, J., Applied Surface Science, 390, 2016, 60-67
https://doi.org/10.1016/j.apsusc.2016.08.016

[813] Luo, M., Shen, Y.H., Yin, T.L., JETP Letters, 104[8], 2016, 557-562
https://doi.org/10.1134/S0021364016200029

[814] Liu, M., Chen, Q., Huang, Y., Cao, C., He, Y., Superlattices and Microstructures, 100, 2016, 131-141
https://doi.org/10.1016/j.spmi.2016.09.014

[815] Luo, M., Shen, Y.H., Yin, T.L., Japanese Journal of Applied Physics, 56[1], 2017, 015201
https://doi.org/10.7567/JJAP.56.015201

[816] Sun, M., Wang, S., Du, Y., Yu, J., Tang, W., Applied Surface Science, 389, 2016, 594-600
https://doi.org/10.1016/j.apsusc.2016.07.091

[817] Du, J., Xia, C., An, Y., Wang, T., Jia, Y., Journal of Materials Science, 51[20], 2016, 9504-9513
https://doi.org/10.1007/s10853-016-0194-z

[818] Zeraati, M., Vaez Allaei, S.M., Abdolhosseini Sarsari, I., Pourfath, M., Donadio, D., Physical Review B, 93[8] 2016, 085424
https://doi.org/10.1103/PhysRevB.93.085424

[819] Kou, L., Chen, C., Smith, S.C., Journal of Physical Chemistry Letters, 6[14] 2015, 2794-2805
https://doi.org/10.1021/acs.jpclett.5b01094

[820] Balendhran, S., Walia, S., Nili, H., Sriram, S., Bhaskaran, M., Small, 11[6] 2015, 640-652
https://doi.org/10.1002/smll.201402041

[821] Bagheri, S., Mansouri, N., Aghaie, E., International Journal of Hydrogen Energy, 41[7] 2016, 4085-4095
https://doi.org/10.1016/j.ijhydene.2016.01.034

[822] Jing, Y., Zhang, X., Zhou, Z., Wiley Interdisciplinary Reviews - Computational Molecular Science, 6[1] 2016, 5-19
https://doi.org/10.1002/wcms.1234

[823] Fukuoka, S., Taen, T., Osada, T., Journal of the Physical Society of Japan, 84[12] 2015, 121004
https://doi.org/10.7566/JPSJ.84.121004

[824] Ribeiro-Soares, J., Almeida, R.M., Cançado, L.G., Dresselhaus, M.S., Jorio, A., Physical Review B, 91[20] 2015, 205421
https://doi.org/10.1103/PhysRevB.91.205421

[825] Köpf, M., Eckstein, N., Pfister, D., Grotz, C., Krüger, I., Greiwe, M., Hansen, T., Kohlmann, H., Nilges, T., Journal of Crystal Growth, 405, 2014, 6-10 *https://doi.org/10.1016/j.jcrysgro.2014.07.029*

[826] Lu, W., Nan, H., Hong, J., Chen, Y., Zhu, C., Liang, Z., Ma, X., Ni, Z., Jin, C., Zhang, Z., Nano Research, 7[6] 2014, 853-859 *https://doi.org/10.1007/s12274-014-0446-7*

[827] Serrano-Ruiz, M., Caporali, M., Ienco, A., Piazza, V., Heun, S., Peruzzini, M., Advanced Materials Interfaces, 3[3] 2016, 1500441 *https://doi.org/10.1002/admi.201500441*

[828] Brent, J.R., Savjani, N., Lewis, E.A., Haigh, S.J., Lewis, D.J., O'Brien, P., Chemical Communications, 50[87] 2014, 13338-13341 *https://doi.org/10.1039/C4CC05752J*

[829] Sresht, V., Pádua, A.A.H., Blankschtein, D., ACS Nano, 9[8] 2015, 8255-8268 *https://doi.org/10.1021/acsnano.5b02683*

[830] Mu, Y., Si, M.S., EPL, 112[3] 2015, 37003 *https://doi.org/10.1209/0295-5075/112/37003*

[831] Wang, L., Sofer, Z., Pumera, M., ChemElectroChem, 2[3] 2015, 324-327 *https://doi.org/10.1002/celc.201402363*

[832] Feng, Y., Zhou, J., Du, Y., Miao, F., Duan, C.G., Wang, B., Wan, X., Journal of Physics - Condensed Matter, 27[18] 2015, 185302 *https://doi.org/10.1088/0953-8984/27/18/185302*

[833] Pei, J., Gai, X., Yang, J., Wang, X., Yu, Z., Choi, D.Y., Luther-Davies, B., Lu, Y., Nature Communications, 7, 2016, 10450 *https://doi.org/10.1038/ncomms10450*

[834] Zeng, J., Cui, P., Zhang, Z., Physical Review Letters, 118[4], 2017, 046101 *https://doi.org/10.1103/PhysRevLett.118.046101*

[835] Guan, J., Zhu, Z., Tománek, D., ACS Nano, 8[12] 2014, 12763-12768 *https://doi.org/10.1021/nn5059248*

[836] Liu, J., Guo, Y., Zhang, S., Wang, Q., Kawazoe, Y., Jena, P., Journal of Physical Chemistry C, 119[43] 2015, 24674-24680 *https://doi.org/10.1021/acs.jpcc.5b08593*

[837] Guan, J., Zhu, Z., Tománek, D., Physical Review Letters, 113[4] 2014, 046804 *https://doi.org/10.1103/PhysRevLett.113.046804*

[838] Feng, J.R., Wang, G.C., RSC Advances, 6[27] 2016, 22277-22284 *https://doi.org/10.1039/C5RA25361F*

[839] Wu, M., Fu, H., Zhou, L., Yao, K., Zeng, X.C., Nano Letters, 15[5] 2015, 3557-3562
https://doi.org/10.1021/acs.nanolett.5b01041

[840] Zhao, T., He, C.Y., Ma, S.Y., Zhang, K.W., Peng, X.Y., Xie, G.F., Zhong, J.X., Journal of Physics - Condensed Matter, 27[26] 2015, 265301
https://doi.org/10.1088/0953-8984/27/26/265301

[841] Wang, H., Li, X., Liu, Z., Yang, J., Physical Chemistry Chemical Physics, 19[3], 2017, 2402-2408
https://doi.org/10.1039/C6CP07944J

[842] Sha, Z.D., Pei, Q.X., Ding, Z., Jiang, J.W., Zhang, Y.W., Journal of Physics D, 48[39] 2015, 395303
https://doi.org/10.1088/0022-3727/48/39/395303

[843] Hao, F., Chen, X., Journal of Applied Physics, 118[23] 2015, 234304
https://doi.org/10.1063/1.4937920

[844] Liu, N., Hong, J., Pidaparti, R., Wang, X., Nanoscale, 8[10] 2016, 5728-5736
https://doi.org/10.1039/C5NR08682E

[845] Sorkin, V., Zhang, Y.W., Nanotechnology, 27[39], 2016, 395701
https://doi.org/10.1088/0957-4484/27/39/395701

[846] Sorkin, V., Zhang, Y.W., 2D Materials, 2[3] 2015, 035007

[847] Xu, F., Ma, H., Lei, S., Sun, J., Chen, J., Ge, B., Zhu, Y., Sun, L., Nanoscale, 8[28], 2016, 13603-13610
https://doi.org/10.1039/C6NR02487D

[848] Kou, L., Ma, Y., Smith, S.C., Chen, C., Journal of Physical Chemistry Letters, 6[9] 2015, 1509-1513
https://doi.org/10.1021/acs.jpclett.5b00522

[849] Wang, G., Loh, G.C., Pandey, R., Karna, S.P., Nanotechnology, 27[5] 2015, 055701
https://doi.org/10.1088/0957-4484/27/5/055701

[850] Hu, T., Han, Y., Dong, J., Nanotechnology, 25[45] 2014, 455703
https://doi.org/10.1088/0957-4484/25/45/455703

[851] Wang, L., Kutana, A., Zou, X., Yakobson, B.I., Nanoscale, 7[21] 2015, 9746-9751
https://doi.org/10.1039/C5NR00355E

[852] Ding, Y., Wang, Y., Shi, L., Xu, Z., Ni, J., Physica Status Solidi R, 8[11] 2014, 939-942
https://doi.org/10.1002/pssr.201409385

[853] Fei, R., Yang, L., Applied Physics Letters, 105[8] 2014, 083120
https://doi.org/10.1063/1.4894273

[854] Elahi, M., Khaliji, K., Tabatabaei, S.M., Pourfath, M., Asgari, R., Physical Review B, 91[11] 2015, 115412
https://doi.org/10.1103/PhysRevB.91.115412

[855] Sorkin, V., Zhang, Y.W., Nanotechnology, 26[23] 2015, 235707
https://doi.org/10.1088/0957-4484/26/23/235707

[856] Verma, D., Hourahine, B., Frauenheim, T., James, R.D., Dumitric, Ă.T., Physical Review B, 94[12], 2016, 121404
https://doi.org/10.1103/PhysRevB.94.121404

[857] Wang, H., Li, X., Li, P., Yang, J., Nanoscale, 9[2], 2017, 850-855
https://doi.org/10.1039/C6NR08550D

[858] Li, W., Yang, Y., Zhang, G., Zhang, Y.W., Nano Letters, 15[3] 2015, 1691-1697
https://doi.org/10.1021/nl504336h

[859] Zhang, R., Wu, X., Yang, J., Nanoscale, 8[7] 2016, 4001-4006
https://doi.org/10.1039/C5NR06856H

[860] Wang, D., Guo, G.C., Wei, X.L., Liu, L.M., Zhao, S.J., Journal of Power Sources, 302, 2016, 215-222
https://doi.org/10.1016/j.jpowsour.2015.10.059

[861] Zhang, C., Yu, M., Anderson, G., Dharmasena, R.R., Sumanasekera, G., Nanotechnology, 28[7], 2017, 075401
https://doi.org/10.1088/1361-6528/aa52ac

[862] Liu, X., Wen, Y., Chen, Z., Shan, B., Chen, R., Physical Chemistry Chemical Physics, 17[25] 2015, 16398-16404
https://doi.org/10.1039/C5CP02419F

[863] Zhang, S., Li, C., Xiao Guo, Z., Cho, J.H., Su, W.S., Jia, Y., Nanotechnology, 26[29] 2015, 295402
https://doi.org/10.1088/0957-4484/26/29/295402

[864] Zhu, Z., Li, C., Yu, W., Chang, D., Sun, Q., Jia, Y., Applied Physics Letters, 105[11] 2014, 11310
https://doi.org/10.1063/1.4895924

[865] Du, Y., Liu, H., Xu, B., Sheng, L., Yin, J., Duan, C.G., Wan, X., Scientific Reports, 5, 2015, 8921
https://doi.org/10.1038/srep08921

[866] Zhai, C., Dai, X., Li, W., Ma, Y., Wang, T., Tang, Y., Superlattices and Microstructures, 101, 49-56
https://doi.org/10.1016/j.spmi.2016.10.090

[867] Sui, X., Si, C., Shao, B., Zou, X., Wu, J., Gu, B.L., Duan, W., Journal of Physical Chemistry C, 119[18] 2015, 10059-10063
https://doi.org/10.1021/jp5129468

[868] Seixas, L., Carvalho, A., Castro Neto, A.H., Physical Review B, 91[15] 2015, 155138
https://doi.org/10.1103/PhysRevB.91.155138

[869] Khan, I., Hong, J., New Journal of Physics, 17, 2015, 023056
https://doi.org/10.1088/1367-2630/17/2/023056

[870] Zheng, H., Yang, H., Wang, H., Du, X., Yan, Y., Journal of Magnetism and Magnetic Materials, 408, 2016, 121-126
https://doi.org/10.1016/j.jmmm.2016.02.014

[871] Zheng, H., Zhang, J., Yang, B., Du, X., Yan, Y., Physical Chemistry Chemical Physics, 17[25] 2015, 16341-16350
https://doi.org/10.1039/C5CP00916B

[872] Yang, L., Mi, W., Wang, X., Journal of Alloys and Compounds, 662, 2016, 528-533
https://doi.org/10.1016/j.jallcom.2015.12.095

[873] Boukhvalov, D.W., Physical Chemistry Chemical Physics, 17[40] 2015, 27210-27216
https://doi.org/10.1039/C5CP05071E

[874] Ziletti, A., Huang, S.M., Coker, D.F., Lin, H., Physical Review B, 92[8] 2015, 085423
https://doi.org/10.1103/PhysRevB.92.085423

[875] Hashmi, A., Farooq, M.U., Hong, J., Journal of Physical Chemistry Letters, 7[4] 2016, 647-652
https://doi.org/10.1021/acs.jpclett.5b02600

[876] Farooq, M.U., Hashmi, A., Hong, J., ACS Applied Materials and Interfaces, 7[26] 2015, 14423-14430
https://doi.org/10.1021/acsami.5b03284

[877] Chen, M., Yu, Z., Xie, Y., Wang, Y., Applied Physics Letters, 109[14], 2016, 142409
https://doi.org/10.1063/1.4964463

[878] Tan, X.Y., Wang, J.H., Zhu, Y.Y., Zuo, A.Y., Jin, K.X., Acta Physica Sinica, 63[20] 2014, 207301

[879] Guo, H., Lu, N., Dai, J., Wu, X., Zeng, X.C., Journal of Physical Chemistry C, 118[25] 2014, 14051-14059
https://doi.org/10.1021/jp505257g

[880] Ezawa, M., New Journal of Physics, 16, 2014, 115004
https://doi.org/10.1088/1367-2630/16/11/115004

[881] Morgan Stewart, H., Shevlin, S.A., Catlow, C.R.A., Guo, Z.X., Nano Letters, 15[3] 2015, 2006-2010
https://doi.org/10.1021/nl504861w

[882] Cai, Y., Dr., Ke, Q., Dr., Zhang, G., Feng, Y.P., Prof., Shenoy, V.B., Prof., Zhang, Y.W., Advanced Functional Materials, 25[15] 2015, 2230-2236
https://doi.org/10.1002/adfm.201404294

[883] Li, P., Appelbaum, I., Physical Review B, 90[11] 2014, 115439
https://doi.org/10.1103/PhysRevB.90.115439

[884] Woomer, A.H., Farnsworth, T.W., Hu, J., Wells, R.A., Donley, C.L., Warren, S.C., ACS Nano, 9[9] 2015, 8869-8884
https://doi.org/10.1021/acsnano.5b02599

[885] Peng, X., Wei, Q., Materials Research Express, 1[4] 2015, 045041
https://doi.org/10.1088/2053-1591/1/4/045041

[886] Carvalho, A., Rodin, A.S., Castro Neto, A.H., EPL, 108[4] 2014, 47005
https://doi.org/10.1209/0295-5075/108/47005

[887] Li, W., Zhang, G., Zhang, Y.W., Journal of Physical Chemistry C, 118[38] 2014, 22368-22372
https://doi.org/10.1021/jp506996a

[888] Liu, H., Neal, A.T., Zhu, Z., Luo, Z., Xu, X., Tománek, D., Ye, P.D., ACS Nano, 8[4] 2014, 4033-4041
https://doi.org/10.1021/nn501226z

[889] Dai, J., Zeng, X.C., Journal of Physical Chemistry Letters, 5[7] 2014, 1289-1293
https://doi.org/10.1021/jz500409m

[890] Han, X., Morgan Stewart, H., Shevlin, S.A., Catlow, C.R.A., Guo, Z.X., Nano Letters, 14[8] 2014, 4607-4614
https://doi.org/10.1021/nl501658d

[891] Peng, X., Wei, Q., Copple, A., Physical Review B, 90[8] 2014, 085402
https://doi.org/10.1103/PhysRevB.90.085402

[892] Wang, C., Xia, Q., Nie, Y., Guo, G., Journal of Applied Physics, 117[12] 2015, 124302
https://doi.org/10.1063/1.4916254

[893] Osada, T., Journal of the Physical Society of Japan, 84[1] 2015, 013703
https://doi.org/10.7566/JPSJ.84.013703

[894] Xie, J., Si, M.S., Yang, D.Z., Zhang, Z.Y., Xue, D.S., Journal of Applied Physics, 116[7] 2014, 073704
https://doi.org/10.1063/1.4893589

[895] Xu, L.C., Song, X.J., Yang, Z., Cao, L., Liu, R.P., Li, X.Y., Applied Surface Science, 324, 2015, 640-644
https://doi.org/10.1016/j.apsusc.2014.10.166

[896] Hu, T., Hong, J., Journal of Applied Physics, 118[5] 2015, 054301
https://doi.org/10.1063/1.4927848

[897] Peng, X., Copple, A., Wei, Q., Journal of Applied Physics, 116[14] 2014, 144301
https://doi.org/10.1063/1.4897461

[898] Sun, M., Tang, W., Ren, Q., Wang, S.K., Yu, J., Du, Y., Applied Surface Science, 356, 2015, 110-114
https://doi.org/10.1016/j.apsusc.2015.08.009

[899] Ziletti, A., Carvalho, A., Trevisanutto, P.E., Campbell, D.K., Coker, D.F., Castro Neto, A.H., Physical Review B, 91[8] 2015, 085407
https://doi.org/10.1103/PhysRevB.91.085407

[900] Sa, B., Li, Y.L., Sun, Z., Qi, J., Wen, C., Wu, B., Nanotechnology, 26[21] 2015, 215205
https://doi.org/10.1088/0957-4484/26/21/215205

[901] Wang, V., Liu, Y.C., Kawazoe, Y., Geng, W.T., Journal of Physical Chemistry Letters, 6[24] 2015, 4876-4883
https://doi.org/10.1021/acs.jpclett.5b02047

[902] Owens, F.J., Solid State Communications, 223, 2015, 37-39
https://doi.org/10.1016/j.ssc.2015.09.007

[903] Kong, L.J., Liu, G.H., Zhang, Y.J., RSC Advances, 6[13] 2016, 10919-10929
https://doi.org/10.1039/C5RA22004A

[904] Hashmi, A., Hong, J., Journal of Physical Chemistry C, 119[17] 2015, 9198-9204
https://doi.org/10.1021/jp511574n

[905] Yu, W., Zhu, Z., Niu, C.Y., Li, C., Cho, J.H., Jia, Y., Physical Chemistry Chemical Physics, 17[25] 2015, 16351-16358
https://doi.org/10.1039/C5CP01732G

[906] Son, J., Hashmi, A., Hong, J., Current Applied Physics, 16[5] 2016, 506-514
https://doi.org/10.1016/j.cap.2016.02.003

[907] Boukhvalov, D.W., Rudenko, A.N., Prishchenko, D.A., Mazurenko, V.G., Katsnelson, M.I., Physical Chemistry Chemical Physics, 17[23] 2015, 15209-15217
https://doi.org/10.1039/C5CP01901J

[908] Hu, T., Dong, J., Physical Review B, 92[6] 2015, 064114
https://doi.org/10.1103/PhysRevB.92.064114

[909] Ding, B., Chen, W., Tang, Z., Zhang, J., Journal of Physical Chemistry C, 120[4] 2016, 2149-2158
https://doi.org/10.1021/acs.jpcc.5b09159

[910] Li, R., Tao, Y., Liu, Z.L., Tan, Y., Physica E, 88, 2017, 174-181
https://doi.org/10.1016/j.physe.2016.12.014

[911] Xu, R., Zhang, S., Wang, F., Yang, J., Wang, Z., Pei, J., Myint, Y.W., Xing, B., Yu, Z., Fu, L., Qin, Q., Lu, Y., ACS Nano, 10[2] 2016, 2046-2053
https://doi.org/10.1021/acsnano.5b06193

[912] Prada, E., Alvarez, J.V., Narasimha-Acharya, K.L., Bailen, F.J., Palacios, J.J., Physical Review B, 91[24] 2015, 245421
https://doi.org/10.1103/PhysRevB.91.245421

[913] Duan, H., Yang, M., Wang, R., Physica E, 81, 2016, 177-181
https://doi.org/10.1016/j.physe.2016.03.011

[914] Mohammadi, Y., Nia, B.A., Superlattices and Microstructures, 89, 2016, 204-215
https://doi.org/10.1016/j.spmi.2015.10.049

[915] Ong, Z.Y., Zhang, G., Zhang, Y.W., Journal of Applied Physics, 116[21] 2014, 214505
https://doi.org/10.1063/1.4902545

[916] Zhang, X., Li, Q., Xu, B., Wan, B., Yin, J., Wan, X.G., Physics Letters A, 380[4] 2016, 614-620
https://doi.org/10.1016/j.physleta.2015.11.019

[917] Ostahie, B., Aldea, A., Physical Review B, 93[7] 2016, 075408
https://doi.org/10.1103/PhysRevB.93.075408

[918] Keshtan, M.A., Esmaeilzadeh, M., Journal of Physics D, 48[48] 2015, 485301
https://doi.org/10.1088/0022-3727/48/48/485301

[919] Yang, M., Duan, H.J., Wang, R.Q., JETP Letters, 102[9] 2015, 610-615
https://doi.org/10.1134/S0021364015210134

[920] Li, Y., Wei, Z., Li, J., Applied Physics Letters, 107[11] 2015, 112103
https://doi.org/10.1063/1.4931119

[921] Liang, L., Wang, J., Lin, W., Sumpter, B.G., Meunier, V., Pan, M., Nano Letters, 14[11] 2014, 6400-6406
https://doi.org/10.1021/nl502892t

[922] Hu, W., Lin, L., Yang, C., Physical Chemistry Chemical Physics, 17[47] 2015, 31397-31404
https://doi.org/10.1039/C5CP00333D

[923] Zhang, T., Lin, J.H., Yu, Y.M., Chen, X.R., Liu, W.M., Scientific Reports, 5, 2015, 13927
https://doi.org/10.1038/srep13927

[924] Taghizadeh Sisakht, E., Zare, M.H., Fazileh, F., Physical Review B, 91[8] 2015, 085409
https://doi.org/10.1103/PhysRevB.91.085409

[925] Zhang, R., Zhou, X.Y., Zhang, D., Lou, W.K., Zhai, F., Chang, K., 2D Materials, 2[4] 2015, 045012

[926] Zhou, X.Y., Zhang, R., Sun, J.P., Zou, Y.L., Zhang, D., Lou, W.K., Cheng, F., Zhou, G.H., Zhai, F., Chang, K., Scientific Reports, 5, 2015, 12295
https://doi.org/10.1038/srep12295

[927] Tahir, M., Vasilopoulos, P., Peeters, F.M., Physical Review B, 92[4] 2015, 045420
https://doi.org/10.1103/PhysRevB.92.045420

[928] Ghazaryan, A., Chakraborty, T., Physical Review B, 92[16] 2015, 165409
https://doi.org/10.1103/PhysRevB.92.165409

[929] Pereira, J.M., Katsnelson, M.I., Physical Review B, 92[7] 2015, 075437
https://doi.org/10.1103/PhysRevB.92.075437

[930] Lew Yan Voon, L.C., Lopez-Bezanilla, A., Wang, J., Zhang, Y., Willatzen, M., New Journal of Physics, 17, 2015, 025004
https://doi.org/10.1088/1367-2630/17/2/025004

[931] Yu, S., Zhu, H., Eshun, K., Arab, A., Badwan, A., Li, Q., Journal of Applied Physics, 118[16] 2015, 164306
https://doi.org/10.1063/1.4934852

[932] Arabieh, M., Azar, Y.T., Applied Surface Science, 396, 2017, 1411-1419
https://doi.org/10.1016/j.apsusc.2016.11.175

[933] He, Y., Xia, F., Shao, Z., Zhao, J., Jie, J., Journal of Physical Chemistry Letters, 6[23] 2015, 4701-4710
https://doi.org/10.1021/acs.jpclett.5b01920

[934] Lu, J., Wu, J., Carvalho, A., Ziletti, A., Liu, H., Tan, J., Chen, Y., Castro Neto, A.H., Özyilmaz, B., Sow, C.H., ACS Nano, 9[10] 2015, 10411-10421
https://doi.org/10.1021/acsnano.5b04623

[935] Yang, T., Dong, B., Wang, J., Zhang, Z., Guan, J., Kuntz, K., Warren, S.C., Tománek, D., Physical Review B, 92[12] 2015, 125412
https://doi.org/10.1103/PhysRevB.92.125412

[936] Cai, Y., Ke, Q., Zhang, G., Zhang, Y.W., Journal of Physical Chemistry C, 119[6] 2015, 3102-3110
https://doi.org/10.1021/jp510863p

[937] Ziletti, A., Carvalho, A., Campbell, D.K., Coker, D.F., Castro Neto, A.H., Physical Review Letters, 114[4] 2015, 046801
https://doi.org/10.1103/PhysRevLett.114.046801

[938] Dai, J., Zeng, X.C., RSC Advances, 4[89] 2014, 48017-48021
https://doi.org/10.1039/C4RA02850C

[939] Jing, Y., Tang, Q., He, P., Zhou, Z., Shen, P., Nanotechnology, 26[9] 2015, 095201
https://doi.org/10.1088/0957-4484/26/9/095201

[940] Wang, G., Pandey, R., Karna, S.P., Nanoscale, 7[2] 2015, 524-531
https://doi.org/10.1039/C4NR05384B

[941] Ding, Y., Wang, Y., Journal of Physical Chemistry C, 119[19] 2015, 10610-10622
https://doi.org/10.1021/jp5114152

[942] Babar, R., Kabir, M., Journal of Physical Chemistry C, 120[27], 2016, 14991-15000
https://doi.org/10.1021/acs.jpcc.6b05069

[943] Kulish, V.V., Malyi, O.I., Persson, C., Wu, P., Physical Chemistry Chemical Physics, 17[2] 2015, 992-1000
https://doi.org/10.1039/C4CP03890H

[944] Hu, T., Hong, J., Journal of Physical Chemistry C, 119[15] 2015, 8199-8207
https://doi.org/10.1021/acs.jpcc.5b01300

[945] Zhang, R., Li, B., Yang, J., Journal of Physical Chemistry C, 119[5] 2015, 2871-2878
https://doi.org/10.1021/jp5116564

[946] Shao, L., Ye, H., Yelongwu, W., Yinxiao, D., Ding, P., Zeng, F., Yuan, Q., Materials Research Express, 3[2] 2016, 025013
https://doi.org/10.1088/2053-1591/3/2/025013

[947] Zhang, Z.Y., Si, M.S., Peng, S.L., Zhang, F., Wang, Y.H., Xue, D.S., Journal of Solid State Chemistry, 231, 2015, 64-69
https://doi.org/10.1016/j.jssc.2015.07.043

[948] Ling, Z.P., Ang, K.W., APL Materials, 3[12] 2015, 126104
https://doi.org/10.1063/1.4937468

[949] Xu, R., Yang, J., Zhu, Y., Yan, H., Pei, J., Myint, Y.W., Zhang, S., Lu, Y., Nanoscale, 8[1] 2016, 129-135
https://doi.org/10.1039/C5NR04366B

[950] Lin, J.H., Zhang, H., Cheng, X.L., Frontiers of Physics, 10[4] 2015
https://doi.org/10.1007/s11467-015-0468-y

[951] Tran, V., Yang, L., Physical Review B, 89[24] 2014, 245407
https://doi.org/10.1103/PhysRevB.89.245407

[952] Zhang, S., Yang, J., Xu, R., Wang, F., Li, W., Ghufran, M., Zhang, Y.W., Yu, Z., Zhang, G., Qin, Q., Lu, Y., ACS Nano, 8[9] 2014, 9590-9596
https://doi.org/10.1021/nn503893j

[953] Seixas, L., Rodin, A.S., Carvalho, A., Castro Neto, A.H., Physical Review B, 91[11] 2015, 115437
https://doi.org/10.1103/PhysRevB.91.115437

[954] Tokár, K., Derian, R., Brndiar, J., Štich, I., Journal of Applied Physics, 120[19], 2016, 194305
https://doi.org/10.1063/1.4968009

[955] Srivastava, A., Khan, M.S., Gupta, S.K., Pandey, R., Applied Surface Science, 356, 2015, 881-887
https://doi.org/10.1016/j.apsusc.2015.08.109

[956] Umar Farooq, M., Hashmi, A., Hong, J., Scientific Reports, 5, 2015, 12482
https://doi.org/10.1038/srep12482

[957] Wang, V., Kawazoe, Y., Geng, W.T., Physical Review B, 91[4] 2015, 045433
https://doi.org/10.1103/PhysRevB.91.045433

[958] Hu, T., Dong, J., Nanotechnology, 26[6] 2015, 065705
https://doi.org/10.1088/0957-4484/26/6/065705

[959] Hu, W., Yang, J., Journal of Physical Chemistry C, 119[35] 2015, 20474-20480
https://doi.org/10.1021/acs.jpcc.5b06077

[960] Cai, Y., Ke, Q., Zhang, G., Yakobson, B.I., Zhang, Y.W., Journal of the American Chemical Society, 138[32], 2016, 10199-10206
https://doi.org/10.1021/jacs.6b04926

[961] Sha, Z.D., Pei, Q.X., Zhang, Y.Y., Zhang, Y.W., Nanotechnology, 27[31], 2016, 315704
https://doi.org/10.1088/0957-4484/27/31/315704

[962] Hao, F., Chen, X., Journal of Applied Physics, 120[16], 2016, 165104
https://doi.org/10.1063/1.4966167

[963] Srivastava, P., Hembram, K.P.S.S., Mizuseki, H., Lee, K.R., Han, S.S., Kim, S., Journal of Physical Chemistry C, 119[12] 2015, 6530-6538
https://doi.org/10.1021/jp5110938

[964] Wang, G., Pandey, R., Karna, S.P., Applied Physics Letters, 106[17] 2015, 173104
https://doi.org/10.1063/1.4919389

[965] Chintalapati, S., Shen, L., Xiong, Q., Feng, Y.P., Applied Physics Letters, 107[7] 2015, 072401
https://doi.org/10.1063/1.4928754

[966] Lopez-Bezanilla, A., Physical Review B, 93[3] 2016, 035433
https://doi.org/10.1103/PhysRevB.93.035433

[967] Chen, W.H., Chen, I.C., Cheng, H.C., Yu, C.F., Journal of Materials Science, 52[6], 2017, 3225-3232
https://doi.org/10.1007/s10853-016-0611-3

[968] Li, X., Ma, L., Wang, D., Zeng, X.C., Wu, X., Yang, J., Nanoscale, 8[41], 2016, 17801-17808
https://doi.org/10.1039/C6NR05414E

[969] Jang, W., Kang, K., Soon, A., Nanoscale, 7[45] 2015, 19073-19079
https://doi.org/10.1039/C5NR05605E

[970] Aierken, Y., Leenaerts, O., Peeters, F.M., Physical Review B, 92[10] 2015, 104104
https://doi.org/10.1103/PhysRevB.92.104104

[971] Zhu, Z.L., Yu, W.Y., Ren, X.Y., Sun, Q., Jia, Y., EPL, 109[4] 2015, 47003
https://doi.org/10.1209/0295-5075/109/47003

[972] Sorkin, V., Zhang, Y.W., Nanotechnology, 28[7], 2017, 075704
https://doi.org/10.1088/1361-6528/aa537b

[973] Mehboudi, M., Utt, K., Terrones, H., Harriss, E.O., SanJuan, A.A.P., Barraza-Lopez, S., Proceedings of the National Academy of Sciences of the United States of America, 112[19] 2015, 5888-5892
https://doi.org/10.1073/pnas.1500633112

[974] Hu, W., Wang, T., Yang, J., Journal of Materials Chemistry C, 3[18] 2015, 4756-4761
https://doi.org/10.1039/C5TC00759C

[975] Padilha, J.E., Fazzio, A., Da Silva, A.J.R., Physical Review Letters, 114[6] 2015, 066803
https://doi.org/10.1103/PhysRevLett.114.066803

[976] Hashmi, A., Farooq, U., Hong, J., Current Applied Physics, 16[3] 2016, 318-323
https://doi.org/10.1016/j.cap.2015.12.019

[977] Tian, X., Liu, L., Du, Y., Gu, J., Xu, J.B., Yakobson, B.I., Physical Chemistry Chemical Physics, 17[47] 2015, 31685-31692
https://doi.org/10.1039/C5CP05443E

[978] Hong, Y., Zhang, J., Zeng, X.C., Nanoscale, 8[46], 2016, 19211-19218
https://doi.org/10.1039/C6NR07977F

[979] Cai, Y., Zhang, G., Zhang, Y.W., Journal of Physical Chemistry C, 119[24] 2015, 13929-13936
https://doi.org/10.1021/acs.jpcc.5b02634

[980] Hu, T., Hong, J., ACS Applied Materials and Interfaces, 7[42] 2015, 23489-23495
https://doi.org/10.1021/acsami.5b05694

[981] Gillgren, N., Wickramaratne, D., Shi, Y., Espiritu, T., Yang, J., Hu, J., Wei, J., Liu, X., Mao, Z., Watanabe, K., Taniguchi, T., Bockrath, M., Barlas, Y., Lake, R.K., Lau, C.N., 2D Materials, 2[1] 2015, A1

[982] You, B., Wang, X., Zheng, Z., Mi, W., Physical Chemistry Chemical Physics, 18[10] 2016, 7381-7388
https://doi.org/10.1039/C5CP07585H

[983] Shao, D.F., Lu, W.J., Lv, H.Y., Sun, Y.P., EPL, 108[6] 2014, 67004
https://doi.org/10.1209/0295-5075/108/67004

[984] Huang, G.Q., Xing, Z.W., Chinese Physics B, 25[2] 2016, 027402
https://doi.org/10.1088/1674-1056/25/2/027402

[985] Huang, G.Q., Xing, Z.W., Xing, D.Y., Applied Physics Letters, 106[11] 2015, 113107
https://doi.org/10.1063/1.4916100

[986] Ge, Y., Wan, W., Yang, F., Yao, Y., New Journal of Physics, 17 2015, 035008
https://doi.org/10.1088/1367-2630/17/3/035008

[987] Jain, A., McGaughey, A.J.H., Scientific Reports, 5, 2015, 8501
https://doi.org/10.1038/srep08501

[988] Zhu, L., Zhang, G., Li, B., Physical Review B, 90[21] 2014, 214302
https://doi.org/10.1103/PhysRevB.90.214302

[989] Xu, W., Zhu, L., Cai, Y., Zhang, G., Li, B., Journal of Applied Physics, 117[21] 2015, 214308
https://doi.org/10.1063/1.4922118

[990] Ong, Z.Y., Cai, Y., Zhang, G., Zhang, Y.W., Journal of Physical Chemistry C, 118[43] 2014, 25272-25277
https://doi.org/10.1021/jp5079357

[991] Zhang, Y.Y., Pei, Q.X., Jiang, J.W., Wei, N., Zhang, Y.W., Nanoscale, 8[1] 2016, 483-491
https://doi.org/10.1039/C5NR05451F

[992] Peng, X.F., Chen, K.Q., Scientific Reports, 5, 2015, 16215
https://doi.org/10.1038/srep16215

[993] Hong, Y., Zhang, J., Huang, X., Zeng, X.C., Nanoscale, 7[44] 2015, 18716-18724
https://doi.org/10.1039/C5NR03577E

[994] Liu, T.H., Chang, C.C., Nanoscale, 7[24] 2015, 10648-10654
https://doi.org/10.1039/C5NR01821H

[995] Hu, W., Lin, L., Yang, C., Dai, J., Yang, J., Nano Letters, 16[3] 2016, 1675-1682
https://doi.org/10.1021/acs.nanolett.5b04593

[996] Cai, Y., Zhang, G., Zhang, Y.W., Scientific Reports, 4, 2014, 6677
https://doi.org/10.1038/srep06677

[997] Kou, L., Frauenheim, T., Chen, C., Journal of Physical Chemistry Letters, 5[15] 2014, 2675-2681
https://doi.org/10.1021/jz501188k

[998] Ray, S.J., Sensors and Actuators, B, 222, 2016, 492-498
https://doi.org/10.1016/j.snb.2015.08.039

[999] Cui, S., Pu, H., Wells, S.A., Wen, Z., Mao, S., Chang, J., Hersam, M.C., Chen, J., Nature Communications, 6, 2015, 8632
https://doi.org/10.1038/ncomms9632

[1000] Suvansinpan, N., Hussain, F., Zhang, G., Chiu, C.H., Cai, Y., Zhang, Y.W., Nanotechnology, 27[6] 2016, 065708
https://doi.org/10.1088/0957-4484/27/6/065708

[1001] Rubio-Pereda, P., Cocoletzi, G.H., Journal of Molecular Modeling, 23[2], 2017, 49
https://doi.org/10.1007/s00894-017-3225-z

[1002] Mahabal, M.S., Deshpande, M.D., Hussain, T., Ahuja, R., Journal of Physical Chemistry C, 120[36], 2016, 20428-20436
https://doi.org/10.1021/acs.jpcc.6b06791

[1003] Rahman, M.Z., Kwong, C.W., Davey, K., Qiao, S.Z., Energy and Environmental Science, 9[3] 2016, 709-728
https://doi.org/10.1039/C5EE03732H

[1004] Sa, B., Li, Y.L., Qi, J., Ahuja, R., Sun, Z., Journal of Physical Chemistry C, 118[46] 2014, 26560-26568
https://doi.org/10.1021/jp508618t

[1005] Chen, O.P., Lin, Y.J., Cao, W.Z., Chang, C.T., Materials Letters, 190, 2017, 280-282
https://doi.org/10.1016/j.matlet.2017.01.030

[1006] Guo, G.C., Wei, X.L., Wang, D., Luo, Y., Liu, L.M., Journal of Materials Chemistry A, 3[21] 2015, 11246-11252
https://doi.org/10.1039/C5TA01661D

[1007] Yao, Q., Huang, C., Yuan, Y., Liu, Y., Liu, S., Deng, K., Kan, E., Journal of Physical Chemistry C, 119[12] 2015, 6923-6928
https://doi.org/10.1021/acs.jpcc.5b02130

[1008] Cao, J., Shi, J., Hu, Y., Wu, M., Ouyang, C., Xu, B., Applied Surface Science, 392, 2017, 88-94
https://doi.org/10.1016/j.apsusc.2016.09.004

[1009] Zhao, S., Kang, W., Xue, J., Journal of Materials Chemistry A, 2[44] 2014, 19046-19052
https://doi.org/10.1039/C4TA04368E

[1010] Guo, G.C., Wang, D., Wei, X.L., Zhang, Q., Liu, H., Lau, W.M., Liu, L.M., Journal of Physical Chemistry Letters, 6[24] 2015, 5002-5008
https://doi.org/10.1021/acs.jpclett.5b02513

[1011] Kulish, V.V., Malyi, O.I., Persson, C., Wu, P., Physical Chemistry Chemical Physics, 17[21] 2015, 13921-13928
https://doi.org/10.1039/C5CP01502B

[1012] Sun, J., Lee, H.W., Pasta, M., Yuan, H., Zheng, G., Sun, Y., Li, Y., Cui, Y., Nature Nanotechnology, 10[11] 2015, 980-985
https://doi.org/10.1038/nnano.2015.194

[1013] Yu, Z.G., Zhang, Y.W., Yakobson, B.I., Nano Energy, 23, 2016, 34-39
https://doi.org/10.1016/j.nanoen.2016.03.010

[1014] Fei, R., Faghaninia, A., Soklaski, R., Yan, J.A., Lo, C., Yang, L., Nano Letters, 14[11] 2014, 6393-6399
https://doi.org/10.1021/nl502865s

[1015] Zhang, J., Liu, H.J., Cheng, L., Wei, J., Liang, J.H., Fan, D.D., Shi, J., Tang, X.F., Zhang, Q.J., Scientific Reports, 4, 2014, 6452
https://doi.org/10.1038/srep06452

[1016] Konabe, S., Yamamoto, T., Applied Physics Express, 8[1] 2015, 015202
https://doi.org/10.7567/APEX.8.015202

[1017] Lv, H.Y., Lu, W.J., Shao, D.F., Sun, Y.P., Physical Review B, 90[8] 2014, 085433
https://doi.org/10.1103/PhysRevB.90.085433

[1018] Liao, B., Zhou, J., Qiu, B., Dresselhaus, M.S., Chen, G., Physical Review B, 91[23] 2015, 235419
https://doi.org/10.1103/PhysRevB.91.235419

[1019] Zare, M., Rameshti, B.Z., Ghamsari, F.G., Asgari, R., Physical Review B, 95[4], 2017, 045422
https://doi.org/10.1103/PhysRevB.95.045422

[1020] Liu, Y.S., Zhang, X., Yang, X.F., Hong, X.K., Feng, J.F., Si, M.S., Wang, X.F., Physical Chemistry Chemical Physics, 17[16] 2015, 10462-10467
https://doi.org/10.1039/C5CP00391A

[1021] Yin, D., Han, G., Yoon, Y., IEEE Electron Device Letters, 36[9] 2015, 7159025, 978-980

[1022] Chang, J., Hobbs, C., Applied Physics Letters, 106[8] 2015, 083509
https://doi.org/10.1063/1.4913842

[1023] Wu, Q., Shen, L., Yang, M., Cai, Y., Huang, Z., Feng, Y.P., Physical Review B, 92[3] 2015, 035436
https://doi.org/10.1103/PhysRevB.92.035436

[1024] Kim, J.S., Liu, Y., Zhu, W., Kim, S., Wu, D., Tao, L., Dodabalapur, A., Lai, K., Akinwande, D., Scientific Reports, 5, 2015, 8989
https://doi.org/10.1038/srep08989

[1025] Liu, H., Neal, A.T., Si, M., Du, Y., Ye, P.D., IEEE Electron Device Letters, 35[7] 2014, 6823630, 795-797

[1026] Luo, X., Rahbarihagh, Y., Hwang, J.C.M., Liu, H., Du, Y., Ye, P.D., IEEE Electron Device Letters, 35[12] 2014, 6936890, 1314-1316

[1027] Anugrah, Y., Robbins, M.C., Crowell, P.A., Koester, S.J., Applied Physics Letters, 106[10] 2015, 103108
https://doi.org/10.1063/1.4914978

[1028] Das, S., Zhang, W., Demarteau, M., Hoffmann, A., Dubey, M., Roelofs, A., Nano Letters, 14[10] 2014, 5733-5739
https://doi.org/10.1021/nl5025535

[1029] Wan, R., Cao, X., Guo, J., Applied Physics Letters, 105[16] 2014, 163511
https://doi.org/10.1063/1.4900410

[1030] Chang, J., Journal of Applied Physics, 117[21] 2015, 214502
https://doi.org/10.1063/1.4921806

[1031] Mukhopadhyay, A., Banerjee, L., Sengupta, A., Rahaman, H., Journal of Applied Physics, 118[22] 2015, 224501
https://doi.org/10.1063/1.4937148

[1032] Cao, X., Guo, J., IEEE Transactions on Electron Devices, 62[2] 2015, 6987321, 659-665

[1033] Das, S., Demarteau, M., Roelofs, A., ACS Nano, 8[11] 2014, 11730-11738
https://doi.org/10.1021/nn505868h

[1034] Liu, Q., Zhang, X., Abdalla, L.B., Fazzio, A., Zunger, A., Nano Letters, 15[2] 2015, 1222-1228
https://doi.org/10.1021/nl5043769

[1035] Zhang, C., Xiang, G., Lan, M., Tang, Z., Deng, L., Zhang, X., RSC Advances, 5[50] 2015, 40358-40362
https://doi.org/10.1039/C5RA04056F

[1036] Fan, Z.Q., Sun, W.Y., Jiang, X.W., Luo, J.W., Li, S.S., Organic Electronics, 44, 2017, 20-24
https://doi.org/10.1016/j.orgel.2017.02.002

[1037] Du, J., Zhang, M., Guo, Z., Chen, J., Zhu, X., Hu, G., Peng, P., Zheng, Z., Zhang, H., Scientific Reports, 7, 2017, 42357
https://doi.org/10.1038/srep42357

[1038] Jiang, Z.T., Liang, F.X., Zhang, X.D., Physics Letters A, 381[4], 373-378
https://doi.org/10.1016/j.physleta.2016.11.006

[1039] Zhou, B., Zhou, B., Zhou, X., Zhou, G., Journal of Physics D, 50[4], 2017, 045106
https://doi.org/10.1088/1361-6463/aa52b5

[1040] Zhang, W., Huynh, T., Xiu, P., Zhou, B., Ye, C., Luan, B., Zhou, R., Carbon, 94, 2015, 895-902
https://doi.org/10.1016/j.carbon.2015.07.075

[1041] Walker, E.S., Na, S.R., Jung, D., March, S.D., Kim, J.S., Trivedi, T., Li, W., Tao, L., Lee, M.L., Liechti, K.M., Akinwande, D., Bank, S.R., Nano Letters, 16[11], 2016, 6931-6938
https://doi.org/10.1021/acs.nanolett.6b02931

[1042] Aktürk, E., Aktürk, O.Ü., Ciraci, S., Physical Review B, 94[1], 2016, 014155
https://doi.org/10.1103/PhysRevB.94.014115

[1043] Ji, W.X., Zhang, C.W., Ding, M., Zhang, B.M., Li, P., Li, F., Ren, M.J., Wang, P.J., Zhang, R.W., Hu, S.J., Yan, S.S., New Journal of Physics, 18[8], 2016, 083002
https://doi.org/10.1088/1367-2630/18/8/083002

[1044] Mannix, A.J., Zhou, X.F., Kiraly, B., Wood, J.D., Alducin, D., Myers, B.D., Liu, X., Fisher, B.L., Santiago, U., Guest, J.R., Yacaman, M.J., Ponce, A., Oganov, A.R., Hersam, M.C., Guisinger, N.P., Science, 350[6267] 2015, 1513-1516
https://doi.org/10.1126/science.aad1080

[1045] Piazza, Z.A., Hu, H.S., Li, W.L., Zhao, Y.F., Li, J., Wang, L.S., Nature Communications, 5, 2014, 3113
https://doi.org/10.1038/ncomms4113

[1046] Li, W.L., Chen, Q., Tian, W.J., Bai, H., Zhao, Y.F., Hu, H.S., Li, J., Zhai, H.J., Li, S.D., Wang, L.S., Journal of the American Chemical Society, 136[35] 2014, 12257-12260
https://doi.org/10.1021/ja507235s

[1047] Li, W.L., Zhao, Y.F., Hu, H.S., Li, J., Wang, L.S., Angewandte Chemie – International, 53[22] 2014, 5540-5545

[1048] Zabolotskiy, A.D., Lozovik, Y.E., Physical Review B, 94[16], 2016, 165403 *https://doi.org/10.1103/PhysRevB.94.165403*

[1049] Lopez-Bezanilla, A., Littlewood, P.B., Physical Review B, 93[24], 2016, 241405 *https://doi.org/10.1103/PhysRevB.93.241405*

[1050] Xu, S., Zhao, Y., Liao, J., Yang, X., Xu, H., Nano Research, 9[9], 2016, 2616-2622 *https://doi.org/10.1007/s12274-016-1148-0*

[1051] Yuan, J., Zhang, L.W., Liew, K.M., RSC Advances, 5[91] 2015, 74399-74407 *https://doi.org/10.1039/C5RA14939H*

[1052] Wang, H., Li, Q., Gao, Y., Miao, F., Zhou, X.F., Wan, X.G., New Journal of Physics, 18[7], 2016, 073016 *https://doi.org/10.1088/1367-2630/18/7/073016*

[1053] Le, M.Q., Mortazavi, B., Rabczuk, T., Nanotechnology, 27[44], 2016, 445709 *https://doi.org/10.1088/0957-4484/27/44/445709*

[1054] Pang, Z., Qian, X., Wei, Y., Yang, R., EPL, 116[3], 2016, 36001 *https://doi.org/10.1209/0295-5075/116/36001*

[1055] Yuan, J., Yu, N., Xue, K., Miao, X., RSC Advances, 7[14], 2017, 8654-8660 *https://doi.org/10.1039/C6RA28454J*

[1056] Zhang, Z., Mannix, A.J., Hu, Z., Kiraly, B., Guisinger, N.P., Hersam, M.C., Yakobson, B.I., Nano Letters, 16[10], 2016, 6622-6627 *https://doi.org/10.1021/acs.nanolett.6b03349*

[1057] Yuan, J., Zhang, L.W., Liew, K.M., Current Nanoscience, 12[5], 2016, 636-644 *https://doi.org/10.2174/1573413712666160517165813*

[1058] Gao, M., Li, Q.Z., Yan, X.W., Wang, J., Physical Review B, 95[2], 2017, 024505 *https://doi.org/10.1103/PhysRevB.95.024505*

[1059] Xiao, R.C., Shao, D.F., Lu, W.J., Lv, H.Y., Li, J.Y., Sun, Y.P., Applied Physics Letters, 109[12], 2016, 122604 *https://doi.org/10.1063/1.4963179*

[1060] Li, J.Y., Lv, H.Y., Lu, W.J., Shao, D.F., Xiao, R.C., Sun, Y.P., Physics Letters A, 380[46], 2016, 3928-3931 *https://doi.org/10.1016/j.physleta.2016.09.052*

[1061] Jiang, H.R., Lu, Z., Wu, M.C., Ciucci, F., Zhao, T.S., Nano Energy, 23, 2016, 97-104
https://doi.org/10.1016/j.nanoen.2016.03.013

[1062] Zhang, Y., Wu, Z.F., Gao, P.F., Zhang, S.L., Wen, Y.H., ACS Applied Materials and Interfaces, 8[34], 2016, 22175-22181
https://doi.org/10.1021/acsami.6b05747

[1063] Mortazavi, B., Dianat, A., Rahaman, O., Cuniberti, G., Rabczuk, T., Journal of Power Sources, 329, 2016, 456-461
https://doi.org/10.1016/j.jpowsour.2016.08.109

[1064] Zhang, X., Hu, J., Cheng, Y., Yang, H.Y., Yao, Y., Yang, S.A., Nanoscale, 8[33], 2016, 15340-15347
https://doi.org/10.1039/C6NR04186H

[1065] Rao, D., Zhang, L., Meng, Z., Zhang, X., Wang, Y., Qiao, G., Shen, X., Xia, H., Liu, J., Lu, R., Journal of Materials Chemistry A, 5[5], 2017, 2328-2338
https://doi.org/10.1039/C6TA09730H

[1066] Liu, C.S., Wang, X., Ye, X.J., Yan, X., Zeng, Z., Journal of Chemical Physics, 141[19] 2014, 194306
https://doi.org/10.1063/1.4902062

[1067] Shahbazi Kootenaei, A., Ansari, G., Physics Letters A, 380[34], 2016, 2664-2668
https://doi.org/10.1016/j.physleta.2016.06.016

[1068] Rostami, Z., Soleymanabadi, H., Journal of Molecular Modeling, 22[4] 2016, 70
https://doi.org/10.1007/s00894-016-2954-8

[1069] Rastgou, A., Soleymanabadi, H., Bodaghi, A., Microelectronic Engineering, 169, 2017, 9-15
https://doi.org/10.1016/j.mee.2016.11.012

Keywords

(13 x 13).. 20, 24
(2 x 2)................................... 20, 35, 191
(2√3 x 2√3) 19, 22, 26- 29, 89
(23 x 23)............................... 20, 26, 101
2-dimensional material . 3, 152, 180, 192-193, 215, 252-254, 273-274, 278
(3 x 3). 20-22, 26-27, 35, 39, 53, 84, 102, 112, 122, 151, 161, 191
(4 x 4) 19-29, 37, 39, 53, 84, 89, 115, 122
(7 x 7)................................... 20, 26, 114
ab initio..2, 11, 14-15, 19, 24, 35, 48, 62, 91, 101, 109, 120-21, 124, 131, 136, 138, 148, 150, 160, 165, 184, 192, 214-216, 225, 238, 260
Ag(001).. 3
Ag(110)............... 2-3, 21, 24-25, 30, 114
Ag(111). 3, 14, 16, 19-32, 37, 39, 40, 72-73, 77, 84, 89, 92, 101, 112-116, 122-126, 129, 142, 147, 154, 158, 160, 182, 273-277
Al_2O_3(00•1)................................ 40, 161
AlN 22, 121, 154, 191
ambipolar 3, 256, 263, 266-267
analogue...... 2, 5, 23, 54, 73, 76, 92, 113, 133, 151, 157
angle-resolved photo-emission16, 18, 22, 92, 115, 180
anisotropic .. 2, 13, 90, 93, 114, 159, 165, 196, 201, 204, 212-213, 216-219, 221-222, 227, 231, 235-240, 247-248, 250, 252, 254-255, 257-259, 263-265, 268, 272, 274-275
annealing... 6, 19, 26, 112, 130, 149, 159, 188, 247, 266
antimonene................................. 195-202
armchair 5-18, 30, 40, 43, 57, 68, 70, 75-76, 81, 83, 85-87, 90, 95, 97-98, 100, 103-105, 108, 111, 123, 125, 129, 131-139, 155, 170, 175, 177, 181-183, 190, 195, 204-206,212, 217-221, 225-232, 235-241, 247-254, 257-259, 262-266, 268, 274
arsenene 197, 204-211
asymmetry.............. 2, 190, 199, 237, 267
auxetic behavior........................... 13, 114
Bader charge 28, 144, 147, 153, 185
band structure... 1, 3-4, 18, 23, 26-29, 34, 36, 38, 40, 42-43, 52, 54, 62, 70, 75, 80-86, 90, 94-97, 101, 103-105, 111, 124, 131, 140, 145, 151, 154, 156, 158, 161, 173, 179, 181, 184, 188, 191-192, 220, 224, 226-228, 230,234, 236, 240-241, 246-248, 253-254, 256, 259, 268-269
band-gap.. 1, 2, 4-6, 11, 13-18, 31-34, 40, 43, 55-57, 61, 64, 66, 68-69, 71, 73-74, 79, 80, 82-83, 85-114,119, 122, 124, 128-137, 141-142, 147, 150, 152-154, 156-158, 160-161, 164, 166-167, 169-170, 172-178, 181, 183-184, 186, 188-197, 199, 201, 204-207, 210, 213-214, 216-217, 219-220, 223, 225-240, 242-243, 246-253, 255, 259-261, 264, 266, 268, 271
barrier-less 9, 116
battery 145-146, 221, 263
Bernal.. 10
biaxial strain...... 9, 12, 20, 101, 105, 111, 157, 190, 193, 201, 204-205, 207, 211, 220, 223, 231, 257
bilayer ... 9, 13, 17, 18, 21, 24, 26, 28, 65, 76-78, 80, 83, 85, 91, 93, 96, 99, 104, 111, 122, 134, 138, 151, 195, 198, 201, 204, 214, 219, 227, 231, 239-240, 247, 254, 256, 259, 265, 268, 271, 274-275
bismuthene 271

BN.... 18, 29, 31-32, 38, 95-96, 104, 124, 126, 158, 178, 190-191, 197, 210, 225, 245-246, 255
boat structure 13
Boltzmann transport ... 46, 48, 76, 87, 96, 176, 192, 200, 212, 257, 258
bond-angle 16, 71, 105, 130, 216
bond-length............................ 10, 25, 131
bond-stretching 11, 155
bond-weakening......................... 155, 193
borophene 272-279
bridge-... 17
Brillouin zone 3, 8, 16, 19, 21-22, 29, 30-31, 33, 36, 38, 53, 70, 79, 84, 91, 95, 101, 104-105, 140, 156, 157, 173, 181, 192, 220, 227
brittle fracture 12, 274
buckling 2, 4, 13-16, 20-21, 25, 27-28, 31, 36, 39, 46, 57, 71-72, 82, 84, 86, 89, 94, 101, 103, 108, 112, 113, 123, 131, 133, 137, 147, 149, 151-153, 157, 159, 170, 173, 176, 181, 184, 195, 201, 219, 272
CaF_2(111).. 33
carbon ..2, 5, 7, 10, 14-15, 21, 31, 66, 68, 70, 76, 89, 92, 107-108, 113, 116, 121, 128, 145, 147, 152, 157, 166, 181, 199, 205-210, 224, 226, 232-233, 236, 241, 251, 253, 272, 278, 296, 361
$CaSi_2$ 18
C-H 5, 68
chain........... 5, 26, 68, 104, 128, 166, 228
charge transport 3
charge-reorganisation 28
chemical functionalization...... 11, 80, 98, 107, 146, 155, 176, 255
chirality..10-11, 29, 47, 63, 74, 105, 146, 182, 216, 258
chlorine-passivated 33
clusters 4, 14, 124, 128, 143, 147, 159, 167, 272-273, 278
compressive stress........................ 12, 156
conduction...... 14, 36-38, 49, 56, 99-101, 106, 139, 141, 157, 199, 204, 208, 227, 231-232, 239-240, 246, 248, 252, 254, 260, 264, 267-268
corrugated structure 7-8
Coulomb..33, 41, 45, 53, 59, 78, 91, 147, 163, 214
coupling-constant.................................. 4
covalent... 5, 9, 13, 18, 26, 28, 32, 39, 68, 89, 92, 104, 117, 122, 147, 154, 160, 163, 169, 184, 193, 196, 234, 243, 255, 260
crack propagation.................................. 6
critical strain 7, 11-12, 220, 265, 274
C-Si 5, 68
Cu(111) 38, 215
D_{6h} 4, 146, 152, 156
defected......20, 27, 41, 86, 124-126, 129, 132, 245, 251, 261
defect-free 10, 131, 221
defects 6, 7, 10, 12, 16, 40, 46, 86, 91-92, 113, 115, 122-125, 127, 129-131, 133, 141-143, 150, 168, 170-171, 188, 199, 215, 221, 224-245, 249-255, 266, 275
density functional theory4, 5, 7, 9, 11-12, 15, 17-18, 20-21, 24-30, 35- 36, 38-39, 42, 51-52, 66, 69, 73-76, 81, 83-84, 86, 88, 91, 97-98, 101, 103, 106-107, 109-113, 116-117, 119, 122, 126, 128-129, 132, 134-135, 141-142, 145, 147, 150, 153-155, 160, 163, 165-166, 168, 170, 173, 175-176, 180-182, 184, 186, 197, 201, 204-205, 226-230, 232-234, 238, 241-243, 247- 248, 252, 260, 264, 269, 274-275, 277-279
diamond 7, 25-27

diboride 5, 36, 89
diffusion. 15, 18, 126, 132, 145-147, 176, 182, 188, 221-222, 247, 250-251, 262-263, 277-278
dimer 5, 26, 89, 119, 141, 147
Dirac cone.. 2, 3, 9, 11, 14, 18, 20-22, 24, 27, 29-33, 37-40, 52-53, 57, 66, 85, 87, 90, 92, 94-95, 97, 102-103, 105, 119, 133, 135, 140, 149-150, 154, 158, 161, 165, 173, 177-178, 181, 190-191, 210, 220, 273
Dirac point 17, 19, 21, 31-32, 37, 40, 53-54, 58, 63, 82-83, 85, 86, 88, 90, 95, 97, 100-101, 106, 117, 123, 136, 142, 156-158, 160-161, 164-165, 169-170, 173, 175-176, 184, 190, 192, 208, 254, 273
Dirac states 11, 26, 40, 155, 161
divacancy ... 125-126, 143, 170-171, 187-188, 199-200, 249-252
dumb-bell 28, 46, 107-109, 151, 165, 181, 189, 191
electrode 255, 262, 268, 277
electronegativity . 5, 32, 68, 96, 161, 166, 251, 260
energy-harvesting 264
environment-dependent 10
epitaxial growth 3, 14, 73, 182, 196, 215, 216
equiaxial tensile strain 7
Fermi level 5, 9-11, 15, 18-21, 25-28, 31, 36-38, 41, 51, 57, 59, 66-68, 70, 77-79, 81, 83-84, 86-87, 90, 92, 100-108, 110, 126, 140, 148, 150, 154, 156, 157, 159, 163, 170, 174, 177, 191, 195, 197, 207-208, 210, 223, 232, 239-240, 246, 249, 256-257, 265, 267, 269, 273
Fermi velocity 2, 9, 16, 19, 22, 31, 34, 85-86, 106, 151-152, 178, 181, 190-191, 254
fermions .4, 19, 21, 28, 61-63, 74, 78, 85, 92, 104, 150-151, 159, 228, 254, 273
ferrimagnetic 43
ferromagnetic 41-43, 50-53, 60-70, 77, 82, 87, 91, 100, 102, 105-109, 111-113, 119, 131, 138, 140, 154, 163, 170, 172, 174, 177, 179, 189-190, 211, 222, 224-225, 233, 237, 250, 276
field effect transistor 2-3, 82, 133-136, 138, 157, 176-178, 266-267, 270
field-tunable 4, 99
fifth-order elastic constants 8
figure of merit 40
first-principles calculations. 5, 28, 36, 37, 39-40, 46-48, 54, 66, 78, 81, 92, 114, 120, 124, 129, 133, 140, 143-144, 146, 161, 164, 176, 178, 180, 182-184, 192-193, 196, 203, 211, 217, 232, 245, 247, 252, 257-258, 262
flip-flop motion 21, 26
fluorinated 13, 70, 109, 113, 132
fracture strain 6, 131
fracture strength6, 12, 131, 182, 217, 251
free-standing ... 12, 15, 17, 19, 23, 27, 30, 32, 34, 38-39, 47, 49, 80, 97, 102-104, 110, 114, 125-126, 129, 145-146, 148-150, 153-155, 158, 160-161, 168, 171, 173, 182, 184, 191-192, 196, 220, 238, 271, 276-277
GaAs 32, 160-161, 206
gallium . 31, 117, 160, 173, 207, 210, 234
germanene ... 2, 19, 39, 54-55, 67, 76, 80, 149-163, 165-179, 181, 183-184, 193, 197
giant magnetoresistance ... 41, 50-52, 112, 178, 189
grain boundaries 6, 7, 16, 62, 123, 188, 252-253
grain size 6

graphene-like 1, 22, 24, 129, 184, 189, 192, 272
graphite . 2, 14, 22, 29, 39, 104, 146, 149, 221, 277-278
grazing-incidence X-ray diffraction 30
Grimme method 40
group-IV 1, 31, 46, 71, 86, 109, 139, 165, 172-273, 179, 183-184, 189, 208
Grüneisen parameter...... 11, 13, 155, 220
halogen.............................. 5, 13, 66, 109
Harrison bonding-orbital 13
hexasilabenzene 4
hole-doped 11, 155
hollow- 17
honeycomb structure 4, 10, 14, 16, 20-25, 27-28, 40, 55, 83, 94, 109, 113, 139, 150, 159, 160-161, 165, 174, 183-184, 186, 189, 195, 204, 238
H-terminated 5, 68
hybridization 1, 5, 9, 13, 15, 20-21, 23, 26-29, 36-37, 71-72, 80, 83-84, 89-90, 101, 109-110, 115, 117, 119, 124, 126, 141, 145, 147, 149, 155, 158, 164, 173, 196-198, 207, 211-212, 217, 225, 234-235, 278
hydrogenated 4, 17-18, 31, 34, 42, 62, 67, 69-70, 85, 91, 93, 98, 101-102, 106, 110-113, 122, 128, 134, 152, 159, 166-167, 175-176, 191, 210, 228-229, 234, 271
hydrogen-functionalized 6
hydroxyl 16, 114, 167, 232, 235
in-plane stress 7
interface-engineering 18
ion bombardment 12
Ir(111) 3, 14, 37
iridium 5, 37, 136, 184
islands 22, 24, 30, 114, 119, 149-150
isotope 46, 49
Jahn-Teller 4-5, 66, 82, 152
Janus silicene 5
K̄' point.. 22
Kagome lattice 39, 181
Kane-Mele 32, 59, 65, 71, 189
Keating model...................................... 13
Kleinman internal displacement 7
K-point 17, 19, 38, 54, 79, 84, 89, 95, 100-101, 104, 112, 181
Lagrangian 13
Landau level sequence 23
Landau-Peierls-Mermin-Wagner 2
lattice mismatch 19, 22, 35, 103, 133, 135, 150, 157, 254
Lennard-Jones 2
lithiation 82, 154, 221, 262, 276
lithium-adsorbed 12
lithography-patterned........................... 19
longitudinal acoustical modes................ 8
Lorentzian shapes 51
low-buckle 2, 7-8, 40, 46, 54, 77, 103, 105, 124, 126, 132, 150-151, 158, 161, 171, 180
low-energy electron diffraction 16, 20, 25-27, 36, 114
magnetic field . 23, 28, 52, 58, 60, 64, 69, 71, 74, 79, 80, 95, 139, 140, 156, 237, 239, 240, 256
magnetoresistance ..41, 43, 51-53, 57, 64, 105, 112, 179, 189, 226
massless Dirac fermions . 4, 19, 151, 159, 273
mini-gap 64, 75
mirror-symmetric 21, 27, 133
mobility 2, 13, 34, 74, 83, 87-88, 90, 102-103, 114, 122, 130, 133, 135-136, 158, 160, 175-176, 178-179, 197, 205, 213,

216-217, 221-222, 227, 229, 236-237, 241-242, 250, 252, 262-263, 266, 268
moiré pattern 275
molecular dynamics simulation 7, 11, 14-15, 43, 48-50, 121, 123-125, 144, 148, 181, 192, 196, 215, 243, 255, 258, 270, 277-278
monovacancy..... 124-126, 185-186, 188, 199, 207, 250-252
Monte Carlo 47, 138, 152, 179
MoS_2 ... 3, 36, 88, 92, 103, 126, 133, 175, 181, 221, 227, 229, 245-246, 248, 256, 259, 266, 268, 277
$MoSe_2$ 34, 217, 256
multilayer..... 16, 21-22, 26, 73, 108-109, 115, 132, 165, 206-207, 214, 226-227, 242, 271
multi-level 4
nanocapacitor 17
nanocomposite 197
nanodisk 33, 178
nano-electronics3, 40, 102, 161, 167, 177
nanoribbon ... 5, 10-12, 14, 30, 41, 43-45, 51, 53, 56, 59, 64, 68, 70, 75-76, 79, 81, 87, 98, 108, 125, 131, 133-135, 138-141, 157, 172, 177, 179, 191, 193, 195, 223, 228, 232, 235, 237, 240, 258, 264, 266, 269-270, 274
nanosheet 10, 31, 40, 102, 111, 170, 260, 275, 279
nearest-neighbor distance 25, 159
$NiSi_2(111)$ 34
non-covalent binding 18
non-linear behavior 10, 13, 213
n-type doping 17, 102, 106-107, 136, 209, 245, 265
OH-termination 16, 235
one-dimensional 10, 16, 24, 30, 102, 114, 123, 133, 137, 146, 188, 223, 228, 233-234, 236, 255, 262, 275
opto-electronics; 6
oscillation behavior 5, 189
oxygen-terminated 16, 246
Pb(111) 38
peeling 39
Peierls-Feynman-Bogoliubov 33
phenyl-modified 39
phonon dispersion . 5, 8, 35, 48, 138, 181, 218, 244, 257, 275
phonon soft modes 8
phosphorene195-196, 206, 214-217, 219-259, 261-271, 277
piezoelectric 17, 96
pile-up model 6
Poisson ratio. 8-11, 13-14, 105, 114, 131, 155, 182, 183, 205, 216, 218-219, 221, 274
polarization .. 2, 31, 41-42, 51-52, 56, 58-59, 61-64, 67, 69, 72-74, 82-83, 87, 89-90, 95-97, 102, 105-106, 113, 122, 130-131, 134, 138-140, 145, 154, 163, 170, 174, 177, 196, 198-199, 205, 236, 252, 254, 278
polycrystalline 6, 16
pristine . 10, 12, 18-19, 24, 26, 30, 48, 50, 53, 62, 67-69, 72, 76, 85-89, 98, 101-102, 106, 108-110, 122-128, 131-132, 141-143, 145-147, 166-167, 174-175, 177, 197, 206-207, 210, 214, 222, 228, 232, 234-235, 241-242, 245, 249, 251, 255, 260-264, 276, 278
projector-augmented 5, 87
pseudo-Jahn-Teller 4, 5, 82
pseudomorphic 19
Pt(111) 14, 39, 151

p-type doping.....106-107, 136, 156, 170, 201, 208
puckering 4, 5, 143, 251
pulling direction.................................... 8
pyramidal 16
quantum spin Hall effect2, 19, 53-54, 79-80, 150, 153, 175, 178, 189, 190, 196, 207, 211, 269
quantum well 19, 38, 139, 236
quantum-spin 4, 53, 97, 239
quasi-particle ...29, -, 63, 70, 86, 99, 173, 204, 234
Raman spectroscopy..... 20, 22, 115, 119, 196, 214
reconstructed surface 16
Rh(111)... 15
rhombic phase............................... 21, 27
ribbons 3-4, 16, 19, 41, 75, 81-82, 90-91, 112, 131, 140, 171, 181, 189, 191-192, 225, 229, 233, 238
ring-unit 4
Ru(00•1)... 14
scanning tunnelling microscopy3, 12, 14, 16, 20-21, 24-25, 27-30, 35, 37-38, 89, 92, 112, 115, 119, 123, 150, 159, 180, 248, 275
second-order elastic constants 8, 183
Seebeck coefficient. 41, 46, 51, 105, 264, 265
semiconduction 82, 245
semi-metal . 10, 13, 17, 53, 55, 66-68, 79, 89-90, 92, 93, 96, 100, 104, 106, 111, 114, 120, 130, 131, 133, 136, 146-147, 154, 164, 166, 174, 185, 192, 207
Si(111). 12, 16, 21, 26, 27, 29-30, 33-34, 36-37, 71, 102, 114-115, 121-122
Si_3Pt 39
Si_6.. 35, 146
SiC .. 15, 18, 31, 109, 122, 147, 165, 189, 245
SiC(00•1) 31, 122
Si-H 5, 32, 75
silicane . 4-7, 18, 85-86, 90, 93, 101, 104-105, 110, 112-113, 137, 144-145
silicene 1-2, 4-27, 29-80, 82-147, 149-150, 152, 156, 158, 165-166, 168, 171, 174, 178, 181, 183-184, 192-193, 195, 208, 221, 241, 268
silicon-based ... 4, 82, 116, 119, 122, 150, 208
silicone 5
simulations2, 6, 10-12, 14-15, 19, 21, 34, 38, 43, 46, 48-50, 73, 82, 109, 111, 116, 121-122, 124-125, 133, 136, 138, 144, 148, 150, 152, 155, 179, 182, 193, 196, 201-202, 208, 215, 217, 220, 243, 252, 255, 258, 265, 267, 270, 274, 276, 278
Si-Si bonds 3, 5, 7, 127, 144
six-membered........................ 4, 143, 146
skew 16
solidification 15, 121
sp^2.....1, 3, 5, 9, 14-15, 19, 21, 23, 47, 72, 83, 105-106, 109, 119, 124, 137, 169, 181, 183, 251
sp^3 2, 5, 15, 16, 20, 23, 35, 71, 80, 83, 94, 105-106, 109, 110, 116, 124, 126, 141, 146, 169, 181, 183, 216-217, 251
spin-filter......50-51, 60, 65, 79, 106, 112, 138, 140, 179, 265
spin-orbit ... 2, 4, 8, 22, 32-33, 41, 46, 54-61, 64-66, 69, 71, 74-75, 78-79, 83-84, 86, 89, 91, 94, 96-97, 99, 102, 120, 131, 134-135, 138, 153, 156, 157, 159, 161, 172-173, 179,-181, 184, 189-191, 199, 202, 205-206, 209, 228, 267, 271
spintronics ... 63, 119, 137, 139, 150, 163, 178, 196

spontaneous buckling 21
standalone 39
stanene 54, 165, 180-193
stiffness. 6-8, 13, 182, 192, 221, 263, 274
Stillinger-Weber potential 48
Stone-Wales defect..... 16, 127-131, 142-143, 168, 170, 188, 199, 221, 252
strain-rate 6, 12, 218
stress-concentrating sites 6
stress-strain relationship 7, 183, 251
stripes 2, 3, 35
sub-lattice.... 4, 18, 23, 33, 42, 53, 55, 59-60, 66, 68, 71-72, 77, 79, 88-89, 93, 97, 99, 101-103, 105, 112, 116, 119-120, 133, 135, 136, 152, 156, 159, 174, 178, 273
substrate 1, 2, 5, 12, 14, 19-26, 28-31, 33-35, 36-40, 49, 53, 66, 73, 84, 87-90, 93, 101-103, 105-107, 109, 112, 114-116, 122, 127, 135, 137, 142, 144, 149-151, 153, 158-159, 161, 174, 177-178, 182-183, 190-191, 194, 210, 217, 246, 275, 277
sulfide 31
superstructure...... 25, 27, 29, 37, 53, 102, 112, 151
surfactant 26, 196
switchable polar bilayer....................... 17
Tersoff potential 47, 131
tetramers 39
thermal conductivity...... 46-49, 172, 192, 200, 205, 212, 257, 258, 264, 265
thermal expansion........ 12, 155, 252, 274
thermal stability..... 15, 18, 125, 132, 247
thermal transport 18, 46, 48-50, 171, 192, 258
thermoelectric..... 18, 40-46, 51, 105, 264
thermopower...... 42, 44-45, 87, 105, 108, 172
third-order elastic constants................ 13
three-dimensional...... 7, 14, 93, 125, 134, 147, 152, 195, 246
tight-binding..... 4, 8-9, 33, 54, 61, 76, 79, 92, 96, 101, 108, 120, 156, 172, 193, 202, 218, 220, 237, 239-240, 270, 273
tight-binding calculations 9
tinene... 181
top- 17, 26, 157
topological insulator .. 32-33, 41, 52-55, 59, 68, 71, 74, 76, 79, 81, 91, 94, 97, 99-100, 157, 178, 183, 189-190, 195, 201, 207, 211, 268, 271
transistor....... 2-3, 89, 134-136, 138, 157, 176-177, 260, 266, 269
trigonal warping............................ 29, 81
tunability 2
tunable direct gap................................ 13
two-dimensional. 1-3, 6-7, 10, 13-15, 19, 22-24, 35, 38-42, 48-49, 54-55, 59, 62, 66-67, 69-72, 75, 77-78, 82, 84, 87, 93-94, 96-97, 100-101, 106, 109, 110-111, 114-116, 119, 123, 126, 133, 135, 137, 145-147, 149-151, 153-155, 157-159, 163, 170, 173-176, 181, 183, 192-193, 195, 201, 204, 207, 210, 213-217, 224-226, 228, 232-233, 236, 246, 248, 251, 253-259, 266, 268, 272
uniaxial tensile deformation 6
uniaxial tensile strain 7, 65, 105, 111, 205, 218-219, 236, 250
vacancy 10, 46, 66, 88, 106, 124, 125-129, 131-132, 166, 169-170, 188, 200, 221, 249- 250, 252, 255, 262, 272-275
valence orbital................................. 3, 21
valleytronics...................... 2, 53, 63, 140

Van der Waals 13, 31, 49, 158, 191, 246, 261, 278
vibronic 4
washboard 147, 195, 199, 271
wavelength. 2, 47, 72, 171, 192, 240, 275
wetting 22, 26
zero-frequency 35
zig-zag ... 5-11, 13, 16, 18, 33, 40-41, 43-45, 47, 50-52, 56, 59, 61, 64-65, 67-68, 70, 75-76, 79, 81-82, 87, 90-91, 95-96, 104-109, 112-114, 123, 126, 128-129, 131-133, 135, 137-138, 140, 145, 155, 159, 167, 175, 177-178, 182-183, 189-190, 205-206, 212, 217-222, 225-233, 235-239, 247-254, 257-259, 262-266, 268-270, 274
ZrB_2(00•1)............3, 12, 34-36, 115, 121
ZrC(111) 3, 35, 36
β-Si_3N_4(00•1) 34
Γ-point..... 17, 89, 92, 101, 106, 191, 205, 225, 240
π* .3, 21, 22, 36, 69, 81, 85, 86, 173, 177
π-electron 18, 62, 74, 84, 89

www.ingramcontent.com/pod-product-compliance
Ingram Content Group UK Ltd.
Pitfield, Milton Keynes, MK11 3LW, UK
UKHW020658041026
15590UKWH00001B/352

9 781945 291302